# Wavelength Division Multiple Access Optical Networks

For a complete listing of the *Artech House Optoelectronics Library*,
turn to the back of this book.

# Wavelength Division Multiple Access Optical Networks

Andrea Borella
Giovanni Cancellieri
Franco Chiaraluce

Artech House
Boston • London

**Library of Congress Cataloging-in-Publication Data**
Borella, Andrea
Wavelength division multiple access optical networks / Andrea Borella, Giovanni Cancellieri, Franco Chiaraluce.
p. cm. — (Artech House optoelectronics library)
Includes bibliographical references and index.
ISBN 0-89006-657-4 (alk. paper)
1. Optical communications. 2. Multiplexing. 3. Light—Wavelength.
I. Cancellieri, Giovanni. II. Chiaraluce, Franco. III. Title. IV. Series.
TK5103.59.B67 1998
621.382'7—dc21 98-2922
CIP

**British Library Cataloguing in Publication Data**
Borella, Andrea
Wavelength division multiple access optical networks. –
(Artech House optoelectronics library)
1.Optical communications 2. Radio waves
I. Title II. Cancellieri, G. (Giovanni) III. Chiaraluce, Franco
621.3'827
ISBN 0890066574

**Cover design by Joseph Sherman and Deborah Dutton Design**


**685 Canton Street**
**Norwood, MA 02062**



International Standard Book Number: 0-89006-657-4
Library of Congress Catalog Card Number: 98-2922

10 9 8 7 6 5 4 3 2 1

# *Contents*

# *Preface*

In September 1994, when we had just returned from the International Conference on Computer Communications and Networks (IC$^3$N) held in San Francisco, Mark Walsh, acquisitions editor of Artech House, suggested that we write a book on wavelength division multiplexing (WDM) optical networks. Our first reaction to this proposal was one of worried satisfaction. We felt satisfaction because of the recognition of our achievements in this important field, which involves many valuable researchers all over the world. We felt worried because we promptly realized, based on previous similar experiences, that to write a book on optical networking would be very exacting.

Actually, what is true when writing a book on any technical subject becomes more amplified in the case of optical networks. Research on lightwave interconnections is evolving so quickly, and the number of printed works on the subject are multiplying so rapidly, that it seems impossible to summarize in a few hundred pages the vast quantity of new ideas that have emerged and are continuously emerging.

We feel, and the preparation of the book has reinforced our belief, that it is wise to stop periodically to summarize the situation. Producing a text that is an overview of the field is useful both for the researchers, who must struggle daily with the jungle of published papers, and for the graduate and undergraduate students, who need a concise text that reviews the current body of knowledge in the optical network field.

Because of the inclusive nature of the subject matter, a point of view is necessary, and we have adopted one that classifies wavelength division multiple access (WDMA) networks on the basis of whether wavelength conversion is or is not requested to establish a connection. This explains the structure of the book where, apart from Chapter 1, which introduces general concepts and briefly reviews components, Chapter 2 is devoted to single-hop networks, Chapter 3 to multihop networks, and Chapter 4 discusses a combination of both systems. Though our aim is to explain the subject clearly, the more technical discussions require some basic knowledge for full comprehension. Where the text may become difficult for the reader to follow—mainly in the discussion of protocols and routing—we have inserted some very short summaries (in the form of keypoints) that should help the reader to choose rapidly if the section is of interest or not, for a specific application.

Obviously, many other approaches are possible, in principle, and we recognize that readers may not find in this book discussion of all the issues they may desire. Our goal was not to write a comprehensive text on WDM networks. As we mentioned above, in our view, this is impossible with a restricted number of pages. On the contrary, we will be satisfied if this book helps scientists and technicians to deepen their understanding of optical networks, both theoretically and practically. We hope that other books will soon continue the discussion set forth in the current work.

We wish to thank our wives and children for having patiently awaited the conclusion of this job. Particularly Andrea's daughters (Caterina and Chiara) and Franco's daughters (Giulia and Martina) for whom the production of the book has certainly meant giving up many hours of recreation with their fathers. But they are very young and will forget soon. So we are sure that they will not hate optical networks.

Giovanni's children (Elisabetta and Daniele) deserve separate mention. They have seen their father work on half a dozen books, and perhaps they think that writing books is one of the pleasures of life. In that case they should be potential candidates for authoring a text on a future generation of optical networks.

Finally, we would like to thank the whole Artech staff, and particularly Julie Lancashire and Susanna Taggart. They have followed the project with competence and constancy, helping us to surmount all the difficulties we encountered.

# Chapter 1

# Optical Networking

## 1.1 BRIEF HISTORY OF OPTICAL COMMUNICATIONS

A brief outline of the main steps that characterized the evolution of technological and engineering aspects of optical communications in recent years can provide some keys for interpreting the growing interest that is presently devoted to all-optical networks. By *all-optical network*, technicians usually mean a telecommunication network where the signals travel, from their source up to their destination, in the form of optical guided waves. This solution for the fixed worldwide telecommunication network appears to be the most attractive, and its practical realization, by proper merging of several pre-existent national or continental networks, is deferred, without doubt, to the more or less distant future, depending on which solution will finally be preferred.

Silica-based multimode fibers were first proposed for long-haul signal transmission in 1966 [1], on the basis of the prediction that the quartz of which they were made would exhibit, in subsequent years, a reduction in its attenuation due to absorption and scattering, from more than 100 dB/km to about 20 dB/km. In 1974, an experimental value of 2 dB/km was announced with an operating wavelength of 0.85 μm [2]. Two years later an attenuation as low as 0.5 dB/km was reached at a wavelength of 1.3 μm [3]. In 1979, the value of 0.2 dB/km was finally achieved at 1.55 μm [4]. In parallel, after the early observations of laser action in semiconductors [5–7], the technology of laser diode optical sources showed continuous improvement, thus providing efficient power injection as well into fiber cores whose diameters were becoming smaller and smaller.

In this direction the fiber technology had been progressively driven by the need for obtaining single-mode operation [8], to avoid multimode signal distortion [9]. The remaining cause of signal distortion, which is due to chromatic dispersion [10], was faced by reducing the spectral linewidth of laser emission up to a few megahertz [11] and by operating at a wavelength characterized by zero second-order chromatic dispersion [12]. The former result made coherent modulation of laser light of practical interest [13], although at present this technique often remains unattractive with respect to the traditional intensity modulation, because of the requirement of more sophisticated demodulation procedures which are usually based on heterodyning. Coherent modulation, on the other hand, could become very attractive if a true frequency division multiplexing (FDM) of the optical carrier could be employed in the future.

When a set of signals travel the same fiber, all intensity modulated, avoiding interference among themselves due to different wavelengths, it is preferable to adopt the name wavelength division multiplexing (WDM) [14]. These signals can be separated, at the output fiber end, by optical filtering, without the necessity of heterodyning, since their wavelength separation is on the order of 1 nm. Transmission systems of this type were proposed early to overcome the bandwidth limitation due to chromatic dispersion on a single modulated wavelength, for point-to-point connections [15]. Nevertheless, their use appeared almost immediately to be inefficient, mainly because of the number of regeneration circuits needed, which equals the number of wavelengths. A modification in this perspective has been recently produced by the advent of optical amplifiers, which operate on many wavelengths at the same time. Furthermore, the possibility of routing different wavelengths independently one from another, within an all-optical network, contributes to make WDM as the most attractive technique to be employed today, as we will further discuss in the following.

Coming back to the relatively short history of optical communications, during the early 1970s, circuit and quantum noise in the photodetection process were accurately evaluated [16], in order to fix nearly ultimate limits of sensitivity in direct detection of an optical signal. *Sensitivity* is usually defined as the minimum received optical power required for ensuring a prefixed quality of the signal. Circuit noise can be circumvented owing to the effect of the local oscillator in heterodyne detection of a coherently modulated wave [13]. Nevertheless this expedient leads to a practical improvement of few decibels in sensitivity with respect to the best performance of a system based on intensity modulation and avalanche direct detection. Also these considerations contributed to making coherent modulation not yet competitive with traditional intensity modulation.

During the 1980s the fraction of long-haul telecommunication traffic carried on point-to-point single-mode fiber connections progressively grew in the transport networks of all the most advanced nations. About 200 km was achieved as the maximum limit for regeneration spacing, mainly due to attenuation, in wide bandwidth links, whose bit rate had reached the extraordinary order of 10 Gbps [17]. Submarine optical cables were laid down, sometimes along transoceanic runs [18], thus raising the interest in guided intercontinental communications, and were almost completely abandoned after the advent of satellites.

Nevertheless one of the most significant results of the late 1980s certainly is the construction of erbium-doped fiber amplifiers (EDFAs) [19]. They allow direct amplification of the optical signal, independently of its modulation format. By their use, the maximum limit in regeneration spacing is enlarged up to a value by far exceeding the circumference of the Earth. At present, amplification spacing is set from 50 to 200 km, depending on the overall length and capacity of the link, because the power noise increases linearly with the number of amplification sections, but, on the other hand, the greater the optical gain the higher the amplifier noise figure [20].

Today's telecommunication transport networks in most nations in the world where digital transmissions are dominant are nearly totally made up of optical connections. They contain single-mode optical fibers, already installed, in an amount often overabundant compared to actual needs. This is due to the fact that the costs of laying down cable result in a considerable fraction of the total costs, and this investment has to be paid off over many years. The problem that public and private operators now have to face is what is the best use of such an inheritance, taking into account the expected growth in telecommunication traffic due to traditional services, but also the possible advent of new services. The solution to this problem is to be found not individually, but rather in a worldwide context, because of the increasing amount of international information transfer that characterizes modern life.

In parallel, local area networks (LANs) [21] are being realized mainly by means of optical fiber connections, whose reliability is very high because of their complete immunity to any kind of electromagnetic interference. The reliability requirement becomes more and more stringent as the bit rate increases since, correspondingly, the amount of carried information generally increases as well. Furthermore, optical cables for indoor applications appear lighter and more flexible than their copper counterpart. In this framework, some passive all-fiber components, like couplers, splitters, and wavelength-selective branches, have become mature for high-reliability applications.

A different aspect, also not completely investigated at present, because of the uncertainty in the definition of future telecommunication services, is that regarding the access network [22–25]. This includes the segment of telecommunication infrastructure between a private user and the core public network, and is expected to replace actual subscriber tails, made of traditional copper twisted pairs. Two main solutions appear possible: one based on copper coaxial cables, and mainly oriented to broadband distribution of analog signals, the other completely optical, digital, and oriented to provide service on demand. Obviously the latter solution appears more attractive on a long time perspective.

The research on all-optical effects and on devices able to support them has been developed with great effort, either in laboratories or in factories, with the objective of maintaining the signals at optical level over the longest path possible in their transport section. Therefore, space optical switches, wavelength converters, optical delay lines, or even true memories have been proposed under the embracing denomination of photonic devices [26]. They perform, by direct processing of light signals, functions that are well established in electric signal processing at the nodes of a traditional transport network. An intermediate solution, probably better achievable in the next few years, could be based on electro-optical components, where the transmission optical signals are processed under the action of electric control signals.

In light of all these facts, it is clear how the interest of public and private operators is devoted to employ at best the installed optical fibers of the existent transport network for ensuring the largest variety of possible future service. This will mean addition of proper equipment, able to generate, combine, switch, select, and detect light signals, at the access terminations and in the nodes of this network. Taking into account the still mature technology of EDFAs and the improvement in the performance obtainable by electro-optical and photonic devices for signal processing in the widest sense of the word, a WDM solution appears to be the best suited one. Wavelength division multiple access (WDMA) optical networks are thus being intensively studied and experimented [27,28], in view of defining a reference model, to which all operators should adapt.

In the following sections of the present introductory chapter, these subjects will be treated, thus gradually introducing the reader into the rather involved world of WDMA networks. Section 1.2 stresses the main features of these networks. In Section 1.3 their classification is presented, which is responsible for the organization of the entire book. Section 1.4 describes some practical limits, deriving from fiber and passive component attenuation, signal distortion produced by chromatic dispersion in each modulated wavelength, intermodulation among the wavelengths due to undesired nonlinear effects and wavelength crosstalk. In Sections 1.5 and 1.6 passive components and active devices are treated, respectively. Section 1.7 is expressly devoted to EDFAs. Finally, in Section 1.8 some traffic aspects are reviewed, in order to make reading of the subsequent chapters more straightforward.

## 1.2 MAIN FEATURES OF WDMA OPTICAL NETWORKS

In a WDMA optical network a wavelength is used to connect two nodes of the network, thus forming a one-way channel. This channel can cross intermediate nodes, without changing the wavelength that supports it. Although traditionally there is a difference between access nodes and transit nodes, the former ones being representative of users and the latter ones of sites where the channels are switched, in WDMA networks this distinction is not necessary. In fact, even peripheral nodes usually collect the traffic generated by and destined to more than one user. Furthermore a digital transmission organized in data packets, like frame relay or ATM (asynchronous transfer mode) [29], is assumed to be employed among the nodes. This implies the need of buffers, in the nodes, where the packets can be stored, organized, and finally sent to their destination. In conclusion, all the nodes of a WDMA optical network usually operate either as access or as transit nodes.

The above considerations have some consequences on the model that we can adopt for a network of this type. First of all, we have to distinguish between a physical and a virtual topology. The first one consists of the optical fiber connections and of the optical components placed at their input and output ends. The second one, instead, represents the structure of the one-way channels, each obtained by modulating a single wavelength, as stated above, along which the data packets flow from a source node to a destination node. Another consequence regards the possibility that a node transmits and receives data packets at the same time, and it is even able to receive from more than one input channel, as well as to transmit towards more than one output channel. Clearly these data packets refer to different communication sessions, often associated to different pairs of users. Furthermore, we can have locally generated or destined traffic, but also traffic in transit across the considered node. Finally, in a context of this type, it is necessary to abandon the distinction between one-way connection and two-way connection for a pair of nodes, because traffic is the aggregation of various messages, in turn supporting different types of telecommunication services. Therefore, in the following, we will restrict our discussion to one-way connections only.

Although isochronous services, like traditional telephony, still represent a large fraction of the traffic offered to national and international transport networks, and the natural technique for routing telephonic calls remains that based on circuit switching, some other modern services, not isochronous, are rapidly growing. Furthermore, it is a widespread opinion that ATM, especially at high bit rate, will soon be able to also manage large amounts of telephonic traffic. This gives rise to a pronounced orientation toward packet switching, to the point where possible future global networks are planned in such a direction. Observations of this type can support the above-described model for a WDMA network, whose simplicity offers a good opportunity for comparing various possible solutions.

Any type of WDMA network is characterized by the following features:

1. It is intrinsically a high-capacity network;
2. Its structure is modular and scalable, which means that it can grow after aggregation of many similar substructures and that hardware as well as the procedure for managing the network do not change during growth;
3. It is reconfigurable, thus easily bypassing a large variety of failures that can occur in the network elements;
4. It supports clear channels, on which the signals originated by very different services can coexist.

As regards the last feature, we will distinguish between point-to-point, point-to-multipoint, and broadcast transmissions, but only in relation with precise data packets, and not with an entire service, whose characteristics are lost once the packets have been formed. Obviously there is the need of such standardization work, for matching the transmission protocols of the various services with that of the network, but this problem appears solvable. More difficult probably will be to find agreement on recommendations and standardizations regarding the physical layer of WDMA networks themselves. On the other hand, only this step will be able to force a reduction in the costs of commercially available components.

In order to put in action more than one contemporary transmission from the same node, it is necessary to dispose of as many optical sources. The wide bandwidth of the modulation signals and the collimated beam necessary for an efficient power launching into a single-mode fiber input end require that such sources are of laser type. It is then necessary to distinguish between fixed laser sources and tunable laser sources, the latter being characterized by the possibility, at different times, of varying the central wavelength of their emission spectrum. Also in reception, we can distinguish between fixed detection and tunable detection, with the same meaning.

There are practical limits in the total number of wavelengths that can contemporarily travel a fiber. They are due to the components crossed by such wavelengths all together (first of all the fiber itself or the EDFAs), but also to the components necessary for collecting and separating them at the nodes, whose complexity and cost grow with the wavelength number. Referring to the physical topology, it is necessary to take into account this limit. On the other hand, when a certain virtual topology is to be practically realized, it is always possible to separate the wavelengths, thus forming as many parallel one-way channels between two nodes, into more than one fiber, resorting to a form of procedure sometimes called space division multiple access (SDMA). Nevertheless this tool is to be carefully taken into account when an economic comparison among various network solutions must be developed.

Although multiple access to the network is ensured here by the presence of more than one wavelength, it is important to also stress the role of time in the evolution of the state that characterizes a network of this type. In this sense, WDMA must be intended in opposition to other possible access procedures, such as time division multiple access (TDMA) or code division multiple access (CDMA), which, respectively, share the transport medium subdividing time [30] or making use of a properly constructed code [31]. Nevertheless, on many occasions, time division will be considered together with wavelength division, for the description of particular WDMA networks, which take advantage of both the procedures.

A series of practical limits on the modulation bandwidth $B$ for each wavelength and on the overall number of usable wavelengths $W$ contribute to impose an upper bound to the product $W \cdot B$, which theoretically expresses the overall bandwidth available, that in turn approximately coincides with the maximum transmission capacity of the whole network (in bps), being transmission based on a binary intensity modulation format.

Once having stated the maximum capacity, however, the actual number of bits transferred per second through the network depends on the load and on the transmission protocol adopted. Therefore, another figure, called *throughput*, is defined to measure the amount of data that successfully reach the intended destination. Throughput obviously increases with the load, up to a maximum value which depends on the transmission protocol and is further reduced by the overhead inserted for addressing and control. Several examples will be given in the following chapters. Once having reached its maximum value (which corresponds to network saturation), throughput no longer increases, and a further augment of the offered load will cause loss of packets. In practice, it is more usual to use a normalized throughput, defined as the ratio between the true throughput and the transmission rate available. This last parameter can also be referred on average to a single node.

The time delay necessary for a packet to reach its destination is an important parameter too for describing the performance of a network. Sometimes a packet is abandoned when this time is too great with respect to the target imposed by the telecommunication service to be realized, so contributing to the fraction of lost packets. Also, such a parameter depends on the conditions of the traffic. Statistical assumptions about the traffic can be made to give a probabilistic description of time delays (e.g., allowing the determination of their mean value and variance) or sometimes of their entire probability density distribution.

For a given network topology, packet transmission can be based on one of several protocols, and this may originate very different performance, in terms of both the throughput and time delay. Other features of great importance for evaluating network operation efficiency are those regarding its modular and scalable structure and its possibility of being easily reconfigurable, to avoid the effects of faults. As mentioned above, any WDMA network intrinsically enjoys such features, but there are particular topologies and protocols that make it possible to improve certain aspects rather than others.

## 1.3 CLASSIFICATION

A classification of WDMA networks can be made on the following basis:

1. Networks in which the unidirectional connection between two nodes is performed by means of a single one-way channel, that is, one employing a single wavelength;
2. Networks where some connections require that, in an intermediate node, a conversion from one channel to another, that is, a change in the wavelength employed, is performed.

The first type of network will be called *single-hop*, and the second type *multihop*. The above classification regards the virtual topology of the network. In fact various physical topologies can support the same single-hop network, or the same multihop network.

The point is even more important if we consider that the same term is sometimes used to denote a physical topology or a virtual topology. As an example, let us consider the case of a ring, shown in Figure 1.1 for a network with four nodes. We can have a physical ring, where the nodes are connected by an optical fiber in such a way as to form a closed physical path (Figure 1.1(a)), or a virtual ring, where each node has only one channel for transmission (Figure 1.1(b)). A virtual ring is intrinsically multihop (for a number of nodes greater than or equal to three). In Figure 1.1(b), for example, node 1 can send a message directly to node 2 (using wavelength $\lambda_1$), but it needs its message repeated (at $\lambda_2$) if addressed to node 3. A virtual ring can be implemented on a physical ring but, alternatively, the same physical structure shown in Figure 1.1(a) can realize a single-hop network (if each node has sufficient wavelengths to reach directly all the other nodes) or a multihop network based on a different strategy.

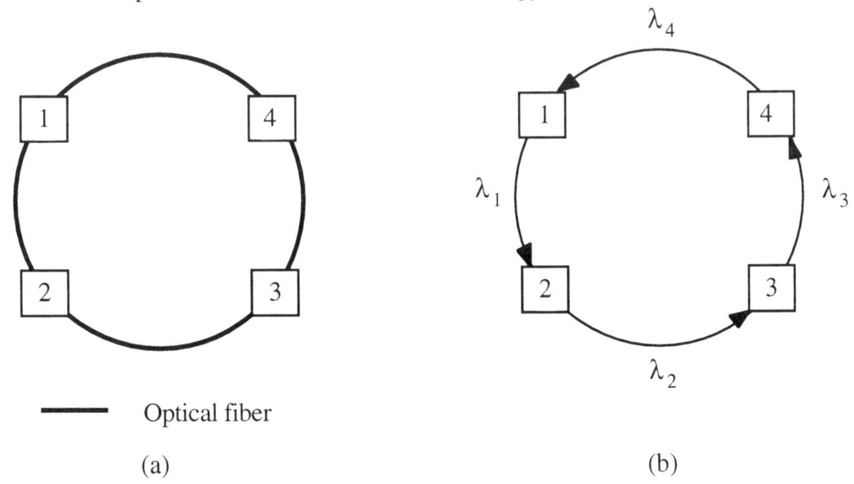


**Figure 1.1** Examples of topologies: (a) physical ring and (b) virtual ring.

Single-hop networks can be organized in two main ways: use of a common medium over which all the nodes can communicate, or use of intermediate devices where the wavelengths are properly routed from an input port to an output port. The first solution is known as *broadcast-and-select*, the second as *wavelength-routing*. Chapter 2 is devoted to the description of single-hop networks, and to a comparison between these two main solutions.

Multihop networks, in turn, are characterized by their own virtual topology, that is by the architecture of the interconnections between any pair of nodes. It takes the form of a graph, in which oriented edges represent one-way channels. A very important parameter is the diameter of such graph, which equals the maximum number of subsequent edges, and hence wavelength hops, necessary for connecting two whichever nodes of the network. Two examples of widely investigated multihop networks are shown in Figure 1.2. On the left we have the so-called *Manhattan street network*, here consisting of 6×6 nodes, organized in rows and columns. It can be seen as a two-dimensional interleaving of as many rings as the sum of the numbers of rows and columns. On the right there is the so-called *shufflenet*, here consisting of 4 + 4 nodes. The nodes are organized in two columns. Two channels leave each node and two channels enter it. Such channels can only connect nodes that belong to adjacent columns. The overall structure can be viewed as a torus.

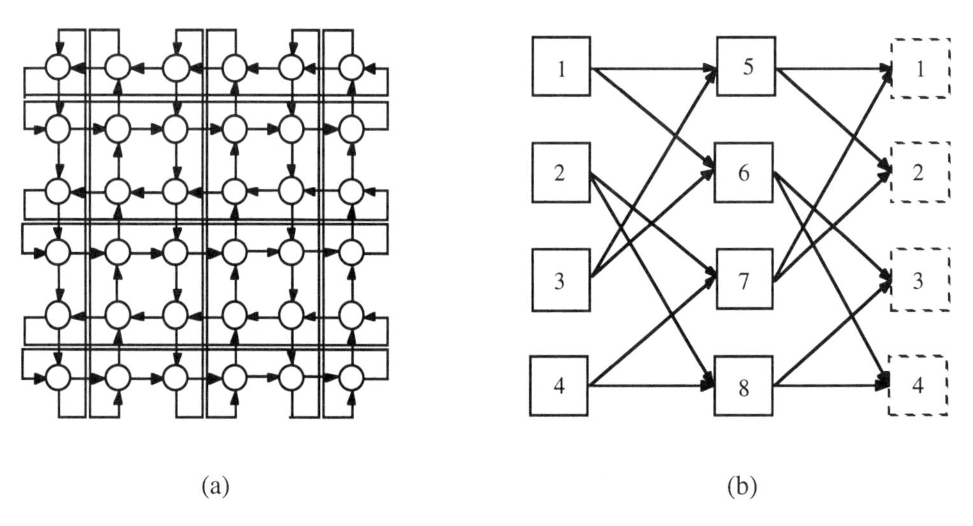


**Figure 1.2** Examples of multihop virtual topologies: (a) 6×6 Manhattan street network and (b) shufflenet with eight nodes.

The diameter of the graphs which represent either Manhattan street networks or shufflenets depends on the number of nodes they contain. In both cases it increases less than linearly with such number. Another important parameter in the description of multihop networks is the average number of hops necessary for routing a packet from its source to its destination with a uniform distribution of the packet traffic over the network. This parameter, which is calculated under the further assumption of an algorithm for routing packets that privileges the path with the lowest number of intermediate nodes to be crossed (shortest path), is called the average distance of the network.

A particular class of multihop networks, which clearly contains the above mentioned two types, is that represented by isotropic graphs, that is, by graphs where any node sees the same distribution of neighboring nodes around it. For the networks of this class, calculation of the diameter and the average distance is easier. Multihop networks will be described in Chapter 3, with particular attention to the two solutions considered here as examples, but also introducing other more involved virtual topologies.

When a network is used for interconnecting nodes which can be geographically grouped, considering also the traffic relations among them, it is often convenient to organize it on more than one level. For example, we can have a first level, or access level, consisting of several single-hop broadcast-and-select networks, able to interconnect nodes with strong traffic relations among them, and a second level in which a backbone network constitutes a sort of core structure for long distance transmissions. An architecture of this type can be easily iterated, leading to a hierarchical organization, whose operation is anyway managed by a global communication protocol. Chapter 4 is devoted to describing this kind of *multilevel networks*.

## 1.4 PRACTICAL LIMITS

There are some practical limits which must be taken into account in the realization of the physical topology of a WDMA network. They are mainly due to:

- Attenuation of the single-mode fiber;
- Attenuation due to passive and active components;
- Signal distortion produced by fiber chromatic dispersion;
- Undesired nonlinear effects of the optical fiber;
- Undesired nonlinear effects in EDFAs;
- Wavelength crosstalk.

Supposing that all the modulated wavelengths are in the neighborhood of 1.55 μm, the power attenuation introduced by a single-mode fiber can be evaluated on the basis of the rate of loss $A = 0.2$ dB/km [32]. To this figure it is necessary to add about 0.5 dB for any splice between two fiber pieces, and 1 dB for any demountable connector between a fiber piece and a device, passive or active, whose input or output port is a single-mode fiber section. Actually, these values can be lowered, by more recent technology, for example, to 0.2 and 0.5 dB, respectively. Anyway, this optimistic assumption will be used seldom in the following chapters (where some design examples will be given), thus preferring to use more standard and cautious solutions.

Passive components, as will be discussed in Section 1.5, except when operating wavelength selection, perform essentially the functions of combining or branching elements (sometimes also called splitters). In the first case losses are mainly due to the nonideal connectors. In the second case, in turn, the loss suffered by each wavelength can be accounted for taking 10 times the decimal logarithm of the number of output ports of the branching element plus an excess loss of 1 dB or more [33], due to intrinsic losses in the constitutive medium and, again, to nonideal coupling.

Wavelength selection in passive components introduces a loss that depends on the particular technique exploited for this optical filtering. Also wavelength-selective space switches can be studied, in regard to their attenuation, in an analogous way, although they are usually classified as active devices, because of the presence of electric or optical control provided by an external signal. To the above-mentioned optical filtering loss, we must here add an additional loss due to the waveguide able to vary its space disposition with respect to the component ports, under the action of the external signal. In Sections 1.5 and 1.6, we will discuss some methods for realizing optical filtering and waveguide spatial switch, and the related optical loss that is necessarily introduced.

Before considering other kinds of practical limits, we can fix typical values for the emission power $P_t$ of any modulated wavelength at the transmitter, and for the correspondent minimum received power $P_r$ in front of the photodiode (receiver sensitivity). Considering a laser source, we can assume $P_t = 0$ dBm. Taking about 1 Gbps as the bit rate of the signal that modulates each wavelength (this seems a quite realistic and even prudent value on the basis of the new trends in technology [34]), and setting at $10^{-8}$ the maximum permitted bit error rate, it is possible to write $P_r = -50$ dBm [8]. Thus, we have an overall attenuation of about 50 dB to distribute among fiber pieces and components. To understand how this figure can be managed in the framework of a design operation, usually called *power budget*, it is advisable to analyze two examples of different situations.

When a broadcast-and-select network has to be realized, a pure trade-off is made among the loss introduced by branching elements (including connectors), that due to fiber pieces and that originated by wavelength selection at the receiver, under the constraint that the sum of these loss contributions does not exceeds 50 dB. If this is impossible, we have to insert some EDFAs, in proper network sections. The number of EDFAs is to be minimized for reducing the overall cost, therefore different solutions for the physical topology, such as a star, a bus, a tree, or even more involved architectures [35], could be preferable for offering a lower number of sections where inserting as many EDFAs, in the same situation of nodes to be served.

With a wavelength-routing network, on the contrary, wavelength-selective space switches are typically employed. These active apparatus include one or more EDFAs in their own structure, so that the output optical signals exhibit power levels which are comparable with those of input optical signals. Here the number of wavelength-selective space switches is directly minimized, for reducing the overall cost, and the power budget appears less constrained.

Nevertheless, with both solutions, too many EDFAs must not be inserted along the same path from the transmitter laser source to the receiver photodiode, in order to avoid amplifier noise accumulation (a mechanism well known also in analog transmissions). Furthermore, nonlinear phenomena add their effects too, in a very involved fashion, as we will discuss more in detail later.

We can distinguish two kinds of single-mode optical fibers as regards their chromatic dispersion $D$: traditional fibers and dispersion-shifted fibers [8]. For the first kind we can assume $D$ on the order of 20 psec/(nm·km), for the second kind this parameter turns out to be an order of smaller magnitude. Such figures can be taken as independent of the disposition of the particular considered wavelength in the spectrum of the overall signal wave traveling the fiber, as long as this spectrum does not exceed the width of some tenths of nanometers.

To reduce the effects of chromatic dispersion, an inherently elegant solution has been recently proposed, which is called the "dispersion management technique" [36]. It consists in alternating long pieces of traditional fiber, having low positive dispersion, with short segments of dispersion compensating fiber, which has a very high negative dispersion. Choosing the lengths correctly, the total positive dispersion accumulated in the first section is almost completely negated by the accumulation of negative dispersion in the second section. At any point of the link there is always some dispersion, but the latter is essentially zero at the end, thus minimizing pulse spreading. At the same time, the fact of having nonnull dispersion along the link is favorable to the inhibition of nonlinear undesired mechanisms.

Nonlinear effects in fibers can be produced by four-wave mixing [37] and stimulated Raman scattering and Brillouin scattering [38]. If EDFAs are also characterized by an emission (output) power on the order of 0 dBm for each amplified wavelength, and the wavelengths are well separated among them (wavelength spacing not smaller than 1 nm), all the above nonlinear effects due to the fiber can be considered negligible [39–41]. A method has even been proposed for designing an optimum channel spacing, which allows one to minimize the degradation due to four-wave mixing [42]. The approach creates nonuniform channel separations for which no four-wave mixing product is superimposed on any of the transmitted channels.

Nonlinear effects in EDFAs are due to the saturation of the overall output emission power. Furthermore we may have the same nonlinear effects described above, because of the fiber structure on which such amplifiers are intrinsically based; their consequences are then negligible on the same conditions mentioned above. Gain saturation, instead, at the present state of the art [43], appears to have negligible effects (among them, wavelength crosstalk turns out to be the most important) only if the overall output emitted power does not exceed the level of about 15 dBm. This limit is assumed for medium-cost devices, and with a not too large number of EDFAs on the same fiber path, since their saturation effects are cumulative. Subdividing equally the above fixed maximum value for the overall emitted power among the wavelengths, each at the suggested output power level of about 0 dBm, results in a maximum number of wavelengths on the order of 30.

Wavelength crosstalk, besides that introduced by gain saturation in EDFAs, is due to wavelength-selective filters at the receivers or at the intermediate space switches. It can limit severely the size of a WDMA network [44], especially when the number of wavelengths is high. Also the costs of these apparatus increase remarkably when increasing this number. From these last considerations, it is probably necessary to reduce the limit for the overall number of wavelengths on the same fiber, previously fixed at about 30. A more realistic value, to allow practical implementations, can be taken as not greater than half the previous value. Since wavelength collection and selection are usually performed on a binary base, the number of 16 wavelengths on the same fiber appears at present as the maximum achievable one even for very complex networks. When the number of wavelengths along the same physical path between two nodes, or between a node and an intermediate apparatus, exceeds this limit, it is however possible to employ more than one fiber along this link, thus using WDMA in conjunction with SDMA.

Sometimes, the passage across EDFAs, wavelength-selective apparatuses, and the fiber pieces themselves give rise to the possibility that different attenuations are suffered by the modulated wavelengths. Even several dBs of penalty may occur between the most attenuated wavelength and the less attenuated one. This fact can be compensated by the introduction of suitable power equalizers, often consisting of passive devices with a well-designed spectral response, in proper sections of the path where a certain ensemble of wavelengths travel together.

## 1.5 PASSIVE COMPONENTS

Among the passive components typically employed in a WDMA network [45], we will briefly describe the following:

- Combiners;
- Branches;
- Directional couplers;
- Wavelength-selective combiners, branches, and directional couplers.

Except for the last category, they are intended as not wavelength-selective, that is characterized by an almost flat response over the whole spectrum of the signal wave.

All these components are preferably constructed by fibers, in order to avoid the loss due to waveguide geometry change at the input and output ports characterizing planar waveguides when interfacing the transmission fibers. Then the components are constituted by a suitable arrangement of single-mode fibers, and the extent of the mode coupling is controlled through the level and features of the overlapping among the fibers. The usage of planar waveguides (with the so-called integrated-optic technology [32]) seems instead recommendable when large-scale production is foreseen, because of the dramatic cost reduction it affords. Furthermore, resorting to lithographic processes sometimes offers additional degrees of freedom to optimize the design.

A combiner exhibits $N$ input ports and one output port. Although some decibels of excess loss should be sometimes budgeted, depending on the specific technological realization, we can say that the optical powers collected from the input ports are almost completely transferred to the output port, independently of the wavelengths modulated on each wave. In this sense, it is necessary to pay attention that the same wavelength does not enter two input ports, at least during the same time interval.

A branch operates with the opposite purpose. It is characterized by one input port and $N$ output ports, among which the optical incoming power is subdivided. Generally such subdivision is uniform, so that the theoretical loss introduced by this device, from its only input port to any output port, is $10 \cdot \log_{10} N$. To such minimum loss, an additional loss of at least 1 dB, due to imperfections of the structure, must typically be added.

A directional coupler has two input ports and two output ports [46]. Generally it exhibits symmetry, either between the two ports on the same side or with respect to an input-output change. In this case, the optical power entering one of its ports is uniformly divided between the two ports on the opposite side. This function is sometimes called *wave splitting*. In particular situations, an unequal separation between the two output ports is required. Considering the power transmission coefficients $\tau_1$ and $\tau_2$ from one input port toward the two output ports on the opposite side, in theory the following equality should hold: $\tau_1 + \tau_2 = 1$; in fact no power must be reflected to the other port on the same side, and no power must be lost within the component. In practice, to the theoretical loss $10 \cdot \log_{10}(1/\tau_1)$ or $10 \cdot \log_{10}(1/\tau_2)$ introduced by this component, it is necessary to add about 1 dB for nonideal behavior in both the above-mentioned requested properties.

Generalization of the concept expressed by a directional coupler is that of a star coupler, where $N$ input ports feed as many output ports. Here the power subdivision among the latter ones is practically always uniform, so that this component can be interpreted as the union of a $N$-ary combiner and a $N$-ary branch. Sometimes the number of input ports and of output ports are different.

In a directional coupler, or in a star coupler, let us focus our attention on the segment of common waveguide between the section where the incoming fiber cores overlap and the section where the outgoing fiber cores individually emerge. Single-mode propagation occurs in such fiber cores, and, in order to limit additional radiation loss, few-mode propagation occurs even along the common waveguide. This mechanism is intrinsically wavelength-selective, and particular care must be devoted to obtain a flat response over the whole spectrum occupied by the various modulated wavelengths.

Wavelength-selective components (combiners, branches, but especially couplers) can intentionally exploit the above mechanism [47]. This principle of operation is efficient when the overall number of different wavelengths is low (at most four). For a higher number of wavelengths, it is necessary to introduce, in the central part of the component, more proper wavelength-selective elements, such as prisms or gratings [48]. In this case, additional loss increases remarkably. Sometimes a single-wavelength filter is to be inserted along a fiber path. It consists of a component, with one input port and one output port, able to select just one wavelength and to eliminate all the other wavelengths. This function is generally performed by including, between two fiber pieces, a prism or a grating, in a proper spatial position with respect to the two fiber cores. In this way, the angular separation among the wavelengths so introduced allows coupling with the output fiber core of the only desired wavelength. Prism-based devices typically exhibit low angles of spatial dispersion, thus resulting in large and rather difficult to produce components. Grating-based devices are therefore preferable (and many examples of commercial solutions recently appeared on the market [49]) though they are affected by the problem of the reflection loss of the grating (> 6 dB), which is extremely sensitive to the steepness of the reflecting sidewalls. Such a problem, however, can be solved by applying an optical phased array as the focusing and dispersive element [50].

In other cases, when the required wavelength selectivity is higher, it is possible to use resonant filters, consisting of proper assembling of dychroic mirrors, Fabry-Perot etalons, or Mach-Zehnder interferometers [32]. Such components are now available in almost all types of micro-optic apparatus.

Fixed filters of the type described above can also be inserted in front of the receiver photodiode to detect the wavelength of interest. Tunable filters can be employed too, and will be described in Section 1.6, being considered as active devices.

Proper combination of more than one passive wavelength-selective component can form a multiport structure where wavelengths are properly routed from one input port to a precise output port. Yet there is a difference between the case of a flat-response component combination (like that considered for a star coupler) and the present combination of wavelength-selective elements. Here it is necessary to organize such elements in subsequent sections, and also the additional losses cumulate section by section.

A very important parameter for evaluating wavelength crosstalk is the extinction ratio that characterizes a wavelength-selective component. For a single-wavelength filter, it consists of the ratio between the power of the desired wavelength and the highest power among those associated to the undesired wavelengths, both measured at output, under a uniform excitation of all the wavelengths at input. For a more involved component, this definition can be extended to any of its output ports, having prefixed a reference situation for wavelength excitation at all its input ports. Typical values of such extinction ratio, in components of medium cost, are on the order of – 45 dB.

When an external signal controls the operations of a combiner, a branch, a coupler, and even a wavelength-selective element, it is preferable to classify the resulting component as an active device. This choice is the one assumed here; therefore such elements, where the optical signal is in some way processed, will be treated in the next section, together with the devices in which the optical signal is generated.

## 1.6 ACTIVE DEVICES AND APPARATUSES

As stressed above, under the definition of active devices we collect all the components where the signal is generated or processed. An active apparatus is, instead, an ensemble of components, passive and active, able to make a certain function.

Light generation at the source node is provided by semiconductor lasers. They can emit a fixed wavelength or be tunable. Tuning, in turn, can be achieved by means of different mechanisms: thermal, mechanical, acousto-optic, and by varying the injection current [48]. Thermal tuning exploits the temperature-dependence of the emitted wavelength in a distributed feedback (DFB) laser. It exhibits a limited tuning range (few nanometers) and is rather slow (some milliseconds). Inserting a laser in an external cavity, and varying mechanically the dimension of such cavity, can provide a wider tuning range, but even a longer response time. It is possible to insert in the cavity, besides the laser, also an acousto-optic tunable filter. This allows one to reduce the response time up to the order of microseconds. Finally the highest tuning speed is reached acting directly on the injection current of the laser. DFB or distributed Bragg reflector (DBR) lasers, having more than one section, with as many injection currents to be adjusted, offer the best performance. In this case, tuning range easily exceeds 10 nm. In order to have laser action at the operation wavelength of about 1.55 μm, it is necessary to construct laser devices with InGaAsP. This is a compound already well established in point-to-point optical transmissions.

Among the active components where the signal is processed, we can mention:

- Tunable filters;

- Optical switches;
- Externally controlled directional couplers.

A passive wavelength-selecting component can be made tunable by varying some of its mechanical parts [51]. So a resonant filter can be tuned acting on the spatial position of a mirror or on the overall length of its etalon. Tuning is faster when acousto-optic or electro-optic effect is exploited.

Optical switches are space matrices, with $N_{in}$ input and $N_{out}$ output ports, able to route the signal entering any input port toward any output port [52]. This function is to be performed independently of the spectral composition of the signal, that is, with a flat overall spectral response. Optical waveguides intersect one another to form a series of branches. An external control signal acts on each branch, varying refractive index or waveguide size, so realizing a two-state elementary switch. The loss introduced on the signal, from the input to the output port, usually increases logarithmically with $N_{out}$, but $N_{in}$ also contributes to the overall loss. Various mechanisms can be exploited by the control signal, which is usually an electric current. Among them, acousto-optic and electro-optic effects are the most commonly used, especially when high switch speed is requested. Recently, also some photonic devices, in which the control signal is also optical, have been proposed [53]. Their main feature is the possibility of self-routing of the signal, by adding proper data on a wavelength reserved to this function.

Let us consider now a directional coupler, in which the power transmission coefficients $\tau_1$ and $\tau_2$ can be varied, under the action of an external control signal. In some network schemes, it is necessary to control such parameters (the one complementary to the other) in an almost continuous way, between 0 and 1. This function can be ensured, for example, by constructing the directional coupler in titanium-diffused lithium niobate, and by exploiting the electro-optic effect [54]. In this way a linear control can be exerted, and any component realized by proper assembling of a certain number of such devices, with even more than two input and output ports, is called *linear combiner-divider* (LCD).

Coming to the description of active apparatus, we can focus our attention on two of them, both wavelength-selective versions of devices already examined: the *wavelength-selective spatial switch* (WSSS), and the *wavelength-selective linear combiner-divider* (WSLCD). They are of basic importance for the practical realization of wavelength-routing single-hop networks, as will be discussed in Chapter 2.

Branches and tunable filters, the latter ones properly controlled by external signals, can be introduced in front of the input ports of a set of optical switches, according to a scheme of the type depicted in Figure 1.3. The combiners, after the output ports, are employed for reconstructing a WDM signal to be sent to its destination. This example refers to the case with four wavelengths on each fiber, and with three input fibers as well as three output fibers. An EDFA for each output fiber is inserted, to compensate the overall loss suffered by the signals in crossing such apparatus. Complete accessibility [55] needs a number of switches equal to that of the wavelengths in each fiber, and, for each switch, a number of tunable filters equal to the number of inbound fibers. The scheme here described corresponds to a WSSS of relatively small capacity. One can easily imagine how complex a WSSS would appear with 16 wavelengths on each fiber, and with more than three fibers.

In WSLCDs, optical switches are replaced by variable couplers. They generally act on wavebands comprehending more adjacent modulated wavelengths, being able to ensure a set of transmission coefficients, whose values depend on the considered waveband. In this sense, it could be more appropriate to speak of waveband selective linear combiners-dividers. A schematic example of possible structure for a WSLCD with three inputs and three outputs, operating on two wavebands, is shown in Figure 1.4 [54].

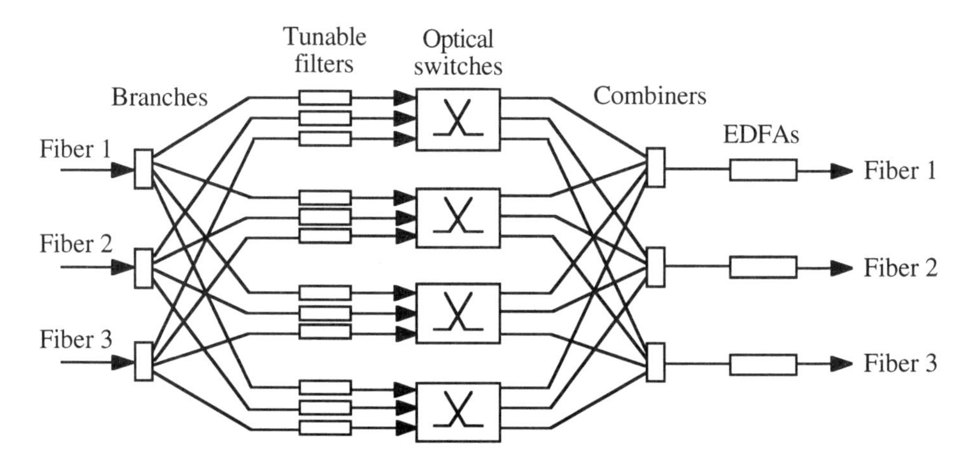


**Figure 1.3** Examples of WSSS with three input fibers, three output fibers, and four wavelengths for each fiber.

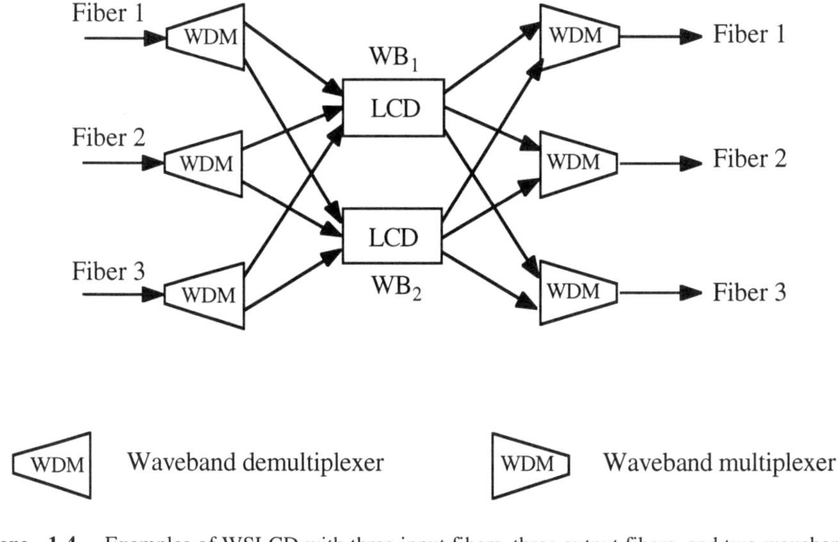


**Figure 1.4** Examples of WSLCD with three input fibers, three output fibers, and two wavebands for each fiber.

Signals arriving at each input fiber are separated by a waveband demultiplexer (e.g., a diffraction grating [56]) in such a way that wavebands $WB_1$ and $WB_2$ can emerge on two separate set of lines and can be routed separately by the two LCDs. The outputs so obtained are then sent to a set of waveband multiplexers, which combine them on the output fibers.

A particular apparatus, to be taken into account for practical realization of multihop networks and, more generally, for increasing the flexibility and the capacity of the network for a fixed set of wavelengths, is the wavelength converter. It transforms an input signal centered at $\lambda_i$ into an output signal centered at $\lambda_j$, with $j \neq i$. The simplest way to achieve wavelength conversion consists in using the received signal to modulate a new tunable source that emits at a different wavelength [57]. Therefore, in this approach, opto-electric and electro-optic conversions are necessary. All optical solutions seem, therefore, more desirable; they can be based on the use of suitably driven optical phase modulators, or can exploit nonlinear processes. Among the most interesting proposals, in this sense, we can mention the adoption of multielectrode DFB or DBR lasers in combination with saturable absorbers [58] or the use of four-wave mixing in semiconductor optical amplifiers [59].

## 1.7 EDFAs

An optical amplifier is usually constructed by properly doping quartz with erbium ions. A single-mode fiber is made from this doped quartz, in which two optical waves simultaneously propagate: the signal that is to be amplified and a pump signal able to provide the necessary power. Amplification is distributed along the length of such fiber, which can vary from a few meters to some hundred meters. EDFAs have been used extensively for either coherently modulated signals or intensity modulated signals since 1990. In a short time, they became the best solution in point-to-point connections, without traffic exchange over distances higher than that of the regeneration spacing in high-capacity digital transmission systems (on the order of 200 km). Here their use is devoted to compensating for the loss introduced by the transmission medium. In an optical network, where the power of a signal is subdivided in a branch, so that, from a single path, different paths are formed, the use of EDFAs is mainly devoted to compensate for splitting loss [19]. In this application, the EDFA's spectral response is to be properly widened, in such a way as to provide a nominally equal amplification to all the modulated wavelengths.

The total power gain, expressed in decibels, that an EDFA is able to produce on the signal traveling it, can be written as

$$G(\lambda) = \int_0^{L_z} \left[\sigma_e(\lambda) n_2(z) - \sigma_a(\lambda) n_1(z)\right] dz \tag{1.1}$$

where $\sigma_e(\lambda)$ and $\sigma_a(\lambda)$ are the emission and absorption cross sections of erbium ions; $n_1(z)$ and $n_2(z)$ are, respectively, the lower- and upper-state population densities at section $z$ along the active fiber; and $L_z$ is its overall length [60]. The wavelength dependence of $\sigma_e(\lambda)$ and $\sigma_a(\lambda)$ is due to the spectroscopic properties of erbium ions in silicate glass. It must be flattened over the whole spectral range occupied by the modulated wavelengths, acting on $n_1(z)$ and $n_2(z)$, which means on the pump signal, the ion concentration, and the length $L_z$. A further degree of freedom consists in the possibility of having pump and signal wave propagating in the same or in the opposite direction. The first solution is called a co-propagating pump wave, the second a counterpropagating pump wave.

Although a very accurate spectral response flattening is ensured by a properly designed single amplifier, the residual differential gain among the wavelengths grows with the number of amplifiers crossed by the signal during its path from source to destination. For example, setting at 1 dB the maximum value of differential gain, and assuming 10 dB as the maximum interval in which a power equalizer can act in order to compensate variations in the levels of the modulated wavelengths at the same section, we have to insert an equalizer not more than any 10 amplifiers. To reduce power level frustration, it is advisable to realize such an equalizer by means of an active device, which turns out to be itself an EDFA [61] with a properly designed spectral response.

Clearly the extension of the nominal spectral range over which EDFAs must operate is a parameter that acts against an easy spectral response flattening in each EDFA, as well as against simple realization of active equalizers. Nevertheless, 3-dB bandwidths that exceed 30 nm are achievable without particular effort [62]. Other features of the amplifiers, such as the overall gain, output saturation power, and noise figure, however, should be taken into account when realizing the flattening of the spectral response [63].

The description of the noise that influences the output power of an EDFA can be very involved [64]. Here we limit ourselves to considering the effects of its main contribution, called amplified spontaneous emission (ASE). It exhibits a continuous power spectrum, and hence the effects produced can be easily accounted for by introducing a properly defined noise figure, similarly to what is usually done for electronic amplifiers with respect to an equivalent thermal noise added by the amplifier itself. ASE noise is independent of the modulated signal, at least when the amplifier output power has intermediate values (between −40 and −10 dBm). In this condition the noise figure $F$ results in a value a little higher than the theoretical minimum, which is fixed at 3 dB. Out of the above range of output power values, $F$ increases. In particular, we are interested in the increase toward the higher values of the output power $P_{out}$. The following rough but practical formula can be assumed between $F$ and $P_{out}$, the first expressed in decibel (dB), the second in dBm—we remind that $P_{out}[\text{dBm}] = 10 \cdot \log_{10} P_{out}[\text{mW}]$—in the case of pump at 980 nm (see below for the discussion of the best choice for the pump wavelength)

$$F \cong 3 + 0.15(P_{out} + 10) \qquad P_{out} \geq -10 \text{ dBm} \tag{1.2}$$

As stressed above, the noise expressed by this figure accumulates when more than one amplifier is inserted along a path from source to destination, therefore a trade-off is to be reached between a lower value for $F$, to reduce directly noise power, and a higher value for $P_{out}$, to increase the length of the repeater section.

Particularly in the case of wavelength-routing networks, the ASE noise generated in the EDFAs can form recirculating loops that lead to amplifier saturation and oscillations

[65]. This could obviously be a serious problem, which must be solved by identifying and eliminating the closed cycles from the network.

Besides the above linear noise effects, there are several causes of nonlinear behavior in an EDFA; they begin in the active fiber, because of the high power levels involved, and continue in the following passive fiber, thus producing additional signal deterioration if not adequately controlled. Among these causes we can list: stimulated Brillouin scattering and Raman scattering, self-phase modulation, cross-phase modulation, four-wave mixing, and soliton formation [37]. The dominance of one over the other, among such causes, depends on different aspects of the transmission, for example, transmission rate, spectral distance between adjacent modulated wavelengths, intensity modulation index. Nevertheless, by increasing the output power level, all these causes produce effects that grow dramatically and interact among themselves in a very complicated fashion. This sometimes suggests fixing a maximum output power $P_{out}$ remarkably smaller than the value that gain saturation should suggest to adopt to avoid traditional intermodulation due to nonlinearity in the input-output characteristic.

Typical values for the most important performance parameters in EDFAs, which are not too sophisticated and hence expensive, are as follows: power gain $G$ from 25 to 35 dB, noise figure of about 6 dB, maximum output power per modulated wavelength $P_{out}/W$ from – 5 dBm to + 5 dBm, nominal width of the spectral response on the order of 30 nm. Another important aspect regards the wavelength and the power level required for the pump wave. The most frequently employed pump wavelength, for EDFAs operating on signal waves at 1.55 μm, is 0.98 μm. At this wavelength, GaAs lasers, with active region made of InGaAs, appear reliable and rather inexpensive. Alternatively it is possible to take a pump wavelength of 1.48 μm, obtained with InGaAsP, the same compound as that employed for constructing laser sources for the signal wave. Apart from material reliability and cost, the choice of the 0.98- or 1.48-μm pump band is also dictated by the need to avoid excited-state absorption (ESA) [66], which is an important limiting factor for the amplification efficiency at different pump bands (e.g., 0.514 or 0.8 μm). In practice, about 7 dB less pump power is required in the 0.98- and 1.48-μm band with respect to the 0.8-μm band, at a parity of gain [62]. In both cases, the pump wave is a continuous wave, whose required power is on the order of 20 dBm for InGaAsP devices, but it can reach 35 dBm for GaAs-InGaAs devices.

The cost of installed EDFAs depends also on the choice of the sections where they are to be inserted. For example, the power feed for generating pump wave could represent a cost for the system, besides a higher probability of failure, if this section is far from the network nodes. A general consideration to minimize costs, as already stressed, is that the network design be oriented to insert EDFAs where a higher number of modulated wavelengths travel the fiber.

## 1.8 TRAFFIC ASPECTS

In general, in wavelength-routing single-hop networks or in multihop networks, several paths exist that can be used to connect any given pair of nodes through the virtual topology. In principle, all of them are suitable for packet delivery, since their ends always coincide in the source and destination nodes, respectively. In practice, instead, the best path is chosen according to a criterion of cost minimization. The cost of a link is not defined univocally [67–71]. One can identify the link cost with its physical length, with the average queueing delay or the packet loss probability experienced on it, with its capacity, and so on. However, once one has chosen the parameter to minimize, the availability of multiple end-to-end paths is exploited to improve the cost/performance ratio. In fact, if two alternative connections can be utilized (two paths are distinct if at least one of their edges does not belong to both the connections) the one characterized by the minimum cumulative cost is chosen. This choice can be made locally by the nodes crossed by traveling packets, or by a centralized supervisor which suitably organizes the network to support the considered transmission. In any case, whichever is the adopted strategy, the latter has to be transferred to all the nodes belonging to the chosen path, in order to make them able to handle the incoming data properly. In fact, when focusing our attention to a single intermediate node, where at least two outputs can be alternatively used to move forward the received packets toward their final destinations, one of them has to be always selected according to the above mentioned strategy that, operatively, organizes and updates the routing tables located at every node. This is the goal of a routing technique, which is not necessary in broadcast-and-select single-hop networks, since they always provide an edge to directly connect any given pair of nodes.

A routing technique is in general based on an algorithm which calculates and selects the best path [72,73]. Such algorithm should be characterized by the following properties that, unfortunately, are not always compatible [74]:

- *Robustness*: with respect to node or link malfunctions and to the dynamic of traffic conditions;
- *Stability*: with respect to short-term variations of network parameters;
- *Fairness*: to equalize the distribution of packet flows in the network and to maintain a uniform allocation of the system resources;
- *Optimality*: to satisfy the design constraints, maximize the network efficiency, and minimize the network costs.

In Section 2.4 and in Chapter 3 many routing schemes will be presented, discussing their characteristics, performance and limits, verified when they are applied in multihop communication systems or in multihop subnetworks included in networks of networks, which are studied in Chapter 4. For the time being, we wish to state that some classification of routing schemes can be done.

First of all, we can group the routing strategies in two classes, based respectively on static routing algorithms and on dynamic, or adaptive, routing algorithms. The static algorithms define the paths on the basis of virtual topology characteristics (number of hops), network extension (physical length of the paths) or a priori estimated traffic conditions (using predictive subalgorithms [68]); adaptive algorithms, on the contrary, since they monitor the network conditions continuously, can modify the routing choices to follow the dynamic of the traffic flows. They have been conceived in many different versions, depending on the structure of the flow control system. In fact, the latter can be centralized in a routing control center (RCC), where all the information concerning the operative conditions of the network is collected, or can be distributed among nodes, where routing schemes are autonomously handled, either using local information concerning adjacent links and nodes, or integrating local information with global information pertaining to the whole network. The limiting case of this class is represented by random, or quasi-random, routing schemes, where any supervisor does not control the packet propagation. Some examples of these different approaches will be given in Chapter 3.

Routing algorithms can be also discriminated according to their objectives; in fact they can be used to manage point-to-point, multicast or broadcast communications. As regards, in particular, the first case, when we have the typical necessity of connecting pairs of remote users, we can distinguish the connection-oriented mode, using virtual circuits, or the connectionless (datagram) mode [75].

In virtual-circuit mode, the network makes one path selection for all the packets of the same communication session and the routing tables are filled by numbers that identify the output link that a node has to use when a packet belonging to a given connection arrives. This selection can sometimes be changed, because of the variation of network conditions; however, in a stationary condition, it marks a fixed sequence of nodes to be crossed by the considered packet flow, which usually represents the shortest path between source and destination. Alternatively, when datagram mode is applied, the network selects a path for each packet individually, so that the routing tables indicate alternate outgoing links, that can be selected randomly, according to a periodic sequence, or by considering their costs in real time at the instant of packet emission.

Virtual circuit mode provides a flow of packets that respects the packet emission sequence whereas datagram mode does not, but the latter is not seriously affected by link failures or topology rearrangement, which may instead affect the virtual circuit.

Multicast and broadcast communications belong to the family of multipoint routing algorithms [69,76], since they concern the connection of a source with many or all the nodes of the networks, respectively. Related techniques have been widely studied [77–82]. In particular, multicasting has been analyzed in the context of single-hop optical networks in [83,84], whereas in [85,86] its application in multihop networks has been focused.

For broadcast connections, usually activated to realize the diffusion of control information, we can adopt the flooding [75] or the spanning tree routing algorithms [74]. The former forces every node to create and emit a copy of each incoming broadcast packet for all its outputs, except that used to receive the packet itself. The latter, on the contrary, organizes the packet propagation along the branches of a tree that partially overlaps the network. Flooding is simple and useful when the emitting node does not know the exact network topology, but it is not efficient because of the production of a high number of copies. In principle, it can be applied, with very high wasting of bandwidth but very high robustness also, to route end-to-end connections. Spanning tree mode is widely considered for either broadcast or multicast communications [67–69], limiting the number of packet transmissions. On the other hand, to identify the best spanning tree, a continuous updating of the databases is requested, containing the information of the network conditions (topology changes, malfunctions, link loads, etc.), that implies a significant amount of bandwidth spent for signaling. Furthermore, a powerful and fast computing device should be available for tree tracing.

The adoption of the so-called depth-first search (DFS) algorithm [87] has been recently proposed [88–91] that represents an interesting example of the simplified multicast routing scheme, since it transforms multipoint connections in point-to-point ones, whose virtual paths can vary dynamically during the communication sessions, depending on the traffic conditions in the network. From a given source, DFS draws a single path on the network, which passes through all the addressed nodes without bifurcation. In this way, the DFS algorithm provides a continuous connection of a set of nodes, through a sort of virtual circuit, realized inside the backbone network, instead of using spanning tree structures. This virtual circuit is activated by the first packets, emitted by source stations at the beginning of every new multipoint communication session, which explore the network without having any information about its global topology. Furthermore, this method, applied in datagram mode, permits one to avoid the traffic congestion present in heavy loaded links and maintains good performance in case of trunk failures and network reconfigurations. Finally, it does not require particular hardware architecture in the switching nodes, such as copy sections, since the stations can manage multipoint connections like point-to-point ones.

Finally, we have to consider that, whichever algorithm is adopted, when the network extension increases, the management of routing problems becomes difficult and expensive. In this case, partitioning the network into several interconnected subdomains is highly recommended, in which hierarchical routing [74] can be applied. Inside a single subnetwork, which can be seen as a cluster of nodes, it operates like one of the previously presented schemes; nevertheless, it is able to control also intercluster communications in a multilevel architecture. This subject is extensively discussed in Chapter 4.

At the end of this section, it could be of some interest to present an overview of the environment in which routing schemes are utilized and of some aspects of their practical application. The traffic conditions characterizing a network, composed by $N$ nodes, can be analytically represented by means of a $N \times N$ traffic intensity matrix $\boldsymbol{T}$, where the $T_{ij}$ element is equal to the average traffic produced by all the virtual connections originated at node $i$ and destined to node $j$ [92]. The network topology can be fitted and optimized to match the constraints indicated by means of the traffic matrix, and therefore allows one to obtain the highest quality-of-service level at the minimum possible cost. Nevertheless, traffic loads can be predicted through statistic models, but they are in any case random variables that can sometimes assume extreme values, very far from their average levels. For this reason, the network resources should be overdimensioned to guarantee the possibility of facing load peaks as well.

Starting from a given condition, represented by matrix $\boldsymbol{T}^{(\mathbf{a})}$, let us consider what happens when a new connection request is presented by one of the users. The RCC, when available, tries to find the best routing choice, which permits one either to find the network resources sufficient to allocate the new call, or to identify the best path for it, according to a prefixed quality criterion [93,94]. Since every choice corresponds to a new traffic matrix $\boldsymbol{T}^{(\mathbf{b})}$, the first goal is reached when $\boldsymbol{T}^{(\mathbf{b})}$ implies average link loads that are never higher than link capacities. If a routing algorithm provides several alternative paths that satisfy this condition, the new connection request is accepted and allocated on the best path. If no path respects the same condition, the access is denied.

Alternatively, especially when the number of refused connections increases, network rearrangement should become convenient. In this case, the network manager has to calculate the connectivity and traffic matrices and the new set of routes, that minimize the largest flow on any link, assuming the presence of the new connections that are waiting to be accepted.

Then, before modifying the network structure, it is necessary to evaluate the new traffic matrix, achievable by accommodating the new connection requests, and verify if its pivot value is now lower than the link capacity. If so, the network can be reconfigured and the new calls accepted; if not, the network should not be changed, since no modification can increase the network throughput.

## REFERENCES


[1] Kao, K. C., and G. A. Hockham, "Dielectric-fibre surface waveguides for optical frequencies," *Proc. IEE*, Vol. 113, 1966, pp. 1151–1158.

[2] French, W. G., J. P. MacChesney, P. B. O'Connor, and G. W. Tasker, "Optical waveguides with very low losses," *Bell Sys. Tech. J.*, Vol. 53, May 1974, pp. 951–954.

[3] Horiguchi, M., and H. Osanai, "Spectral losses of low-OH-content optical fibres," *Electron. Lett.*, Vol. 12, 1976, pp. 310–312.

[4] Miya, T., Y. Terunuma, T. Hosaka, and T. Miyashita, "Ultimate low-loss single-mode fibre at 1.55 μm," *Electron. Lett.*, Vol. 15, 1979, pp. 106–108.

[5] Hall, R. N., G. E. Fenner, J. D. Kingsley, T. J. Soltys, and R. O. Carlson, "Coherent light emission from GaAs junctions," *Phys. Rev. Lett.*, Vol. 9, 1962, pp. 366–368.

[6] Nathan, M. I., W. P. Dumke, G. Burns, F. H. Dill, and G. J. Lasher, "Stimulated emission of radiation from GaAs p-n junctions," *J. Appl. Phys.*, Vol. 1, 1962, pp. 62–64.

[7] Quist, T. M., R. H. Rediker, R. J. Keyes, W. E. Krag, B. Lax, A. L. McWhorter, and H. J. Zeiger, "Semiconductor maser of GaAs," *Appl. Phys. Lett.*, Vol. 9, 1962, pp. 91–92.

[8] Cancellieri, G., *Single-Mode Optical Fibres*, Oxford, England, Pergamon Press, 1991.

[9] Gloge, D., and E. A. J. Marcatili, "Multimode theory of graded-core fibers," *Bell Sys. Tech. J.*, Vol. 52, 1973, pp. 1563–1578.

[10] Marcuse, D., "Pulse distortion in single-mode fibers," *Appl. Opt.*, Vol. 19, 1980, pp. 1653–1660.

[11] Ohtsu, M., *Highly Coherent Semiconductor Lasers*, Norwood, MA, Artech House, 1992.

[12] Nassau, K., "The material dispersion zero in infrared optical waveguide materials," *Bell Sys. Tech. J.*, Vol. 60, 1981, pp. 327–337.

[13] Okoshi, T., and K. Kikuchi, "Heterodyne-type optical fiber communication," *J. Opt. Commun.*, Vol. 2, 1981, pp. 82–88.

[14] Tomlison, W. J., and G. D. Aumiller, "Optical multiplexer for multimode fiber transmission systems," *Appl. Phys. Lett.*, Vol. 30, 1977, pp. 169–171.

[15] Taga, H., S. Yamamoto, M. Mochizuki, and H. Wakabayashi, "2.4 Gb/s 1.55 μm WDM bidirectional optical fiber transmission experiment," *Trans. IEICE*, Vol. E71, No. 10, 1988, pp. 940–942.

[16] Personick, S. D., "Baseband linearity and equalization in fiber optic digital communication systems," *Bell Sys. Tech. J.*, Vol. 52, 1973, pp. 1175–1194.

[17] Miyamoto, Y., T. Kataoka, A. Sano, K. Hagimoto, K. Aida, and Y. Kobayashi, "10 Gbit/s, 280 km nonrepeatered transmission with suppression of modulation instability," *Electron. Lett.*, Vol. 30, No. 10, May 1994, pp. 797–798.

[18] Black, P. W., "Undersea system design constraints," *Proc. SPIE (Fibre Optics '90)*, 1990, Vol. 1314, pp. 112–115.

[19] Giles, C. R., M. Newhouse, J. Wright, and K. Hagimoto, (eds.), "Special issue on system and network application of optical amplifiers," *IEEE/OSA J. Lightwave Technol.*, Vol. 13, No. 5, May 1995, pp. 701–981.

[20] Nakagawa, K., S. Nishi, K. Aida, and E. Yoneda, "Trunk and distribution network application of erbium-doped fiber amplifier," *IEEE/OSA J. Lightwave Technol.*, Vol. 9, No. 2, Feb. 1991, pp. 198–208.

[21] Nasseki, M. M., F. A. Tobagi, and M. E. Marhic, "Fiber optic configurations for local area networks," *IEEE J. Select. Areas Commun.*, Vol. 3, No. 6, 1985, pp. 941–949.

[22] Miki, T., "The potential of photonic networks," *IEEE Commun. Mag.*, Vol. 32, No. 12, Dec. 1994, pp. 23–27.

[23] Warzanskyj, W., and U. Ferrero, "Access network evolution in Europe: A view from EURESCOM," *Proc. ECOC '94*, Florence, Italy, Sept. 1994, pp. 135–142.

[24] Sano, K., and I. Kobayashi, "Access network evolution in Japan," *Proc. ECOC '94*, Florence, Italy, Sept. 1994, pp. 143–149.

[25] Shumate, P. W., "Access network evolution in the United States: Economics and operations drivers," *Proc. ECOC '94*, Florence, Italy, Sept. 1994, pp. 151–158.

[26] van As, H. R., "Photonic terabit/s networks and their key components," *Proc. EFOC&N '93*, The Hague, The Netherlands, June 30- July 2, 1993, pp. 13–20.

[27] Green, P. E., Jr., *Fiber Optics Networks*, Englewood Cliffs, NJ, Prentice Hall, 1993.

[28] Mestdagh, D. J. G., *Fundamentals of Multiaccess Optical Fiber Networks*, Norwood, MA, Artech House, 1995.

[29] de Prycker, M., *Asynchronous Transfer Mode: Solution for Broadband ISDN*, Chichester, England, Hellis Horwood, 1991.

[30] Tucker, R. S., "Optical time-division multiplexing for very high bit-rate transmission," *IEEE/OSA J. Lightwave Technol.*, Vol. 6, No. 11, Nov. 1988, pp. 1737–1749.

[31] Salehi, J. A., "Code division multiple-access techniques in optical fiber networks — Part I: Fundamental principles," *IEEE Trans. Commun.*, Vol. 37, No. 8, 1989, pp. 824–833.

[32] Cancellieri, G., editor, *Single-Mode Optical Fiber Measurement: Characterization and Sensing*, Norwood, MA, Artech House, 1993.

[33] Senior, J. M., *Optical Fiber Communications - Principles and Practice*, 2nd ed., Cambridge, England, Prentice Hall, 1992.

[34] Lee, T.-P., et al., "Multiwavelength DFB laser array transmitters for ONTC reconfigurable optical network testbed," *IEEE/OSA J. Lightwave Technol.*, Vol. 14, No. 6, June 1996, pp. 967–976.

[35] van As, H. R., "Media access techniques: The evolution towards terabit/s LANs and MANs," *Computer Networks and ISDN Systems*, Vol. 26, March 1994, pp. 603–656.

[36] Willner, A. E., "Mining the optical bandwidth for a terabit per second," *IEEE Spectrum*, Vol. 34, No. 4, April 1997, pp. 32–41.

[37] Agrawal, G. P., *Nonlinear Fibre Optics*, San Diego, CA, Academic Press, 1989.

[38] Agrawal, G. P., *Fiber-Optic Communication Systems*, New York, Wiley, 1992.

[39] Chraplyvy, A. R., "Limitations on lightwave communications imposed by optical-fiber nonlinearities," *IEEE/OSA J. Lightwave Technol.*, Vol. 8, No. 10, Oct. 1990, pp. 1548–1557.

[40] Yu, A., and M. J. O'Mahony "Optimisation of wavelength spacing in a WDM transmission system in the presence of fibre nonlinearities," *IEE Proc.-Optoelectron.*, Vol. 142, No. 4, Aug. 1995, pp. 190–196.

[41] Di Pasquale, F., P. Bayvel, and J. E. Midwinter, "Performance limits in amplified dense WDM networks in the presence of four-wave mixing and gain peaking," *Electron. Lett.*, Vol. 31, No. 12, June 1995, pp. 998–1000.

[42] Forghieri, F., R. W. Tkach, and A. R. Chraplyvy, "WDM systems with unequally spaced channels," *IEEE/OSA J. Lightwave Technol.*, Vol. 13, No. 5, May 1995, pp. 889–897.

[43] Willner, A. E., and S.-M. Hwang, "Transmission of many WDM channels through a cascade of EDFAs in long-distance links and ring networks," *IEEE/OSA J. Lightwave Technol.*, Vol. 13, No. 5, May 1995, pp. 802–816.

[44] Takahashi, H., K. Oda, and H. Toba, "Impact of crosstalk in an array-waveguide multiplexer on N×N optical interconnection," *IEEE/OSA J. Lightwave Technol.*, Vol. 14, No. 6, June 1996, pp. 1097–1105.

[45] Ikegami, T., "WDM devices, state of the art," in G. Prati (ed.), *Photonic Networks*, London, Springer-Verlag, 1997, pp. 79–90.
[46] Bricheno, T., and A. Fielding, "Stable low-loss single-mode couplers," *Electron. Lett.*, Vol. 20, No. 6, March 1984, pp. 230–232.
[47] Varshney, R. K., and A. Kumar, "Design of a 4×4 fiber coupler with equal power splitting at two different wavelengths," *Optics Commun.*, Vol. 81, No. 3/4, Feb. 1991, pp. 167–170.
[48] Brackett, C. A., "Dense wavelength division multiplexing networks: principle and applications," *IEEE J. Select. Areas Commun.*, Vol. 8, No. 6, Aug. 1990, pp. 948–964.
[49] Pennings, E., G.-D. Khoe, M. K. Smith, and T. Staring, "Integrated-optic versus micro-optic devices for fiber-optic telecommunication systems: a comparison," *IEEE J. Select. Topics Quantum Electron.*, Vol. 2, No. 2, June 1996, pp. 151–164.
[50] Vellekoop, A. R., and M. K. Smith, "A polarization independent planar wavelength demultiplexer with small dimension," *Proc. ECOISA '89*, Amsterdam, The Netherlands, Sept. 1989, paper D3.
[51] Kobrinski, H., and K. W. Cheung, "Wavelength-tunable optical filter: Applications and technologies," *IEEE Commun. Mag.*, Vol. 27, No. 10, Oct. 1989, pp. 55–63.
[52] Gillner, L., and M. Gustavsson, "Scalability of optical multiwavelength switching networks: Power budget analysis," *IEEE J. Select. Areas Commun.*, Vol. 14, No. 5, June 1996, pp. 952–961.
[53] Thylén, L., G. Karlsson, and O. Nilsson, "Switching technologies for future guided wave optical networks: potentials and limitations of photonics and electronics," *IEEE Commun. Mag.*, Vol. 34, No. 2, Feb. 1996, pp. 106–113.
[54] Stern, T. E., "Linear lightwave networks: How far can they go?," *Proc. GLOBECOM '90*, San Diego, CA, Dec. 1990, pp. 1866–1872.
[55] Beneš, V. E., *Mathematical Theory of Connecting Networks and Telephone Traffic*, New York, Academic Press, 1965.
[56] Mizrahi, V., T. Erdogan, D. J. DiGiovanni, P. J. Lemaire, W. M. MacDonald, S. G. Kosinski, S. Cabot, and J. E. Sipe, "Four channel fibre grating demultiplexer," *Electron. Lett.*, Vol. 30, No. 10, May 1994, pp. 780–781.
[57] Masetti, F., et al., "High speed, high capacity ATM optical switches for future telecommunication transport networks," *IEEE J. Select. Areas Commun.*, Vol. 14, No. 5, June 1996, pp. 979–998.
[58] Kondo, K., M. Kuno, S. Yamakoshi, and K. Wakao, "A tunable wavelength-conversion laser," *IEEE J. Quantum Electron.*, Vol. 28, No. 5, May 1992, pp. 1343–1348.
[59] Jopson, R. M., and R. E. Tench, "Polarization independent phase conjugation of lightwave signals," *Electron. Lett.*, Vol. 29, Dec. 1993, pp. 2216–2217.
[60] Giles, C. R., and E. Desurvire, "Modeling Erbium-doped fiber amplifiers," *IEEE/OSA J. Lightwave Technol.*, Vol. 9, No. 2, Feb. 1991, pp. 271–283.
[61] Goldstein, E. L., L. Eskildsen, V. da Silva, M. Andrejco, and Y. Silberberg, "Inhomogeneously broadened fiber-amplifier cascades for transparent multiwavelength lightwave networks," *IEEE/OSA J. Lightwave Technol.*, Vol. 13, No. 5, May 1995, pp. 782–790.
[62] Bjarklev, A., *Optical Fiber Amplifiers: Design and System Applications*, Norwood, MA, Artech House, 1993.
[63] Delavaux, J.-M. P., and J. A. Nagel, "Multi-stage Erbium-doped fiber amplifier designs," *IEEE/OSA J. Lightwave Technol.*, Vol. 13, No. 5, May 1995, pp. 703–720.
[64] Desurvire, E., *Erbium-Doped Fiber Amplifiers — Principles and Applications*, New York, Wiley, 1994.
[65] Bala, K., and C. A. Brackett, "Cycles in wavelength routed optical networks," *IEEE/OSA J. Lightwave Technol.*, Vol. 14, No. 7, July 1996, pp. 1585–1594.
[66] Lamming, R. I., et al., "Efficient pump wavelengths of Erbium-doped fibre optical amplifier," *Electron. Lett.*, Vol. 25, No. 1, Jan. 1989, pp. 12–14.
[67] Tanaka, Y., and P. C. Huang, "Multiple destination routing algorithms," *IEICE Trans. Fundamentals*, Vol. E76-B, No. 5, May 1993, pp. 544–552.
[68] Huang, P. C., and Y. Tanaka, "Multicast routing based on predicted traffic statistics," *IEICE Trans. Fundamentals*, Vol. E77-B, No. 10, Oct. 1994, pp. 1188–1193.

[69] Bharath-Kumar, K., and J. M. Jaffe, "Routing to multiple destinations in computer networks," *IEEE Trans. Commun.*, Vol. COM-31, No. 3, March 1983, pp. 343–351.
[70] Simha, R., and B. Narahari, "Single path routing with delay considerations," *Computer Networks and ISDN Systems*, Vol. 24, June 1992, pp. 405–419.
[71] Maxemchuk, N. F., "Dispersity routing in high-speed networks," *Computer Networks and ISDN Systems*, Vol. 25, Jan. 1993, pp. 645–661.
[72] Papadimitriou, C. H., *Combinatorial Optimization*, Englewood Cliffs, NJ, Prentice Hall, 1982.
[73] Schwartz, M., and T. E. Stern, "Routing techniques used in computer communication networks," *IEEE Trans. Commun.*, Vol. COM-28, No. 4, April 1980, pp. 539–552.
[74] Walrand, J., *Communication Networks: A First Course*, Aksen Associates, 1991.
[75] Tanenbaum, A. S., *Computer Networks*, Englewood Cliffs, NJ, Prentice Hall, 1986.
[76] Lee, K. J., A. Gersht, and A. Friedman, "Multipoint connection routeing," *Int. J. Digital and Analog Commun. Sys.*, Vol. 3, 1990, pp. 177–186.
[77] Thaler, D. G., and C. V. Ravishankar, "Distributed center-location algorithms," *IEEE J. Select. Areas Commun.*, Vol. 15, No. 3, April 1997, pp. 291–303.
[78] Billhartz, T., J. B. Cain, E. Farrey-Goudreau, D. Fieg, and S. G. Batsell, "Performance and resource cost comparison for the CBT and PIM multicast routing protocols," *IEEE J. Select. Areas Commun.*, Vol. 15, No. 3, April 1997, pp. 304–315.
[79] Salama, H. F., D. S. Reeves, and Y. Viniotis, "Evaluation of multicast routing algorithms for real-time communication on high-speed networks," *IEEE J. Select. Areas Commun.*, Vol. 15, No. 3, April 1997, pp. 332–345.
[80] Roukas, G. N., and I. Baldine, "Multicast routing with end-to-end delay and delay variation constraints," *IEEE J. Select. Areas Commun.*, Vol. 15, No. 3, April 1997, pp. 346–356.
[81] Maxemchuk, N. F., "Video distribution on multicast networks," *IEEE J. Select. Areas Commun.*, Vol. 15, No. 3, April 1997, pp. 357–372.
[82] Shaikh, A., and K. Shin, "Destination-driven routing for low-cost multicast," *IEEE J. Select. Areas Commun.*, Vol. 15, No. 3, April 1997, pp. 373–381.
[83] Borella, M. S., and B. Mukherjee, "A reservation based multicasting protocol for WDM local lightwave networks," *Proc. ICC '95*, Seattle, WA, June 1995, pp. 1277–1281.
[84] Borella, M. S., and B. Mukherjee, "Limits of multicasting in a packet-switched WDM single-hop local lightwave network," *J. High Speed Networks*, Vol. 4, No. 2, 1995, pp. 155–167.
[85] Tridandapani, S. B., and J. S. Meditch, "Supporting multipoint connections in multi-hop WDM optical networks," *J. High Speed Networks*, Vol. 4, No. 2, 1995, pp. 169–188.
[86] Tridandapani, S. B., and B. Mukherjee, "Channel sharing in multi-hop WDM lightwave networks: Realization and performance of multicast traffic," *IEEE J. Select. Areas Commun.*, Vol. 15, No. 3, April 1997, pp. 488–500.
[87] Cormen, T. H., C. E. Leiserson, and R. L. Rivest, *Introduction to Algorithms*, Boston, MA, Massachusetts Institute of Technology, 1990.
[88] Borella, A., and F. Meschini, "Depth-first search algorithm for multipoint connection routing in extensive packet switching networks," *Proc. SITA '93*, Kanazawa, Japan, Oct. 19-22, 1993, pp. 57–60.
[89] Borella, A., and G. Cancellieri, "Multicast routing through Depth-First Search Algorithm," *Proc. Networks '96*, Orlando, FL, Jan. 8-10, 1996, pp. 61–65.
[90] Borella, A., and G. Cancellieri, "Performance of an adaptive fault-tolerant routing method for multicast communications," *Proc. NOC '96*, Heidelberg, Germany, June 25-28, 1996, pp. 295–296.
[91] Borella, A., and G. Cancellieri, "Multicast communications in WDM networks: DFS versus MST techniques," *Proc. NOC '97*, Antwerp, Belgium, June 17-20, 1997, Part III, pp. 232–239.
[92] Acampora, A. S., *An Introduction to Broadband Networks*, Plenum Press, New York, 1994.
[93] Logotthetis, M., and G. Kokkinakis, "Network planning based on virtual path bandwidth management," *Int. J. Commun. Sys.*, Vol. 8, No. 2, March-April 1995, pp. 143–153.
[94] Chlamtac, I., A. Farago, and T. Zhang, "Optimizing the system of virtual path," *IEEE/ACM Trans. Networking*, Vol. 2, No. 6, Dec. 1994, pp. 581–587.

# Chapter 2

# Single-Hop Optical Networks

## 2.1 GENERAL CONCEPTS

In a single-hop network, users are connected directly, without having to cross intermediate access nodes. More explicitly, this means that the data packet, once converted into light, reaches the destination without suffering any shifting to electronic frequencies at an intermediate section. Transmission takes place when one of the transmitter lasers of the source node and one of the optical receivers of the destination node are tuned to the same wavelength. All users have access to the complete set of optical wavelengths, and no wavelength conversion takes place inside the network. Actually, looking at the components available at the source and destination nodes, four different situations are possible, namely:

- Fixed transmitter(s) and fixed receivers(s) (FT-FR);
- Tunable transmitter(s) and fixed receiver(s) (TT-FR);
- Fixed transmitter(s) and tunable receiver(s) (FT-TR);
- Tunable transmitter(s) and tunable receiver(s) (TT-TR).

So, the simplest way to specify the structure of each node's Network Interface Unit (NIU) is to assign the four values $i, j, m$, and $n$, which represent the number of fixed transmitters, tunable transmitters, fixed receivers, and tunable receivers, respectively. Accordingly, the notation $FT^i$-$TT^j$-$FR^m$-$TR^n$ is used to identify a node, with the convention of suppressing the parts whose exponent is zero and omitting superscript ($i, j, m$, or $n$) when equal to one. Generally, but not necessarily, the structure of the node is the same all over the network.

An obvious way to conceive single-hop optical connection of $N$ users is through a broadcast-and-select structure of the type depicted in Figure 2.1. There, reference has been made to a physical star topology, but a bus or a tree can be equally, though sometimes less efficiently, employed [1,2].

Typically, the network is requested to provide point-to-point connections and/or many-to-one connections and/or one-to-many (multicast or broadcast) connections. This ability can be achieved, in principle, using a unique kind of transmitter or receiver (tunable or fixed) at each node, on condition that the number of sources and photodiodes is adequate.

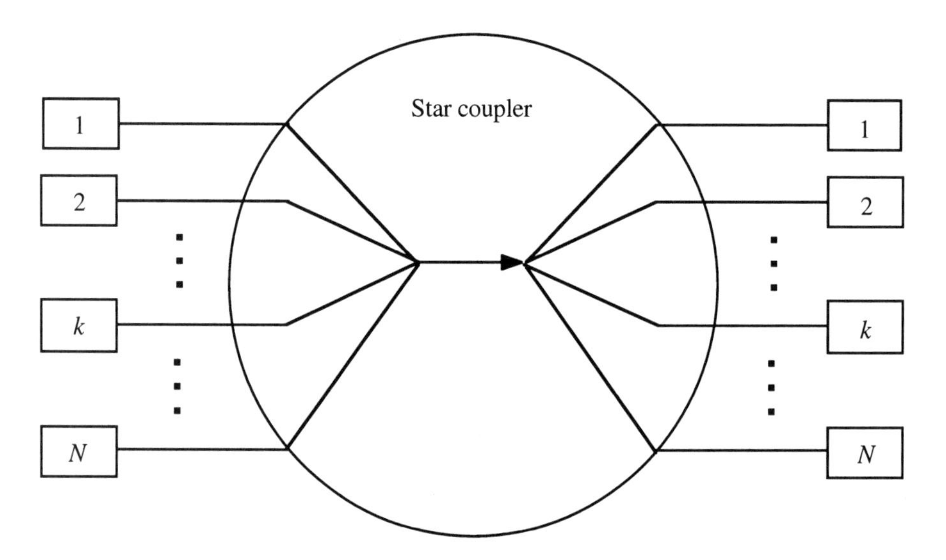


**Figure 2.1** Schematic representation of a broadcast-and-select star network.

Some situations are summarized in Table 2.1, where we have assumed that each tunable transmitter (receiver), in the schemes where it is present, can be tuned to all the wavelengths received (transmitted) by the destination (source) nodes, but only one wavelength at a time. On the other hand, when fixed tuned components are used, each node transmits or receives on a distinct wavelength. The table is not exhaustive (the case where tunable and fixed transmitters and/or tunable and fixed receivers are simultaneously present should be considered) but it includes most of the solutions proposed up to now and experimentally demonstrated. For example, the FT-FR$^N$ system has been implemented in the LAMBDANET [3], a TT-FR solution can be found in the fast optical crossconnect (FOX) demonstration [4], the FT-TR scheme has been proposed in the RAINBOW project [5], and the feasibility of the TT-TR system has been tested in the HYPASS network [6]. The peculiarities of all these architectures will be discussed in the following.

Though introductory, the table is worth some comments. First of all, the case of one fixed transmitter and one fixed receiver at each node has not been considered, since it does not permit full connectivity. Furthermore, the ability of multicasting has been evaluated considering only one active transmission at a time. What we mean is that we cannot be sure that more multicasting transmissions can be simultaneously sustained by the network without collision, so that "yes" in the last column of Table 2.1 simply states that the network has the structural capacity of providing multicasting. This viewpoint is different from the one adopted, in the same table, for discussing the case of point-to-point connections. Here the possibility of having more than one simultaneous transmission from, or to, the same user has been considered, thus emphasizing the risk of collision. For example, in case of TT-FR, it is due to the possibility that two or more sources send a packet, in a given time slot, to the same destination node. The problem can be potentially bypassed by utilizing $N$ fixed receivers at each node but it is evident that this requires a strict pretransmission coordination, in such a way as to employ different wavelengths any time a new connection is established. Moreover, this solution is expensive, requiring a number of components which increases with the network size. Thus, it could be preferable to resolve the contention problem, without increasing the number of receivers but tolerating some packet loss, through a suitable protocol that fixes the connection rules within the network. In general, arbitration of transmission rights can be realized in a preassigned fashion or on the basis of a real time contention. With tunable receivers, however, the adoption of a dedicated control channel for transmission coordination is highly recommended. In case of FT-TR networks, for instance, the receiver must know what wavelength is to be tuned, and at what times. Taking also into account that the tuning time for the transceivers is not null and, on the contrary, generally increases with the range of tunability [7,8], it is easy to understand that the key challenge for single-hop networks really lies in the ability to develop protocols for an efficient coordination of the data transmission. This important topic will be treated in detail in Section 2.2.

**Table 2.1**
Connection Properties Theoretically Achievable by Different Single-Hop Networks

| *Configuration* | *Number of Transmitters and Receivers* | *Point-to-Point Connection* | *Collision or Packet Loss* | *Many-to-One Connection* | *Multicast Connection* |
|---|---|---|---|---|---|
| | $i = m = 1$ | — | — | — | — |
| $FT^i$-$FR^m$ | $i = 1, m > 1$ | Yes (if $m = N$) | No | Yes | Yes |
| | $i > 1, m = 1$ | Yes (if $i = N$) | Yes | No | Yes |
| | $j = m = 1$ | Yes | Yes | No | No |
| $TT^j$-$FR^m$ | $j = 1, m > 1$ | Yes | No (if $m = N$) | Yes | Yes |
| | $j > 1, m = 1$ | Yes | Yes | No | Yes |
| | $i = n = 1$ | Yes | Yes | No | Yes |
| $FT^i$-$TR^n$ | $i = 1, n > 1$ | Yes | No (if $n = N$) | Yes | Yes |
| | $i > 1, n = 1$ | Yes | Yes | No | Yes |
| | $j = n = 1$ | Yes | Yes | No | Yes |
| $TT^j$-$TR^n$ | $j = 1, n > 1$ | Yes | No (if $n = N$) | Yes | Yes |
| | $j > 1, n = 1$ | Yes | Yes | No | Yes |

As an alternative to the broadcast-and-select structure, we can use a wavelength-routing solution. In this case the data pass across a number of passive wavelength-selective elements, where routing is determined by the signal wavelength. The key component is then constituted by a wavelength-selective space switch (WSSS), where the incoming signal is conveyed toward a specific output port, as a function of the wavelength value and under the action of a suitable switch control. A simple example is shown in Figure 2.2: four fibers are used to connect as many users (two transmitters and two receivers). The system globally adopts two wavelengths, generated and selected by tunable transceivers; the WSSS1 selects $\lambda_a$ for connection with $R_1$ and $\lambda_b$ for connection with $R_2$, while the opposite is made by WSSS2. The blocks noted by WDM realize the function of wavelength multiplexing. No contention occurs, the wavelengths being detected by the tunable receiver. The need of tunable transmitters and receivers (or, equivalently, of array of fixed tuned elements) is a drawback of the wavelength-routing solution that, as a counterpart, offers the advantage of a more favorable power budget (splitting losses are avoided) and easy rearrangeability. The latter property is useful to provide dynamics changes in the routing patterns, able to balance the effects of variations in the offered load. A special case of wavelength-routing network is the so-called linear lightwave network (LLN) that will be described in Section 2.4.

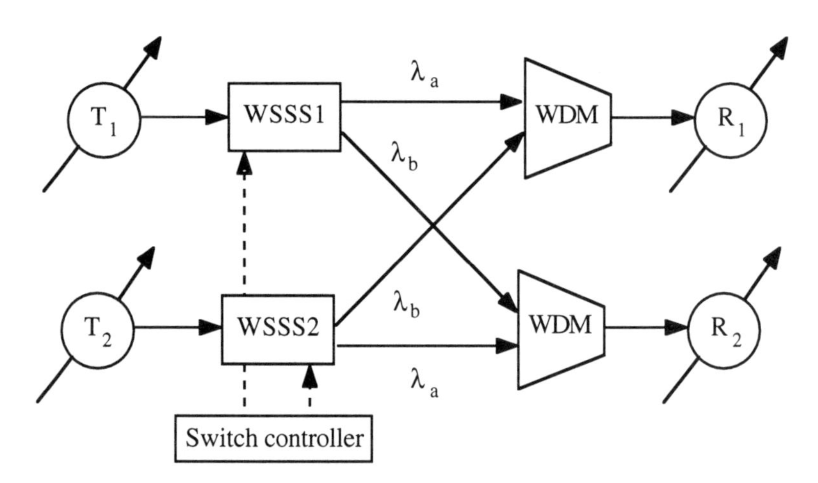


**Figure 2.2** Schematic representation of a wavelength-routing network.

## 2.2 TRANSMISSION PROTOCOLS

A comprehensive survey of access protocols suitable for single-hop networks can be found in [9,10]. Here we highlight the most important results, completed and integrated by some further proposals, which have also appeared in more recent literature. Other details will be given in Section 2.3, which is devoted to the presentation of specific architectures.

As stressed above, the need of defining suitable transmission protocols in a multichannel environment is due to the risk of collisions that potentially exist among packets emitted by different sources. It is usual to distinguish between:

- Channel (or message) collision when two sources try to use the same wavelength for simultaneous transmissions;
- Receiver (or destination) collision when the number of messages simultaneously arriving at a destination node exceeds the receiving capabilities of the node; for example, in case of one receiver at each node, only one message at a time can be successfully detected.

As schematically shown in Figure 2.3, a first subdivision of the transmission protocols is based on the way in which the channel bandwidth is assigned: fixed, on demand, or mixed. Correspondingly, we have three different categories, that is, fixed or semifixed assignment protocols, random access protocols, and hybrid protocols. The random access protocols, in turn, can be roughly divided into reservation, switching, and collision avoidance protocols. From a different point of view, we must distinguish between protocols based on no pretransmission coordination and protocols based on pretransmission coordination. The former encompass most of the fixed assignment and some of the random access protocols; properly speaking, however, instead of no pretransmission coordination, in this case we often have a permanent and prefixed coordination. The remaining random protocols, instead, are based on a dynamic coordination, usually realized through a separate control channel for transferring the control information. This auxiliary channel can be used, for example, to inform a tunable receiver on which wavelength a data packet has been transmitted.

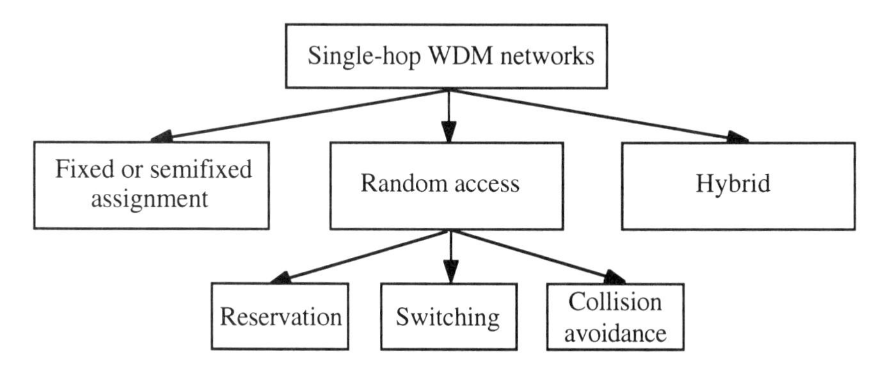


**Figure 2.3** Classification of the protocols for single-hop optical networks.

As qualitatively reasonable, the definition of a transmission protocol cannot be disjointed by the specific $FT^i$-$TT^j$-$FR^m$-$TR^n$ structure of the system to which it is applied. In this sense, the fixed and semifixed assignment protocols we will describe in the next section mainly refer to networks where each node has only one tunable transmitter and one tunable receiver (TT-TR systems). Other structures are obviously possible and will be discussed in the subsequent sections, particularly in the case of random access protocols. Throughout the following discussion, we will denote by $N$ the number of users and by $W$ the number of channels (that is wavelengths) available; the latter will be indicated explicitly as $\lambda_1, \lambda_2, \ldots, \lambda_W$.

With few exceptions, the protocols we will describe are based on a centralized hub architecture, independently of the particular network's physical topology. Where necessary, however, we will refer (sometimes implicitly) to a star topology, though most of the considerations developed can be easily adapted to any other architecture. On the other hand, a linear topology physical network (like a bus) can use protocols conceived ad hoc for it [9]. Examples in this sense are given in [11,12].

> ***Some keypoints.*** In a single-hop network there is a risk of channel or receiver collision. To alleviate the problem suitable transmission protocols must be designed that can employ—or may not employ—pretransmission coordination.

### 2.2.1 Fixed and Semifixed Assignment Protocols

The simplest way to realize single-hop communications consists in combining wavelength division multiplexing with time division multiplexing (TDM); the time is slotted and, inside each slot, some source/destination pairs are allowed to be connected using more than one wavelength. Depending on the number and characteristics of the permissions granted, we can have four different frequency-time division multichannel allocation (FTDMA) protocols [13]: source/destination allocation (S/DA) protocol, destination allocation (DA) protocol, source allocation (SA) protocol, and allocation free (AF) protocol. S/DA is an example of fixed assignment protocol, while DA and SA protocols can be seen as different ways of implementing a semifixed (or partially fixed) assignment scheme. Finally, AF is an example of random access protocol with no pretransmission coordination. S/DA, DA and SA will now be briefly discussed and an example given where possible, with reference, for explicative purposes, to a simple network with $N = 3$ nodes. AF, instead, will be discussed in Section 2.2.2.

In the S/DA protocol, in each time slot $W$ sources are allowed to transmit, using different wavelengths, toward as many distinct destinations. This ensures that neither channel nor receiver collisions occur. The allocation schedule has to foresee the possibility that any pair of nodes is connected. Therefore, calling $L$ the number of slots that ensures that one data packet can be transferred by any source to any destination, it is easy to verify that

$$L = \left\lceil \frac{N \cdot (N-1)}{W} \right\rceil \tag{2.1}$$

having denoted by $\lceil x \rceil$ the next larger integer of $x$. Since $1 \le W \le N$ is set as a rule, we have $N - 1 \le L \le N \cdot (N - 1)$. In reality, the assumption of $W = 1$ (and correspondingly of $L = N \cdot (N - 1)$) is a limit case, only useful for reference purposes. When just one wavelength is adopted, in fact, the system loses its multichannel characteristics. Examples of time slot and channel assignment for the specific case of $N = 3$ are reported in Figure

2.4 assuming $W = 1, 2$, or 3; $T_k$ denotes the $k$th time slot, whereas $i/j$, with $1 \le i, j \le 3$, means that node $i$ has permission to transmit a packet to node $j$ in the specified time slot and to use the assigned wavelength.

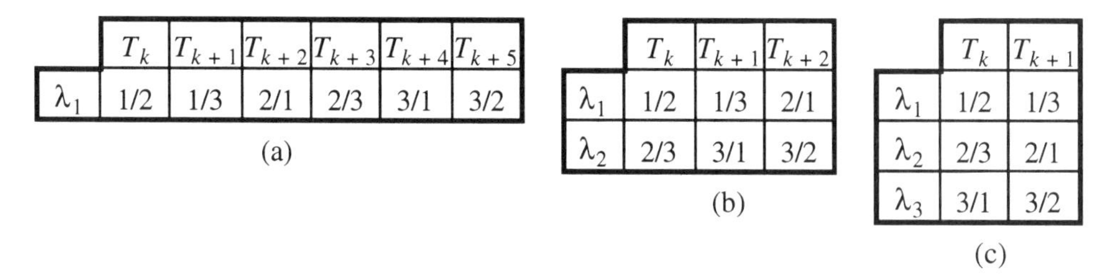

|  | $T_k$ | $T_{k+1}$ | $T_{k+2}$ | $T_{k+3}$ | $T_{k+4}$ | $T_{k+5}$ |
|---|---|---|---|---|---|---|
| $\lambda_1$ | 1/2 | 1/3 | 2/1 | 2/3 | 3/1 | 3/2 |

(a)

|  | $T_k$ | $T_{k+1}$ | $T_{k+2}$ |
|---|---|---|---|
| $\lambda_1$ | 1/2 | 1/3 | 2/1 |
| $\lambda_2$ | 2/3 | 3/1 | 3/2 |

(b)

|  | $T_k$ | $T_{k+1}$ |
|---|---|---|
| $\lambda_1$ | 1/2 | 1/3 |
| $\lambda_2$ | 2/3 | 2/1 |
| $\lambda_3$ | 3/1 | 3/2 |

(c)

**Figure 2.4** Examples of time slot and channel assignment for a network with $N = 3$, under the S/DA protocol: (a) $W = 1$, (b) $W = 2$, and (c) $W = 3$.

The three solutions are equivalent as regards the number of permissions granted, but passing from (a) to (c) in Figure 2.4 the transmission delay decreases at the expense of an increasing number of wavelengths. The best compromise is probably given by (b). As a first approximation the tuning times are here assumed to be zero.

To fix a general criterion for the assignment of time slots and channels to the various connections, it is useful, especially when the network size is large, to divide the time in cycles, each including $N \cdot (N - 1)$ time slots and to define suitable $N \times N$ allocation matrices. The rows of each matrix identify the destination nodes (numbered as $1 \le j \le N$), while the columns identify the source nodes (numbered as $1 \le i \le N$). We can construct one matrix for every channel, so that the $(j, i)$ entry of the $k$th matrix represents the number of the time slot during which node $i$ is allowed to transmit a packet to node $j$ using the wavelength $\lambda_k$. Possible allocation matrices for the case $N = 3$ and $W = 1, 2$, or 3 are reported in Figure 2.5, where the slots have been numbered from 1 to 6.

Entries with $i = j$ are stippled, since transmissions $i/i$ (with $i = 1, 2, 3$) are obviously inhibited. From Figure 2.5 we observe that, in each transmission cycle, any connection $i/j$ is permitted to use, periodically, all the channels available. For example, in (b) node 1 can send a packet to node 2 in the first time slot using $\lambda_1$ and in the fourth time slot using $\lambda_2$.

The figure highlights the simple rules to follow in designing the allocation matrices, that is:

1. No receiver collision occurs;
2. The total number of permissions to transmit in a slot equals the number of available channels;
3. There is only a single allocation of each channel per slot;
4. There is only a single transmission per source node.

As typical of fixed assignment techniques, the S/DA protocol works well when the network operates under high traffic levels (a limit case is that of overload conditions) but it becomes inefficient at light loads, when a great number of slots could be unused.

Furthermore, the protocol does not take into account that the bandwidth requirements may be different from one node to another and, most of all, that they are generally variable with time. Finally, the channel allocation matrices are assigned with reference to a specific number of nodes and need to be reformulated when the network size changes in scale.

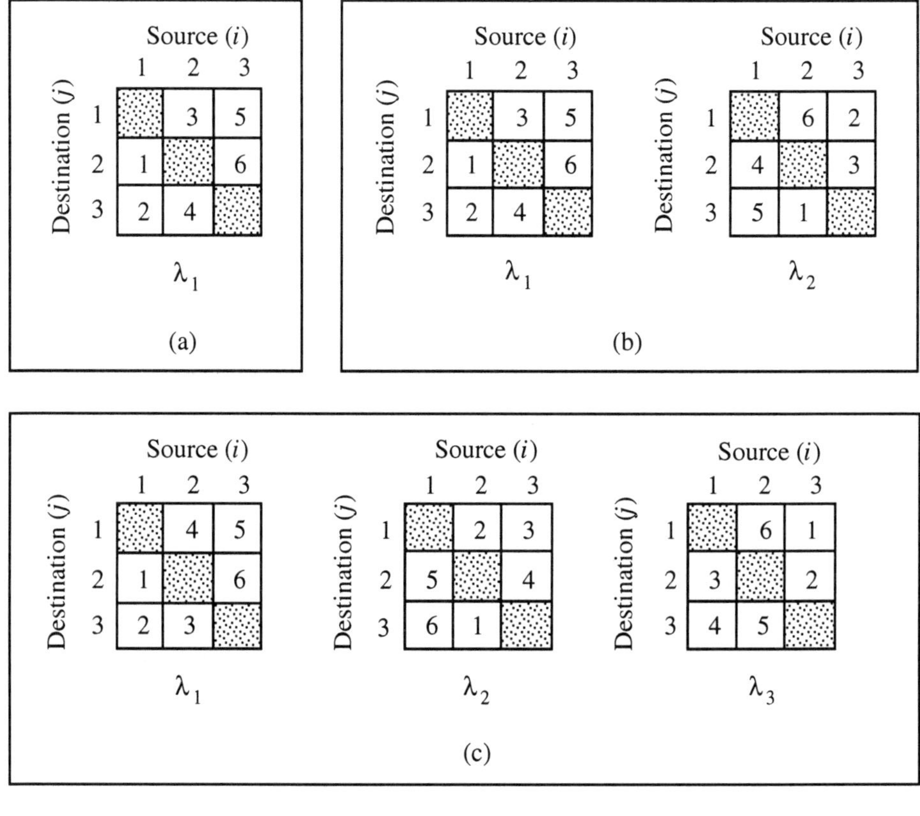


**Figure 2.5** Examples of allocation matrices for a network with $N = 3$, under the S/DA protocol: (a) $W = 1$, (b) $W = 2$, and (c) $W = 3$. Each entry in a matrix shows the slot assigned to the user pair $(i, j)$. Transmissions where $j = i$ are not scheduled (stippled boxes).

Variants of the basic S/DA protocol have been proposed [14,15], where node $i$ is equipped with $t_i$ transmitters and $r_i$ receivers, all of which are tunable over all the available channels. The main merit of these proposals is that they take into account that the transceiver tuning times are not negligible. In [15], for example, two different scheduling algorithms are discussed: the first ensures a minimum packet transmission duration while trying to minimize the tuning time, whereas the second, on the contrary, obtains a schedule with a minimum tuning duration, while trying to minimize the packet transmission time of the schedule. The choice of the best algorithm, for a given system configuration and an expected traffic matrix, is determined by the ratio between the tuning time and the packet transmission time.

In order to optimize the network performance, an organization of the nodes in separate communities (clusters) can be conceived [16], on the basis of the specific traffic flow patterns with other users. In this idea, which anticipates what we will say in Chapter 4 about multilevel optical networks, users within a cluster are connected through their own local WDM star network, whereas users of different clusters can communicate, also in a single-hop fashion, through a separate, more remote, WDM star.

The most obvious way to reduce the packet delay in case of light loads consists in increasing the number of permissions granted in each time slot. This is valid for the DA protocol in which the number of users allowed to transmit in a time slot passes from $W$ (which was the number of permissions for the S/DA protocol) to $N$. Obviously, the DA protocol reduces to S/DA if $W = N$ (see case (c) in Figures 2.4 and 2.5) but when, more frequently, $W < N$ during a time slot, more sources can use the same wavelength for transmission. Since each destination is still required to receive from a fixed channel, this means that receiver collision is avoided but channel collision can occur. An example of slot and channel assignment for the network with three nodes and two channels is reported in Figure 2.6(a): node 1 and node 3 are permitted to transmit a data packet to node 2, in the same time slot and both using $\lambda_1$ (the same occurs in the subsequent time slot for nodes 3 and 2 that are allowed to send a packet to node 1 using $\lambda_2$); for light loads the probability that these two permissions are simultaneously employed is small but the risk of collision obviously increases when the load becomes greater. Comparing Figure 2.6(a) and Figure 2.4(b) we see that the transmission time is reduced by one time slot; this, however, does not mean that the maximum throughput increases, since, in both cases, no more than four packets can be successfully transmitted in the time of two slots.

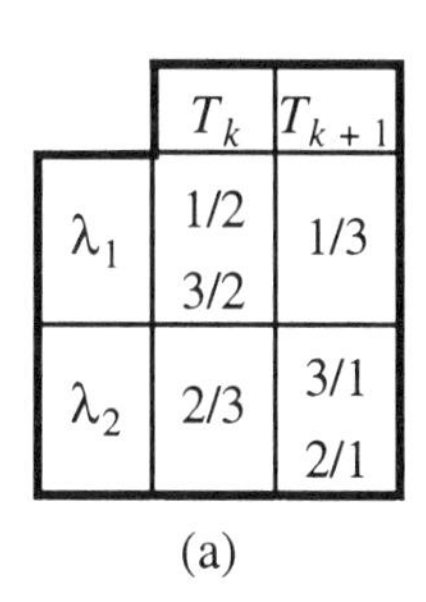


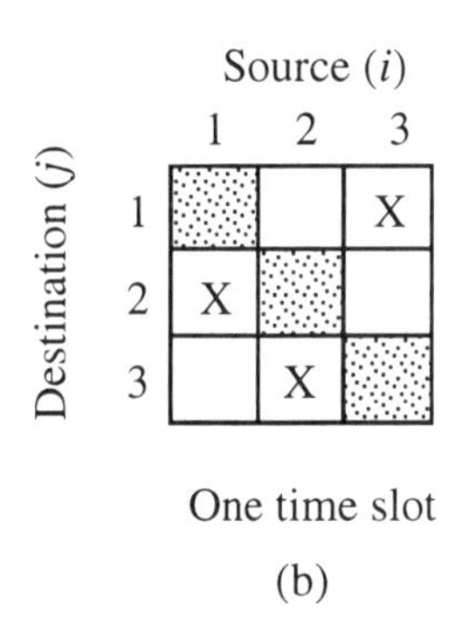


**Figure 2.6** (a) Example of time slot and channel assignment for a network with $N = 3$ and $W = 2$, under the DA protocol and (b) allocation matrix for the same case; an X indicates that any channel can be used.

In more general terms, the DA protocol does not specify the channel on which the generic source has to transmit in a given time slot. In practice, the source having permission and with a packet to transmit randomly selects one of the available channels, the only constraint being to avoid receiver collisions. Thus the allocation matrix can be constructed as shown in Figure 2.6(b), which is relative to a specific time slot, for the usual network with $N = 3$ and $W = 2$. The symbol X for the $(i, j)$ entry of the matrix means that source $i$ is permitted to send a packet to node $j$ using indifferently $\lambda_1$ or $\lambda_2$. Receiver collision is certainly avoided, since, contrary to Figure 2.6(a), each destination receives only one packet at a time. Either the assignment of Figure 2.6(a) or that of Figure 2.6(b) are, however, equally acceptable because both satisfy the rules established by the DA protocol, that is:

1. No receiver collision occurs;
2. The total number of permissions to transmit in a slot equals the number of users;
3. There is only a single transmission per source node.

A further possibility consists in the adoption of the SA protocol, whose philosophy is practically parallel to that of the DA protocol: in this case, in any time slot, $W$ sources can transmit, each on a different channel, toward an arbitrary destination user. This implies that more sources can send a packet, simultaneously and on different wavelengths, to the same destination, thus causing receiver collision. On the contrary, the protocol is such that channel collision is avoided. An example of slot and channel assignment, assuming again $N = 3$ and $W = 2$, is shown in Figure 2.7.

| | $T_k$ | $T_{k+1}$ | $T_{k+2}$ |
|---|---|---|---|
| $\lambda_1$ | 1/2<br>1/3 | 1/2<br>1/3 | 2/1<br>2/3 |
| $\lambda_2$ | 2/1<br>2/3 | 3/1<br>3/2 | 3/1<br>3/2 |

**Figure 2.7** Example of time slot and channel assignment for a network with $N = 3$ and $W = 2$, under the SA protocol.

In general, the rules to satisfy in this protocol are as follows:

1. The number of transmissions in a slot equals the number of channels;
2. A given channel cannot be allocated to more than one source node.

Looking at the performance, receiver collisions are usually considered more destructive than message collisions and, for this reason, we can foresee that the SA protocols will be used less extensively than the S/DA and DA protocols.

***Some keypoints.*** Multiaccess can be managed by combining WDM and TDM. This is the case of some fixed and semifixed assignment protocols, named S/DA, DA, and SA, respectively. Wavelengths and time slots are assigned through suitable allocation matrices. S/DA (i.e., a fixed assignment) eliminates collisions completely, but its efficiency is low at light or variable loads. The packet delay is reduced using DA or SA (semifixed assignment) but channel collisions or receiver collisions may now occur.

### 2.2.2 Random Access Protocols with No Pretransmission Coordination

A large number of random access protocols were conceived, with the aim of providing a more dynamic and efficient sharing of the bandwidth available, especially for relatively light loads. Many of the proposed protocols are not specific for optical networks but have been adapted from previous applications. This is the case of the AF protocol, whose functionality is the same of a multichannel slotted ALOHA [17]. While it seems appropriate to refer to a number of previously valuable books for a full comprehension of ALOHA and the other standard protocols ([17] is probably one of the most widely diffused), we would like to satisfy the curiosity of the reader by reminding that the ALOHA system (which we will largely mention and discuss in the following sections) is so called since it was implemented at the University of Hawaii [18].

In the case of AF the transmission cycle collapses in one time slot, in which we have $N \cdot (N-1)$ permissions to transmit. In practice, in any time slot, any user can send a packet (only one) to any destination using a randomly chosen wavelength. Therefore, in a practical sense, no specific allocation rule is assigned except that of having only one transmission from each source, and channel collisions as well as receiver collisions are possible.

The AF protocol is an example of random access protocol without pretransmission coordination for a TT-TR system. In reality, the possibility of avoiding pretransmission coordination seems more justified when the receiver is fixed, since in this case the only things each transmitter has to know are the wavelengths detectable by the receiver(s) it wishes to communicate and the only ability it must possess is the capacity to tune this set of wavelengths (obviously, this does not prevent conflicts). Therefore, some random access protocols with no pretransmission coordination were also proposed for TT-FR systems.

In [19], for example, two separate schemes were discussed and their performance compared. Both based (similar to AF) on slotted ALOHA, they basically differ on the synchronization boundaries. More precisely, in the first, noted by $SA^{(3)}$ [19], each packet is considered to be of $L$ minislots and synchronization is over minislots (which means that a user is permitted to send a packet at the beginning of the next minislot). In the second protocol, instead, noted by $SA^{(4)}$, there are no minislots, each slot has a length equal to the packet's transmission time and synchronization is over the data packet boundaries (which means that a user is permitted to send a packet at the beginning of the next slot). In terms of throughput, it is proved that $SA^{(4)}$ is better than $SA^{(3)}$. This result is not surprising, since it is due to the reduced vulnerable period of $SA^{(4)}$ with respect to $SA^{(3)}$ ($L$ instead of $2L-1$) and resembles the more general result according to which pure ALOHA has poorer performance than slotted ALOHA [17]. What is interesting, instead, is that $SA^{(3)}$ and $SA^{(4)}$ compare favorably, as regards performance as well as implementation simplicity, with two control-channel-based protocols using slotted ALOHA for the control channel and pure ALOHA for the data channels [20,21]. The latter protocols, which are applied to a TT-TR system, will be further discussed in Section 2.2.3, but this conclusion denies the idea that pretransmission coordination is necessary for improving performance. This is probably true in many cases but cannot be assumed as a general rule.

A TT-FR$^m$ system, with $m$ as a parameter, was considered in [22], and two protocols applied: multichannel slotted ALOHA and random TDMA. The use of multiple, though fixed, receivers at each node allows one to employ tunable transmitters with limited tuning capabilities, so relaxing technological difficulties. Furthermore, in order to consider networks able to support a wide range of applications, the examined configuration was nonhomogeneous, that is, it was characterized by asymmetric node parameters (different packet generation rates and packet destinations distributions), and with a finite number of high-speed buffers with different capacities. Let us denote by $T(i)$ and $R(i)$ the sets of wavelengths over which node $i$ can transmit and receive, respectively. The possibility of connecting two generic nodes, $i$ and $j$ ($i, j = 1, \ldots, N$), in a single hop, follows from the satisfaction of the following relationship:

$$T(i) \cap R(j) \neq \varnothing \qquad \forall i, j \qquad i \neq j \tag{2.2}$$

where $\varnothing$ represents the empty set. As before, time is divided into slots whose duration (fixed) equals the packet transmission time, and all nodes are synchronized to the beginning of the slots. A node is said to be busy when it has at least one packet in its buffer.

In the multichannel slotted ALOHA protocol, at the beginning of each slot, busy node $i$, which has a packet to send to node $j$, transmits with probability $p_i$ on a channel randomly chosen among the $T(i) \cap R(j)$ wavelengths leading to $j$ from $i$. A channel collision occurs when two or more busy nodes try to use the same channel for transmitting at the same time. The duration of a collision is one slot and the collided packets must be retransmitted. On the contrary, a transmission is successful when a single packet is sent via a channel.

In the random TDMA protocol, a slot transmission schedule, called $trans[i]$, is used, which fixes the channel on which node $i$ will be allowed to transmit. At the beginning of each slot, busy node $i$ with $trans[i] = k$ ($1 \leq k \leq W$), is permitted to transmit on channel $k$ to any of the nodes that can receive this wavelength. So, a successful transmission takes place if node $i$ has, in its buffer, at least one packet with such destinations. If more packets are available for transmission on channel $k$, the node randomly selects only one of them; the others must wait for subsequent permissions. If node $i$ is not scheduled for transmission $trans[i] = 0$ is set. The key point of the protocol is the generation of the transmission schedule which, to avoid collisions, must be the same (and known) for all the nodes of the network. In practice, $trans[\cdot]$ is constructed first randomly choosing one wavelength in the set of possible channels, say $k$, and then randomly choosing one node among those able to transmit on such a channel, say $i$. Random channel $k$ and random node $i$ are produced by a random number generator which must be present at any node and must provide the same numbers at any extraction. It is known that this result can be achieved by starting the various (but equal) number generators with the same seed.

The performance of the two protocols was evaluated in [22] through an approximate method, whose efficiency was tested by comparison with simulated results. The measured discrepancies between theory and simulation were of 5%, at most, for the node throughput and 10% for the average node delay. With reference to some typical systems, it was proved that the slotted ALOHA protocol exhibits lower delays for low system loads, whereas the random TDMA protocol results in lower delays for moderate to high system loads. Nevertheless, the final goal of the authors was to propose a procedure that allows one to optimize the design for a required system performance. To make the method meaningful, however, the analytical model has to be combined with a suitable hardware cost function for the transmitters and receivers. For example, it was proved that the best performance is achieved when the number of receivers per node equals the number of wavelengths. This conclusion is obvious, since, correspondingly, the number of nodes competing for a given channel decreases. However, assuming $m = W$, when the number of wavelengths is large the design becomes unfeasible due to the large number of receivers required. Because of its intrinsic complexity, some aspects of the problem are still open.

A further possibility consists in the adoption of a passive star network with a protection against collision (PAC) circuit [23–25]; the system is again of TT-FR type, and the network uses two star couplers, one being the network star and the other a control star coupler. A schematic block diagram of the network is shown in Figure 2.8. Channel collisions are avoided, allowing a node to access the network only if the addressed channel (i.e., the channel the node tries to use for connection with the corresponding destination) is available (i.e., not already busy by another user). Though small, there is a probability that two nodes request simultaneously to use the same available wavelength; in this case, both are denied access to the network. The described behavior is achieved by providing each node with a PAC circuit, located at the central hub, which controls the network access. A PAC circuit verifies the state of the addressed channel utilizing an $n$-bit carrier burst that precedes the packet. The carrier burst is switched through the control star, where it is combined with a fraction of any of the packets coming out of the network star, as well as with the carrier bursts of all the other nodes trying to access the network. The object is to verify if energy is present on the addressed channel, in which case the channel is not available at all. In practice, in the control star, simple optical power measurements are realized, whose result is an electrical signal that controls the optical switch (OS) connecting the input transmission fiber of each node to the network star. The switch is closed if no energy is revealed on the addressed channel, otherwise it is left open (rest position). As an alternative to the scheme of Figure 2.8, when the nodes are all located near the hub, a centralized switch can be used, eliminating the need for the control star and an optical switch for each node.

When a packet cannot access the network, it is reflected back to the input port which can then block its transmission and delay it to a later time. Keeping in mind the distance between the node and the central hub, packet blocking occurs after a feedback propagation delay $\tau$, whose extent influences the network performance. In absence of reflected signal, within $\tau$, the transmission is considered to be successful and the packet, once sent completely, is erased from the node's buffer where it was stored. Because of the presence of noise, the state of the addressed channel is sometimes wrongly estimated. The probability that such an event occurs mainly depends on the length of the $n$-bit carrier burst; it was computed [23] that assuming $n = 20$, the probability of collisions in the network star is on the order of $10^{-14}$, and thus is quite negligible for most applications.

The performance of the PAC optical packet network has been evaluated in [24] under the hypothesis of uniform traffic patterns. It was proved that this network allows to reach rather high values of throughput (normalized to the transmission rate) per channel, typically between 0.4 and 0.5 in the case of geographically distributed applications. This means, for example, that a system with 100 channels, each characterized by a transmission rate of 3 Gbps, can reach a maximum overall throughput between 120 and 150 Gbps. The normalized throughput, however, is further increased, up to exceed 0.8, in the case of centralized switch, where the feedback propagation delay is reduced (if it is normalized to the packet transmission rate, we have $\tau < 1$).

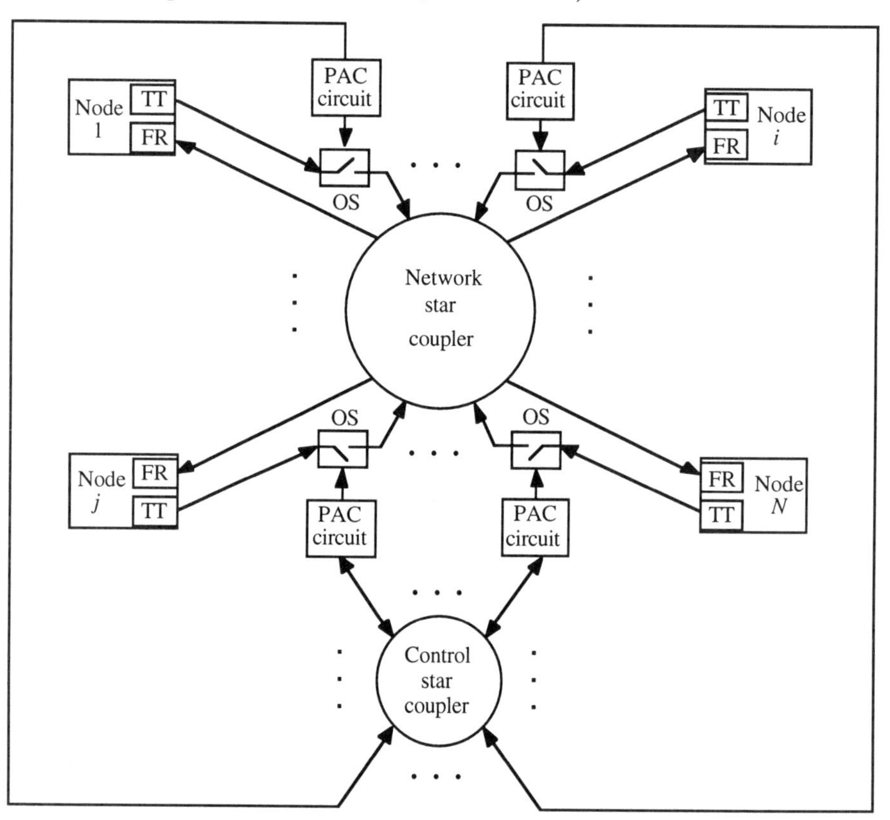


**Figure 2.8** Schematic representation of the PAC optical packet network.

***Some keypoints.*** Random access protocols allow a dynamic and efficient sharing of the available bandwidth, especially in the case of light loads. Many of these protocols are derived from adaptation of slotted ALOHA and do not exploit pretransmission coordination. Channel collisions as well as receiver collisions are possible, and the performance can be improved by reducing the vulnerability period. Collisions can be prevented by using a PAC circuit at each node. This circuit probes in advance the selected wavelength and, if it is occupied, delays the transmission. The structure becomes more involved, requiring a control star coupler germane to the network coupler.

### 2.2.3 Random Access Protocols with Pretransmission Coordination

In the random access protocols described until now, data channels are assigned to the users without any coordination between transmitters and receivers. Under such a strategy, finding optimum solutions is very hard, especially when the traffic exhibits variable dynamics. Moreover, when the number of wavelengths is high, it is recommended to use multiple receivers, which implies augmented costs at each node. For these reasons, the favored choice seems to be the adoption of a TT-TR system, but with the addition of a further control channel for transmission coordination. In particular, the control channel should convey the information about the wavelength the receiver has to tune to, to receive the data packets from other users. Many different protocols have been proposed for such a system. Most of them are based on standard protocols for computer networks, that is, ALOHA, carrier sense multiple access (CSMA), *W*-Server, and so on [17], suitably adapted for application in optical networks. Several examples will be given in this section.

#### *2.2.3.1 Some Basic Random Access Protocols*

Five random access protocols with pretransmission coordination were described in [20], each of them referred to as *X*/*Y*, where *X* is the protocol applied to the control channel and *Y* that applied to the data channels. The analysis neglected the tuning times whereas, as usual, transceivers were assumed to be tunable over the entire wavelength range under consideration. The time was normalized to the duration of a control packet transmission, and data packets with fixed size of $L$ units were considered. Every one of the $W$ data channels was identified by a different wavelength $\lambda_i$, with $i = 1, 2, \ldots, W$, whereas the control channel wavelength was denoted by $\lambda_0$. In general, a control packet contains three basic types of information: the source address, the destination address, and a number identifying the wavelength used for the data channel. The latter, in fact, can be randomly selected by the source for the transmission of its own data packets.

Along with the discussion in [20], let us consider first the case of either the data channels or the control channel based on the ALOHA protocol (so we have an ALOHA/ALOHA protocol). Each node transmits its control packet, on $\lambda_0$, in a random instant and, once such transmission is completed, immediately sends the corresponding data packet on the wavelength $\lambda_i$ as specified in the control packet. The situation is shown in Figure 2.9.

The vulnerability period of the control packet, that is, the time interval within which two control packets should collide, is equal to two time units, extending from $t_0 - 1$ to $t_0 + 1$. The meaning of this notation is, in fact, that we must look at up to one time unit before and one time unit after the instant of emission for the tagged control packet, noted by $t_0$. When collision occurs, it is obviously destructive. It is important to stress that Figure 2.9 refers to the situation as seen at the network hub. From a quantitative point of view, in fact, the packet propagation time should also be considered, which is different for nodes at different distances from the star coupler.

To avoid collision between two or more control packets is a condition necessary but not sufficient for successful transmission of data packets. The tagged data packet, in fact, will certainly collide if another control packet is successfully transmitted over the interval from $t_0 - L$ to $t_0 + L$ and both data packets use the same wavelength $\lambda_i$. Taking this into account, the throughput performance of the protocol was analytically evaluated in [20]. The model, however, ignored the possibility of receiver collisions that, on the contrary, may occur when a control packet is sent to a destination already engaged in the reception of another data packet. In this condition, in fact, the receiver cannot tune to $\lambda_0$ and, even if channel collision is avoided, the corresponding data packet cannot be successfully detected. In this sense, the analysis of [20] is significant only for huge user population, when the effect of receiver collisions becomes negligible.

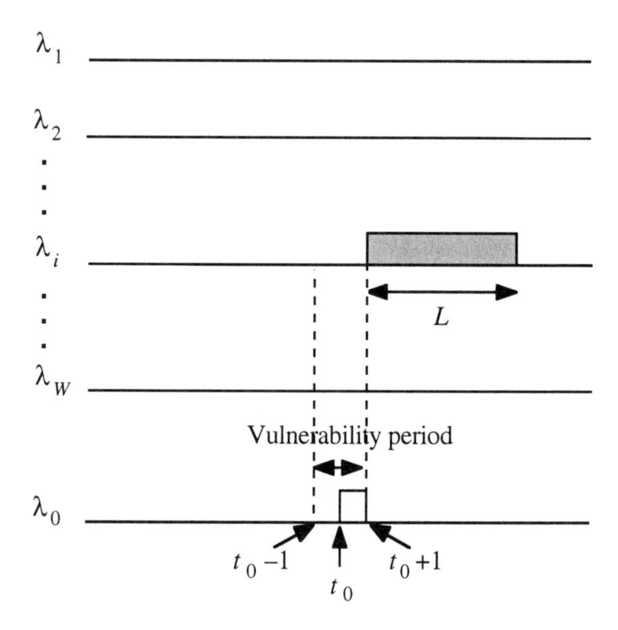


**Figure 2.9** The ALOHA/ALOHA protocol. A node emits a control packet at time $t_0$ and, immediately afterward, transmits the data packet on wavelength $\lambda_i$.

The second protocol proposed in [20] is similar to the previous one except for the slottization of time as regards the control channel; in other words it is a slotted ALOHA/ALOHA protocol. In the third solution, instead, access to the data channels obeys to a carrier sense strategy. In this ALOHA/CSMA protocol, the node that has a packet to transmit senses the data channels up to find an idle wavelength. After that, the user jams the available channel during the time it transmits its control packet and, finally, sends the data packet at the selected wavelength. Since the jamming signal and the control packet are transmitted at the same time, this system really needs two separate transmitters, only one emitting at fixed wavelength $\lambda_0$; in other words, it is an FT-TT-TR system. The ALOHA protocol used for the control channel can be, as above, slotted or not.

The solution proposed for the third protocol is inverted in the fourth one, which is, in fact, CSMA/ALOHA. In this case, the user first randomly chooses a data channel, then verifies the state of the control channel and, if idle, transmits the control packet on it. The data packet is sent immediately after on the selected wavelength. Finally, the last protocol considered in [20] is a CSMA/$W$-server switch protocol: a node monitors the control channel for a period of $L$ time units. In this way, it knows exactly which channels and which receivers are idle for the same time interval. The source transmits its control and data packets only when at least one wavelength is found idle; otherwise, transmission is delayed. Therefore, no channel collision will occur in this protocol.

The analysis of the protocols using CSMA was based on the hypothesis that the carrier sensing provides a practically instantaneous feedback signal (at least in relative terms). It is quite obvious that such hypothesis becomes worse and worse verified when the network speed increases. For very high speeds, in fact, the packet lengths become short and the normalized propagation delay cannot be neglected (it can exceed unity). For this reason, further improvements of the protocols presented in [20] which is mainly focused on the ALOHA and $W$-server switch protocols.

In [21], for example, maintaining unchanged the hardware architecture considered in [20], two improved random access protocols were proposed and their performance discussed. The first modification concerned the slotted ALOHA/ALOHA protocol: permission to transmit a data packet was made conditional on the successful transmission of the corresponding control packet. In this way, contrary to the original protocol, it is possible to avoid the wasting of bandwidth which occurs when a data packet is transmitted on a data channel while its accompanying control packet is lost on the control channel. The same improvement was retained in the second proposal of [21], which was a variant of the CSMA/$W$-server switch protocol, the most efficient, but involved, solution discussed in [20]. Here the CSMA protocol for the control channel was replaced by a much simpler slotted ALOHA, so making less critical the dependence on the normalized propagation time of packets over the optical medium. Apart from the relevant advantage of avoiding carrier sensing, it was shown analytically that either the improved slotted ALOHA/ALOHA protocol or the slotted ALOHA/$W$-server switch protocol have better throughput performance than the corresponding protocols of [20], for typical values of the system parameters.

On the other hand, using the improved slotted ALOHA/ALOHA scheme, when two or more control packets having the same data channel number are successfully transmitted, a channel collision may occur. To overcome this problem, a further improvement was proposed, which is called slotted ALOHA/polite access protocol [26]: a data packet is transmitted only if the control packet is successful and there are no other successful control packets having the same data channel number during the $L - 1$ slots prior to its own control packet transmission. Otherwise, the user repeats the same procedure after a random delay. It is evident that the risk of channel collision is prevented.

In the same paper [26], a variant of the slotted ALOHA/$W$-server switch protocol was also proposed; its characteristics can be summarized as follows. The control channel is divided into frames and any frame is further divided into control slots, one for each user. The control packet contains the receiver address but no data channel number; this means that, contrary to the previous schemes, users do not select the data channels. When all control packets of a frame are received by the users, wavelengths are allocated to the successful control packets according to an assignment algorithm known to everyone in the network. For example, an assignment can be made on a first-in-first-out (FIFO) basis, starting from $\lambda_1$ to $\lambda_W$; the number of successful control packets can be obviously greater than the number of data channels and, in this case, not all the candidates can be served. Immediately after the assignment, the users that are given a data channel can transmit their data packets, whereas the others (who have suffered a collided control packet or with a successful control packet but without an assigned wavelength) repeat the same procedure after a random delay. Owing to the structure of the control channel, this modified protocol is called slotted ALOHA/synchronous $W$-server switch.

Either the slotted ALOHA/polite access protocol or the slotted ALOHA/synchronous $W$-server switch protocol work independently of the network propagation time, and it was shown that they achieve a better performance than the ones in [21]. In particular, it is interesting to remark that the slotted ALOHA/synchronous $W$-server switch protocol has a greater maximum throughput than the slotted ALOHA/$W$-server switch even when the ratio of the network propagation time and the data packet transmission time is supposed to be null if, correspondingly, $W < L < 4W$.

As the propagation delay becomes significant, the maximum throughput of the slotted ALOHA/$W$-server switch protocol decreases, while that of the slotted ALOHA/synchronous $W$-server switch protocol remains the same.

Among the attempts to improve the performance in [20] continuing to use carrier sensing but taking advantage from the conclusions of [21], it is worth mentioning the ALOHA/slotted CSMA protocol described in [27]. A user senses the data channel at the beginning of each CSMA slot; once having found an idle channel, it immediately transmits a control packet that preannounces the employment of such wavelength. As in [21,26], only in the case of successful transmission of the control packet, the user sends the corresponding data packet. It was proved that the ALOHA/slotted CSMA protocol outperforms the ALOHA/CSMA in the case of heavy traffic and large values of packet length. Though the improvement depends on the specific values of the variables involved [27], it was proved that the ALOHA/slotted CSMA protocol achieves a higher throughput and causes much less delay than the original ALOHA/CSMA protocol for a reasonable range of system parameters. Furthermore, it is also relatively cheap since it allows each user to adopt one (tunable) laser and one (tunable) receiver.

As mentioned above, the analysis of [20,21] did not take into account the possibility of receiver collisions. It relied on the observation that the probability of receiver collisions is small in the case of very large (tending to infinity) user population. This mathematical, but unphysical, hypothesis was removed in [28] where a finite user population model was developed for the improved slotted ALOHA/ALOHA protocol of [21]. In the same reference it was proved that the throughput of the system, obviously degraded by the receiver collisions, can have a bimodal behavior, that is, its curve can exhibit two peaks. Really, the same may occur in the case of infinite population, which was also examined in [28], so deriving other interesting features of the protocol. In particular, it was shown that when the control channel is over-dimensioned (which means that it is allocated a fraction of the total bandwidth larger than what is needed) the mean packet delay can have a nonmonotonic characteristic, sometimes decreasing when the load is increased. More precisely, this was found to occur for small values of the mean backoff delay, which is required to resolve packet collisions in the random access protocol [28].

***Some keypoints.*** Pretransmission coordination is realized by adding a control channel to the data channels. The two types of packets can use equal or different protocols; various combinations have been considered based on ALOHA, slotted ALOHA, CSMA, *W*-server switch, and so on. Each of these solutions exhibits advantages and drawbacks, thus lending itself to further refinement and improvement. Analytical models developed for performance evaluation of these protocols should take into account the delays in the propagation of the signals (e.g., noninstantaneous feedback in CSMA) and finite user population. Otherwise, incomplete or overly optimistic conclusions might be drawn.

#### *2.2.3.2 Extended Slotted ALOHA and Reservation ALOHA*

The bimodal throughput characteristics were also confirmed in [29] where six cases of slotted ALOHA, six cases of improved slotted ALOHA and two cases of reservation ALOHA protocols were described. In all the proposed solutions, the time is divided into cycles, each including control packets as well as data packets. The system is TT-TR, which means that no separate fixed tuned transmitter is used for the control channel. On the contrary, all idle users are tuned to $\lambda_0$, for monitoring the control channel.

As the terminology clearly states, the first set of protocols has a relationship with those proposed in [20], where a data packet is transmitted immediately after the corresponding control packet, regardless of whether the control packet transmission was successful or not. The basic principle of the various solutions are summarized in Figure 2.10, where they are numbered from 1 to 6. Both control and data channels are slotted, using the same minislot time reference.

In case 1, the control sequence consists of $W$ minislots, while $L$ denotes the length of the data sequence on each wavelength and each cycle. There are no minislots on the control channel after the $W$th, while the data slots on the data channels start just after the $W$th minislot. Every minislot is strictly preassigned to a different wavelength; therefore, a node that has a packet to transmit in the cycle between ($K$) and ($K$ + 1), and that wishes to use the $i$th channel ($i = 1, 2, \ldots, W$) for such transmission, must send a control packet on the $i$th minislot of the same cycle and then transmit the corresponding data exactly after the $W$th time slot.

Because of the fixed assignment of a control slot for each channel, a successful control packet transmission ensures that the corresponding data packet will be also successful. Retransmission of a given packet will only be implied by an unsuccessful transmission of the corresponding control packet.

This scheme is very simple but, as a counterpart, it is wasteful since, during a cycle, the data channels are unused for the first $W$ minislots and, conversely, the control channel is idle for the last $L$ minislots.

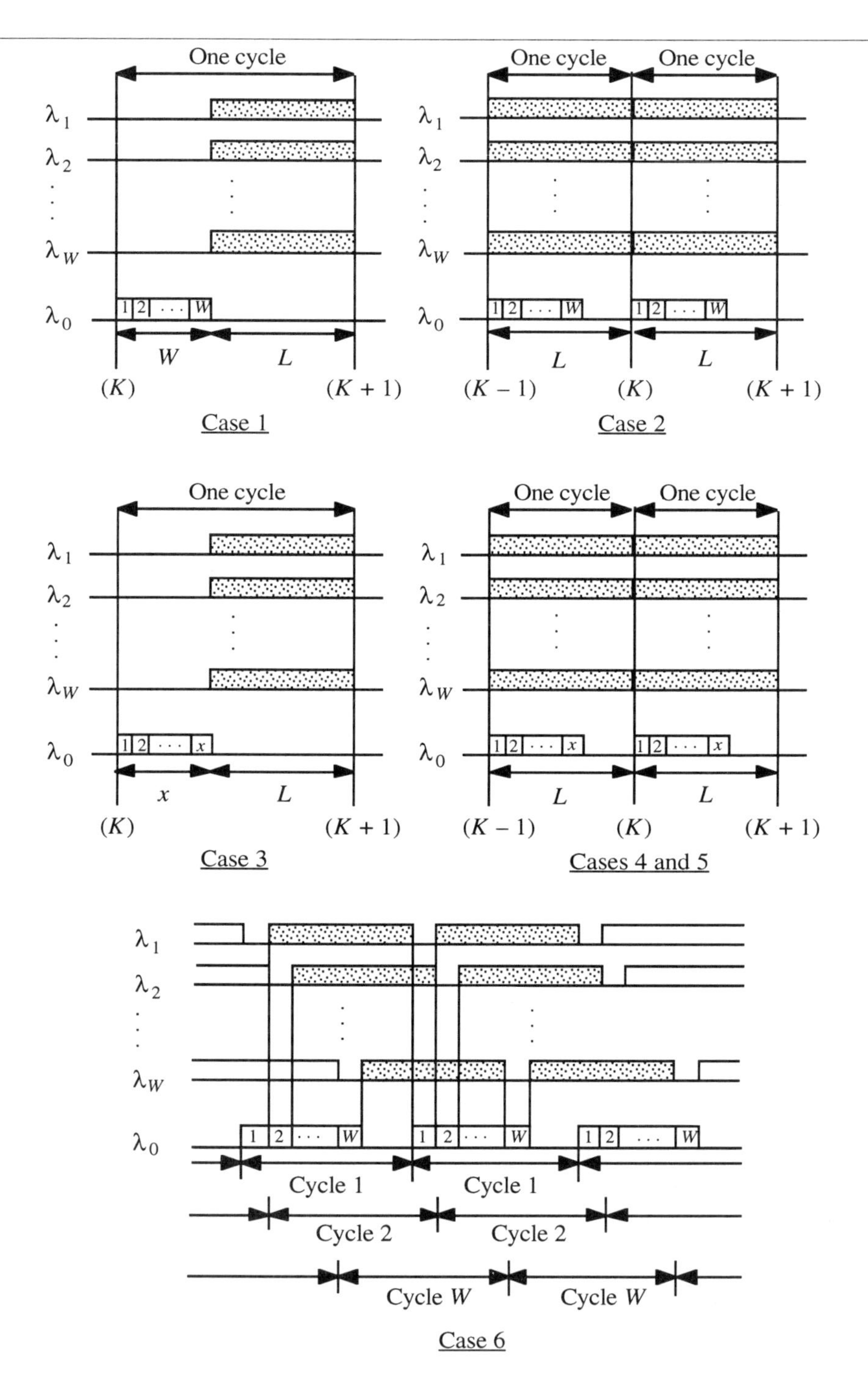


**Figure 2.10** Solutions for the slotted ALOHA protocol [$\lambda_0$ = control channel; $\lambda_i$ ($i$ = 1, . . . ,$W$) = data channels; $L$ = length of a data packet (stippled bar); a control packet is of unit length].

A more efficient solution can be found in case 2; the duration of the cycle here is $L$ minislots (instead of $L + W$, as in the previous case). Because of this partial overlapping of the control sequence and the data sequence, the transmission of corresponding packets must be obviously deferred. In particular, a node that transmits a control packet in the $i$th minislot of the cycle between $(K - 1)$ and $(K)$ will transmit the corresponding data packet in the $i$th channel of the subsequent cycle (i.e., between $(K)$ and $(K + 1)$). In this way, wastage is limited to the last $(L - W)$ minislots on the control channel in each cycle.

Case 3 is similar to case 1, but with the important difference of having $x$ minislots (instead of $W$) for the control sequence, with no rigid preassignment to the channels. In practice, the user wishing to transmit data in the cycle between $(K)$ and $(K + 1)$ randomly selects any of the $x$ minislots for transmitting the control packet and, immediately after the $x$th minislot, begins to transmit the data packet on the $i$th channel. In this case, it is evident that a successful control packet does not necessarily imply a successful data packet, since two or more users may choose the same wavelength in a given cycle (channel collision). Performance comparison between case 3 and case 1, as well as among all the other cases, will be discussed briefly below (see Table 2.2).

Case 4 can be seen as an extension of case 3 where, similar to case 2, the control and data packets are transmitted in two different cycles. Case 5, in turn, is a particularization of case 4, obtained putting $x = L$.

An alternative approach is presented in case 6, where the protocol employs asynchronous cycles on the channels; this means that any channel has its own cycle, having a duration of $L + 1$ minislots. For any channel, the transmission of a data packet follows immediately that of the control packet.

In the above-mentioned slotted ALOHA protocols, the retransmission strategy is not fixed. When a transmission is unsuccessful during the current cycle, the user can select any other wavelength (not necessarily the same chosen earlier) with equal probability and try again the transmission in the next cycle.

The total delay $D$ to transmit a packet from source to destination consists of the sum of three distinct components:

- The average time passing from the packet generation instant and the beginning of the next cycle (when the transmission procedure will start);
- The average retransmission time;
- The average time necessary to transmit a packet.

For all the examined slotted ALOHA protocols, these quantities, and then $D$, can be easily determined analytically [29], together with the average number $S_p$ of successful packets transmitted per cycle on each wavelength. The latter has therefore the meaning of throughput per channel. The results are summarized in Table 2.2, where $G$ represents the average number of control packets transmitted per slot on the control channel, and $t_r$ is the average length of a retransmission period. The latter depends on the backoff strategy adopted; in [29], in particular, $t_r$ was chosen such that the retransmissions take place in the next immediate cycle.

In the second set of six slotted ALOHA protocols presented in [29] a data packet is transmitted only if the corresponding control packet was successful. The total delay and the average number of successful packets transmitted per cycle (throughput per channel) in these modified situations are summarized in Table 2.3 following the same order of Table 2.2.

**Table 2.2**
Total Delay $D$ and Average Number $S_p$ of Successful Packets Transmitted per Cycle with the Slotted ALOHA Protocols of [29]

| | $D$ | $S_p$ |
|---|---|---|
| Case 1 | $(L+W)\cdot e^{G}+\frac{L+W}{2}+\left(e^{G}-1\right)\cdot t_r$ | $\frac{L}{L+W}\cdot G\cdot e^{-G}$ |
| Case 2 | $2L\cdot e^{G}+\frac{L}{2}+\left(e^{G}-1\right)\cdot t_r$ | $G\cdot e^{-G}$ |
| Case 3 | $(L+x)\cdot e^{[G+G(x-1)/W]}+\frac{L+x}{2}+\left\{e^{[G+G(x-1)/W]}-1\right\}\cdot t_r$ | $\frac{L}{L+x}\cdot\frac{Gx}{W}\cdot e^{\{-[1+(x-1)/W]G\}}$ |
| Case 4 | $2L\cdot e^{[G+G(x-1)/W]}+\frac{L}{2}+\left\{e^{[G+G(x-1)/W]}-1\right\}\cdot t_r$ | $\frac{Gx}{W}\cdot e^{\{-[1+(x-1)/W]G\}}$ |
| Case 5 | $2L\cdot e^{[G+G(L-1)/W]}+\frac{L}{2}+\left\{e^{[G+G(L-1)/W]}-1\right\}\cdot t_r$ | $\frac{GL}{W}\cdot e^{\{-[1+(L-1)/W]G\}}$ |
| Case 6 | $(L+1)\cdot e^{G}+\frac{L+1}{2}+\left(e^{G}-1\right)\cdot t_r$ | $\frac{L}{L+1}\cdot G\cdot e^{-G}$ |

Starting from Tables 2.2 and 2.3 a numerical comparison among the slotted ALOHA protocols, improved or not, can be developed. Let us consider, for example, the channel throughput performance. By differentiating, with respect to $G$, the expressions of $S_p$, one can find the maximum value for this parameter as a function of the number of data channels, the length of the data packet and, when applicable, the number of minislots on the control channel. This is shown in Table 2.4. As evident from the last column of Tables 2.2 and 2.3, the throughput performance of case 1, case 2 and case 6 are identical for the slotted ALOHA and the improved slotted ALOHA protocols; thus, in these cases, the results of Table 2.4 refer indifferently to both these schemes. On the contrary, for case 3, case 4, and case 5, only the improved solutions have been considered, in particular assuming $x = W$ for case 3 and case 4.

As stressed above, some of the slotted ALOHA protocols (improved or not) exhibit a wastage of bandwidth. In order to have a better usage, some multicontrol channel protocols were proposed, which extend the control operation over many channels. In [30], for example, the users are subdivided into groups, and each group is assigned its particular and private control slot. A node wishing to send a packet to another node of its own group must transmit a control packet on a minislot of the control slot of the group.

Minislots are assigned on a contention basis. Before its control packet transmission, the source must randomly select one of the data channels, in such a way that it is indicated in the control packet. All groups can access the same available data channels, and this is the reason for a better channel utilization. When communication involves two separate groups, the source becomes a temporary member of the group of the destination node, and it must send its control packet on the control channel of the other group. This protocol is suitable for the interconnection of separate star subnetworks; using active repeater nodes, even wavelength reuse is possible at each subnetwork.

**Table 2.3**
Total Delay $D$ and Average Number $S_p$ of Successful Packets Transmitted per Cycle with the Improved Slotted ALOHA Protocols of [29]

| | $D$ | $S_p$ |
|---|---|---|
| Case 1 | $(L+W)+\frac{L+W}{2}+\left(e^{G}-1\right)\cdot\left(\frac{W+1}{2}+t_r\right)$ | $\frac{L}{L+W}\cdot G\cdot e^{-G}$ |
| Case 2 | $2L+\frac{L}{2}+(e^{G}-1)\cdot\left(\frac{W+1}{2}+t_r\right)$ | $G\cdot e^{-G}$ |
| Case 3 | $(L+x)+\frac{L+x}{2}+\left\{\frac{e^{G}}{\left[1-G\cdot e^{-G}/W\right]^{(x-1)}}-1\right\}\cdot\left(\frac{x+1}{2}+t_r\right)$ | $\frac{L}{L+x}\cdot\frac{Gx}{W}\cdot e^{-G}\cdot\left[1-\frac{G}{W}\cdot e^{-G}\right]^{(x-1)}$ |
| Case 4 | $2L+\frac{L}{2}+\left\{\frac{e^{G}}{\left[1-G\cdot e^{-G}/W\right]^{(x-1)}}-1\right\}\cdot\left(\frac{x+1}{2}+t_r\right)$ | $\frac{Gx}{W}\cdot e^{-G}\cdot\left[1-\frac{G}{W}\cdot e^{-G}\right]^{(x-1)}$ |
| Case 5 | $2L+\frac{L}{2}+\left\{\frac{e^{G}}{\left[1-G\cdot e^{-G}/W\right]^{(L-1)}}-1\right\}\cdot\left(\frac{L+1}{2}+t_r\right)$ | $\frac{GL}{W}\cdot e^{-G}\cdot\left[1-\frac{G}{W}\cdot e^{-G}\right]^{(L-1)}$ |
| Case 6 | $(L+1)+\frac{L+1}{2}+\left(e^{G}-1\right)\cdot\left(1+t_r\right)$ | $\frac{L}{L+1}\cdot G\cdot e^{-G}$ |

In the above multicontrol channel protocol, data channel collisions are obviously possible. In order to avoid them (completely), a different cycle structure was proposed, which is reported in Figure 2.11 [31]. It applies to the case $N = W$, that is, the number of users equals the number of channels; therefore, the system is TT-FR. Each cycle is subdivided into $(N + y + L)$ minislots, $N$ being the length of the control slot (i.e., set equal to the number of users), $y$ the length of an information slot that the destinations employ to inform the sources if their control packet transmissions have been successful or not, and $L$ being the length of the data packet. When the user $i$ has to send a data packet to user $j$, it transmits a control packet in the $i$th minislot on the wavelength $\lambda_j$. At the end of the control slot, user $j$ has received all the connection requests addressed to it, and selects one of them (say that from user $k$) randomly. Then it informs user $k$ that its control packet transmission was successful. Once having received such an acknowledgment, user $k$ sends the data packet to user $j$ in the data slot on $\lambda_j$. Because of its characteristics, the described solution gives an example of switching protocol (and in this sense it is referred to in Figure 2.11).

**Table 2.4**
Maximum Throughput per Channel for the Slotted ALOHA Protocols[1]

| L | W | Case 1 | Case 2 | Case 3 | Case 4 | Case 5 | Case 6 |
|---|---|---|---|---|---|---|---|
| 2 | 2 | 0.184 | 0.368 | 0.150 | 0.300 | 0.300 | 0.245 |
| | 5 | 0.105 | 0.368 | 0.077 | 0.271 | 0.136 | 0.245 |
| | 10 | 0.061 | 0.368 | 0.044 | 0.263 | 0.071 | 0.245 |
| 10 | 2 | 0.307 | 0.368 | 0.250 | 0.300 | 0.387 | 0.334 |
| | 5 | 0.245 | 0.368 | 0.181 | 0.271 | 0.370 | 0.334 |
| | 10 | 0.184 | 0.368 | 0.131 | 0.263 | 0.263 | 0.334 |
| 100 | 2 | 0.361 | 0.368 | 0.294 | 0.300 | 0.370 | 0.364 |
| | 5 | 0.350 | 0.368 | 0.258 | 0.271 | 0.370 | 0.364 |
| | 10 | 0.334 | 0.368 | 0.239 | 0.263 | 0.370 | 0.364 |

*Source*: [29]. © 1991 Institute of Electrical and Electronics Engineers (IEEE).
[1]Cases 3–5 refer to the improved protocols. In case 3 and case 4, $x = W$ has been assumed.

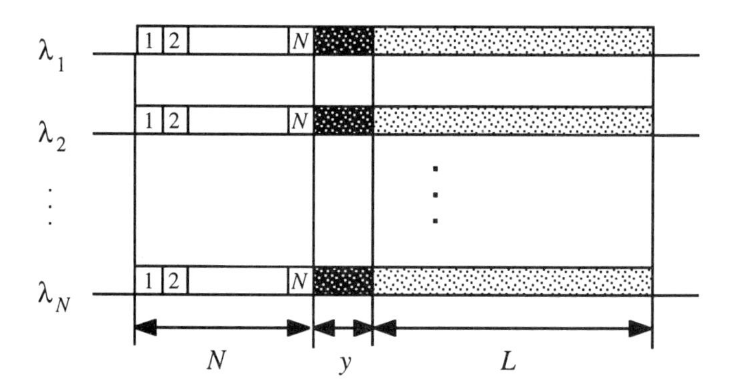


**Figure 2.11** Example of switching protocol; $y$ minislots are used to inform the sources if their control packets have been successful or not.

A fast alternative switching protocol was proposed in [32]. The system is FT$^2$-FR-TR, that is, according with the notation introduced in Section 2.1, any user has two fixed tuned transmitters and two receivers, the first constituted by a fixed tuned filter and the second by a tunable filter. Each user is identified by a different wavelength, so that we have $N = W$ in this case, too. The two transmitters transmit, respectively, on the control channel (common to all the nodes) and on its own data channel. The fixed tuned filter, in turn, receives on the control channel, while the tunable filter spans the set of data channel wavelengths. As in the slotted ALOHA, the time is slotted either on the control channel or on the data channel, and the control slot is here subdivided into $N$ minislots, the $i$th being preassigned, in a deterministic way, to user $i$. Each user always monitors the control channel, and then it knows how many packets are destined to himself and to all the other nodes of the network. This information is used to update, at the end of any control slot, a backlog status matrix $\boldsymbol{B}$, whose generic element $b_{ij}$ indicates the number of packets destined to user $j$ from user $i$. From $\boldsymbol{B}$, another matrix is generated, called $\boldsymbol{F}$, whose $f_{ij}$ element is set to 1 when user $i$ can transmit to $j$, and to 0 otherwise. For this kind of generation, a simple maximum remaining sum (MRS) algorithm is adopted, which can find a suboptimal $\boldsymbol{F}$ in a small number of operations.

Another example of multicontrol channel protocol, instead, can be found in [33]. Each station is provided with a fixed tuned transmitter, a fixed tuned receiver, a rapidly tunable transmitter, and a rapidly tunable receiver (FT-TT-FR-TR system). Each FT and FR is assigned a unique wavelength so that, for a network of $N$ stations, $2N$ wavelengths are needed. The tunable transmitters and fixed receivers are used for control purposes and consequently they are denoted as control transmitters and control receivers, respectively. The fixed tuned transmitters and tunable receivers, in turn, are used for data packets and, accordingly, they are designated as data transmitters and receivers.

Let us consider a station in the network, noted by $A$. On each channel, the time is divided into frames of length $T$. A frame used by $A$'s control receiver is divided into $m$ slots. Since each user has a separate control channel, this provides each user up to $m$ simultaneous connections. More precisely, each connection is assigned one of these $m$ slots at the time of connection setup. A frame used by $A$'s data transmitter is instead divided into $(n + 1)$ slots, with $n \leq m$. $n$ of these slots (say those numbered from 1 to $n$) are used for transmitting data, while in the remaining one, called "status slot", station $A$ transmits the assignment of slots for its control receiver.

The protocol is designed to support three different classes of traffic:

- *Class 1*: Connection-oriented with guaranteed bandwidth; this is the case of video traffic;
- *Class 2*: Connection-oriented but bandwidth not guaranteed; this is the case of file transfer;
- *Class 3*: Datagram (i.e., connectionless) traffic; this is the case of short transactions or control messages.

The detailed description of connection setup, data transfer, acknowledgments, and connection disconnect for all these classes is given in [33]. Summarizing the features of the various classes we can say that:

- For class 1 traffic the packet header is almost always successfully transmitted on a control channel. Transmission is unsuccessful only when another station transmits in a slot previously assigned, being unaware of this. When the header is successfully transmitted, the corresponding data packet is successfully transmitted as well.

- For class 2 traffic the packet header is almost always successfully transmitted on a control channel. In this case, however, two or more stations may transmit to the same destination in the same data slot; only one of the data packets is obviously successful.
- For class 3 traffic the packet header is successfully transmitted on condition that no other station transmits a class 3 header or a connection request or acknowledgment. The data packet is successful if no other station sends a class 1 or class 2 packet in the same data slot to the tagged destination. Anyway, in a particular data slot, even though more class 3 packets are simultaneously addressed to the same destination, only one of them will be successfully received.

As mentioned above, the protocol uses a number of wavelengths ($2N$) which is greater than that required by the other protocols. This number can be reduced imposing that each control channel is shared by many stations. More precisely, all data transmitters are assigned unique wavelengths, but $N/M$ control receivers, grouped in a cluster, are assigned a single wavelength. In this way, the total number of wavelengths becomes $N + M$. A frame on the control channel is still divided into $m$ slots, so that up to $m$ users can establish simultaneous connections with some stations in a cluster. Reducing the number of channels, the amount of processing required increases, since the control receiver at each station must process the transmissions intended for all stations in its cluster. An optimum trade-off exists for any specific application. When $M = 1$, the protocol reduces to the so-called dynamic time-wavelength division multiple access (DT-WDMA) [34], where the total number of wavelengths equals $N + 1$.

Another possible extension of the protocol consists in the reduction from two to one of the number of transmitters and receivers at each station. This result is obtained using a single tunable transmitter and a single tunable receiver both for control and data. The frame, unique at this point, is divided into $m + n + 1$ slots. In the first $m$ slots, the receiver at each station operates as a control receiver and the transmitter as a control transmitter. In the subsequent $n + 1$ slots, instead, the receiver operates as a data receiver and the transmitter as a data transmitter. Similarly to cases 1 and 3 of the slotted ALOHA protocol, the control channels and the data channels cannot be exploited simultaneously, and this implies a reduction in the system efficiency. Nevertheless, the advantage in terms of saving hardware may be preferable to a limited underutilization of the available bandwidth. As a final remark, we can observe that the proposed protocol extensions (reducing the number of channels and reducing the number of transmitters and receivers) are not incompatible and can be easily combined.

As stressed at the beginning of this section, in [29] two cases of reservation ALOHA protocols were also described. Their main goal is to improve the throughput for packet broadcast networks, where a user has to send very long messages (which can be broken into multiple data packets) [35]. The protocol is also suitable for circuit-switched traffic.

The first reservation solution of [29] is an evolution of case 1 in the improved slotted ALOHA protocol. If a user $X$ has successfully transmitted both the control and the data packets, on the wavelength $\lambda_i$, in the cycle between ($K$) and ($K + 1$), then it uses the same data channel for the subsequent slots. To this purpose, it transmits a jam signal on the $i$th minislot up to the end of its session. In this way, the other users are prevented from using $\lambda_i$ during the time it is employed by $X$. Therefore, the node that had the successful transfer of a control packet reserves the channel until it has completed its own transmission.

The second reservation solution is an evolution of case 6 in the improved slotted ALOHA, that is, the usage of $\lambda_i$ obeys the same rules of the previous case, but the protocol is based on asynchronous cycles.

A different contention based reservation protocol is that described in [36]. The system is FT-TT-FR-TR, since each user has one optical transceiver operating at $\lambda_0$ (control channel) and one optical tunable transceiver which can operate at any one of the wavelengths $\lambda_1, \lambda_2 \ldots, \lambda_W$ (data channels, with $W < N$) independently. Any user that has a data packet to send tries to reserve a data channel by sending its reservation packet on the control channel. The slotted ALOHA protocol is used to this purpose, with the control packet containing only the destination address. Reservation is successful if the control packet does not collide with any other, does not have the same destination as those of earlier successful reservations, and the number of currently successful reservations is smaller than $W$. Users always monitor the control channel so that, at the end of a control slot, any user that requested the reservation knows the status of the system and, in particular, if its request has been satisfied or not. In the case of successful reservation, the user transmits the data packet in the next slot, on the wavelength $\lambda_i$ if the number of successful reservations before this one is $i - 1$. The destination node also knows the reservation, and it can tune its receiver to $\lambda_i$ for detecting the arriving data packet. If the reservation fails, the procedure is repeated at the next slot.

A multichannel slot reservation protocol was introduced in [37]: each node has access to one slot per cycle per channel. Cycles are separated by a time interval, whose duration is at least equal to the end-to-end propagation delay; this delay is necessary to permit circulation of the control information required for contention-free slot allocation. The control information is contained in a fixed allocation reservation subframe at the end of a cycle. Each node requires access to all the $W$ available channels, so that the protocol needs $W$ transmitters and receivers, with the appropriate wavelength division multiplexers and demultiplexers.

In [37] the performance advantages of the multichannel protocol were shown in comparison to the single channel protocol ($W = 1$) characterized by the same overall network transmission rate. Both symmetric and asymmetric traffic loading conditions were considered. In the first case, it was shown that the multichannel solution exhibits a significant traffic throughput benefit, the degree of improvement, however, being dependent on the efficiency of the single channel network. More precisely, if the level of efficiency is low (which can occur, for example, because the transmission rate of the single channel network is high) the degree of improvement is greater. The efficiency of the multichannel protocol obviously increases with increasing $W$ but a point exists above which there is a limited benefit from further increasing the number of channels. This optimum number is related to the specific network conditions, such as the end-to-end propagation delay, the transmission rate, and the number of nodes. In any case, in typical conditions [37] one can say that a major performance improvement is achievable just with a small number of channels (e.g., five).

Greater values of $W$ are instead generally required for asymmetric traffic loads; these have the effect of reducing the cycle length, leading to a very substantial deterioration in performance of the single channel protocol. On the contrary, the multichannel protocol reveals to be much less insensitive to this further bandwidth inefficiency, in any case ensuring very high throughputs.

Another simple example of multiple-channel-based-reservation scheme is that presented in [38] where, in a cycle, all channels have a contention slot, followed by a control slot and a data slot. Time slot organization is shown in Figure 2.12: there are $W$ channels, and one of them is also used as control channel ($\lambda_0$). The control slot is divided into $W$ minislots, each of them preassigned to a data channel, while the contention slot consists of $X$ minislots.

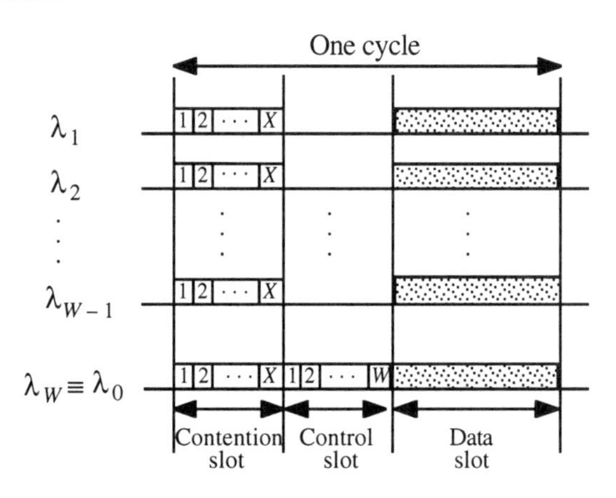


**Figure 2.12** Example of contention-based reservation protocol.

When a user wants to transmit a data packet, he or she has to contend for a randomly selected data channel in the contention slot of that channel. To this purpose, a contention packet is transmitted in one of the randomly selected minislots of the contention slot, using the ALOHA protocol. When the contention is in progress, the contending users monitor the contention slot of the channel of their interest; only the user that succeeds in transmitting a successful contention packet first is allowed to occupy the channel, while the others are inhibited from using it. The winning user transmits a control packet on the control minislot assigned to that data channel, and then sends its data packet in the data slot. The control packet contains source and destination identifiers; once being located in the $i$th minislot of the control slot, it indicates to the destination receiver the channel on which the data packet will be transmitted. Thus, in principle, the control slot could be eliminated using a network where the channels are preassigned to the users; in this case, however, we should have as many channels as the number of users (i.e., $W = N$). As qualitatively reasonable, the performance of this multichannel protocol depends on the number $X$ of minislots on the contention slot; some examples are given in [38].

***Some keypoints.*** Different solutions can be arranged in the framework of the slotted ALOHA protocol. Among them: the number of control packets in each cycle can be as high as the number of channels, or less; the transmission of the control packets can be overlapped to that of the data packets, or not; the cycles on the channels can be synchronous, or asynchronous. Improvement can result by conditioning the transmission of a data packet to the success of the corresponding control packet.
To limit the wastage of bandwidth some multicontrol channel protocols can be conceived that extend the control operation over many channels. Transmission management becomes more involved but also more prone to satisfy the requirements of either specific network organizations or different classes of traffic.
In the case of large file transfer or circuit-switched applications it is better to use protocols based on slot reservation techniques. Generally speaking, each user with a data packet to send attempts to reserve a data channel. This procedure is often contention-based, while the adoption of multichannel slot reservation protocols could be recommended as well, especially in the case of asymmetric traffic loading.

#### *2.2.3.3 Other Collision Avoidance Protocols*

The so-called receiver collision avoidance (RCA) protocol [39] is able to detect and avoid receiver collisions before packets are transmitted. The network consists of one control channel and $W$ data channels; each node has one tunable transmitter and one tunable receiver (TT-TR system). A control packet contains only the destination address and its transmission time is defined as one minislot. The length of a data packet (data slot) is fixed and equal to $L$ minislots; each data slot is further divided into $W$ control slots whose length is then $L/W$. Time on data channels must be synchronized with the time on the control channel.

In order to avoid a receiver collision, a source node has to be able to determine when the destination node is available to receive its packets. Generally speaking, in a structure where each node has only one receiver (as in the present case), the main difficulty arises from the fact that the receiver is responsible for the reception of both control packets and data packets. Thus, when it is tuned on a data channel, the node misses the information transmitted in the control channel, and fails to have a total knowledge on the status of the other nodes at all times. The RCA protocol solves this problem with a simple receiver collision and detection mechanism.

A distinctive feature of the protocol is the management, at each node, of a node activity list (NAL). RCA takes into account also the transceivers' tuning time (equal to $T$ control slots) and the NAL has $2T + W$ entries, one per each control slot, used to record information received on the control channel. Each entry contains a status, which can be either active or quiet. When a successful control packet is received, the status is active and the corresponding NAL entry also contains the destination address, as copied from the control packet. Otherwise the status is quiet. The NAL is updated as long as the receiver is tuned to the control channel; when the node receives on some data channel, instead, the NAL is reset. If all $2T + W$ entries are used up, the first entry is deleted and a new entry is appended to the end of the NAL to record information in the current control slot. Thus, the NAL can contain information on the control channel history throughout the most recent $2T + W$ slots.

The first step of the protocol is the data channel assignment; RCA uses a deterministic algorithm to avoid data channel collisions while maintaining the protocol simple. Precisely, the $W$ data channels are numbered, as usual by $\lambda_1, \lambda_2, \ldots, \lambda_W$; each data slot on the control channel is divided into $W$ control slots and $\lambda_1$ is assigned to the first control slot, $\lambda_2$ to the second and so on. Control packets are transmitted only at the beginning of each control slot; when it happens in the $i$th control slot the transmission of the subsequent data packet will use the channel $\lambda_i$. Because of the deterministic strategy, as stressed above, it is not necessary that the control packet includes the data channel number: synchronization, in fact, ensures that every node can always determine the relative position of a control slot in a data slot , and hence the data channel assigned to it.

Let us denote by $D$ the round-trip propagation delay, in control slots, between any two nodes (assumed to be the same for all the nodes). If a node transmits a control packet at time $t$ and the transmission is successful (i.e., it is received without channel or receiver collisions), the packet will return from the destination node at the time $t + D$. Immediately following this, the sending node will start tuning its transmitter to the assigned channel and will transmit the data packet from time $t + D + 1 + T$ to $t + D + 1 + T + W$. The time has been here expressed in control slots; if $L = W$ the duration of a control slot equals that of a minislot. The destination node will receive the data packet from $t + 2D + 1 + T$ to $t + 2D + 1 + T + W$. This specific scheduling of data packets is called asynchronous transfer on data channel (ATDC). It is evident that it allows the overlapping of one node's tuning time with other nodes' packet transmission time, so reducing the penalty due to large tuning times. It must be noted that the same overlapping mechanism could also have been incorporated in the protocols of [21] where, however, the analysis was developed without considering the propagation delay and, most of all, ignoring receiver collisions.

Coming now to a more detailed description of the packet transmission procedure, let us suppose that node $i$ has to send a data packet to node $j$. Besides the NAL, at each node a reception scheduling queue (RSQ) is also present, to schedule the reception of incoming data packets. Each RSQ's entry has two fields: scheduled reception time and the assigned channel. Scheduled reception time, in particular, is the time the local receiver will start tuning to the data channel for data packet reception. The entries are ordered according to the sequence of arrival times of successful control packets at the local node. Obviously, once a data packet has been received, the top entry is removed. Node $i$ consults the NAL to verify that no control packet to node $j$ has been observed during the past $2T + W$ control slots, and also consults the RSQ, to verify that it has been empty during the same time period. The simultaneous satisfaction of both these conditions allows node $i$ to conclude that no receiver collision is detected, and a new control packet can be emitted. Otherwise the transmission procedure at node $i$ is suspended until the conditions are met.

Before the data transmission can begin, and node $i$ can tune its transmitter to the assigned channel, a delay of $D$ control slots has to be expected, due to the round-trip propagation of the control packet emitted at time $t$. During this period the receiver on node $i$ remains on the control channel, updating its NAL at the beginning of each control slot. If a successful control packet to node $j$ is received between $t + D - (T + W)$ and $t + D - 1$ then node $i$ detects a receiver collision on data packet and the transmission procedure is aborted and restarted. If this is not the case, and at time $t + D$ the control packet is returned without channel collisions, node $i$ starts to tune its transmitter to the assigned data channel from time $t + D + 1$, and after $T$ control slots transmits the data packet. Once the transmission is completed, the node tunes the transmitter back to the control channel.

At the destination node $j$, let us denote by $t_c$ the reception time of the control packet emitted by node $i$. Node $j$ first consults its RSQ to check for receiver collisions on data channel. If the RSQ is empty, it is sure that no receiver collision will happen and the node continues to the next state. The same will be if some entries of the RSQ are used but the scheduled reception time of the last element in the RSQ, $t_r$, is such that $t_c - (t_r - D) > T + W$. If $t_c - (t_r - D) \leq T + W$, instead, then a receiver collision is detected (simultaneously revealed by the sending node through the transmission procedure) and the control packet is ignored. When the control packet is successfully received, the corresponding data packet reception is scheduled at $t_c + D + 1$, and an entry is created in the RSQ at node $j$. Noting by $t_f$ the scheduled reception time of the first element in the RSQ, node $j$ starts tuning its receiver to the assigned data channel at $t_f$, and receives a data packet from $t_f + T$ to $t_f + T + W - 1$. Then, the top element in the RSQ is removed, and node $j$ tunes its receiver back to the control channel or to another data channel if the next scheduled reception time follows immediately.

As regards performance, numerical results show [39] that the maximum throughput achievable by the RCA is about 36% of the total data channel capacity and slightly degrades for increasing transceiver tuning times $T$. This result is satisfactory if compared with other schemes having the same NIU structure [21,29], but not so good in absolute terms. Moreover, previous description clearly shows that the RCA requires a higher amount of protocol processing time.

A different approach to the problem of receiver conflicts is that proposed in QUADRO (QUeueing of Arrivals for Delayed Reception Operation) [40,41]. QUADRO introduces the concept of local conflict resolution at the receiver based on the use of optical delay lines (DLs). Here the system is FT-TR but the same idea of adopting delay lines for solving contentions will also be found in some experimental architectures based on TT-FR systems (see, for example, Section 2.3.9). Similarly, the network has as many users as the number of channels (i.e., $W = N$, which allows to avoid channel collisions) but the utility of QUADRO also remains valid when $W < N$.

In QUADRO, when two or more packets are received in the same time slot, the receiver tunes to one of the wavelengths for reception and, using the DLs, it stores the additional packets, with the object of detecting them in subsequent time slot(s). In practice, each DL delays a packet for one time slot, obtaining a "finite-time" optical buffer, with a reception window for each packet determined by the number of DLs at the receiver. The reception process consists of choosing a specific slot, inside the window, for the reception of each packet. An example of QUADRO implementation is shown in Figure 2.13. The QUADRO receiver consists of $d$ delay lines, $d + 1$ 2×2 optical switches, and one optical combiner. Only the input $I_2$ is used at each switch:

independently for each wavelength, the incoming signal can be transferred to the output $O_2$ (bar state) or to the output $O_1$ (cross state). The default state, for all wavelengths, is the bar state, but this condition can be controlled and changed to reroute some wavelength to the output $O_1$. Outputs $O_1$ of all switches are then combined together using a combiner, with $d + 1$ inputs, whose single output is connected to the tunable filter of the receiver. The receiver controls the switches in a such a way that only one wavelength of only one switch is deflected to one of the outputs $O_1$ at a time. This ensures that only one packet is routed to the tunable filter in each time slot, thus completely avoiding the possibility of conflicts or collisions after the combiner.

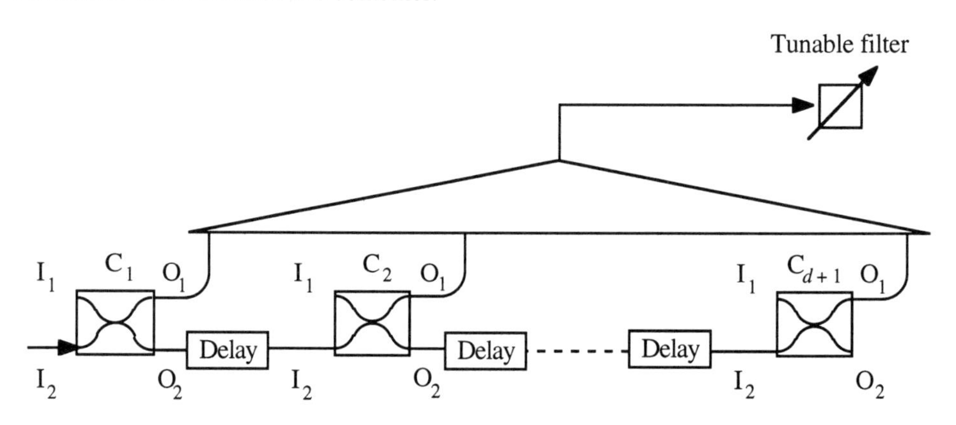


**Figure 2.13** QUADRO receiver block: $C_i$ = 2×2 switch ($i = 1, 2, \ldots, d + 1$). Outputs $O_1$ are transferred to the tunable filter of the receiver through an optical combiner (Delay = optical delay line).

QUADRO allows packets to be received by the tunable filter at arbitrary time slots in the reception window, thus providing a way to optimize the reception process. More precisely, when a packet reaches the destination node, it enters the first switch $C_1$; the receiver can decide to receive the packet in the current slot, in that case rerouting the wavelength carrying the packet to the output $O_1$ of $C_1$ and tuning the filter accordingly. But if a conflict occurs with other packets (or for any other reason), the receiver can delay the reception one slot later, waiting until the packet reaches the second switch $C_2$ and then rerouting its wavelength to the output $O_1$ of $C_2$. Similarly, the receiver can select one of the downstream switches $C_3, C_4, \ldots, C_{d+1}$, to delay packet reception for $2, 3, \ldots, d + 1$ slots, respectively.

It is qualitatively evident that the performance of QUADRO depends on the criterion used to select the slot for reception. Two reception strategies were considered in [40]: a last-in first-out (LIFO) strategy and a first-in first-out (FIFO) strategy. The packet's position in the receiver is defined as the subscript of the switch in which the packet is available; a packet that, in the current slot, is in switch $C_j$ is named to be in position $j$. LIFO selects for reception in the current slot the packet in the leftmost position. If several packets are in the leftmost nonempty position, one of them is chosen deterministically. Such a strategy is easy to implement, but it may exhibit the worst behavior in terms of packet loss, since it selects the packet that has the longest time to spend in the receiver.

Dually, FIFO selects the packet in the rightmost position, for reception in the current slot. If several packets are in the rightmost nonempty position, one of them is chosen deterministically. It is reasonable that such a strategy may have the least packet loss, since it selects the packet that has the shortest time still to spend in the receiver.

In order to understand better the QUADRO operation and the reception strategies adopted, it is useful to consider a simple example (shown in Figure 2.14), focusing on a specific configuration of packet arrivals.

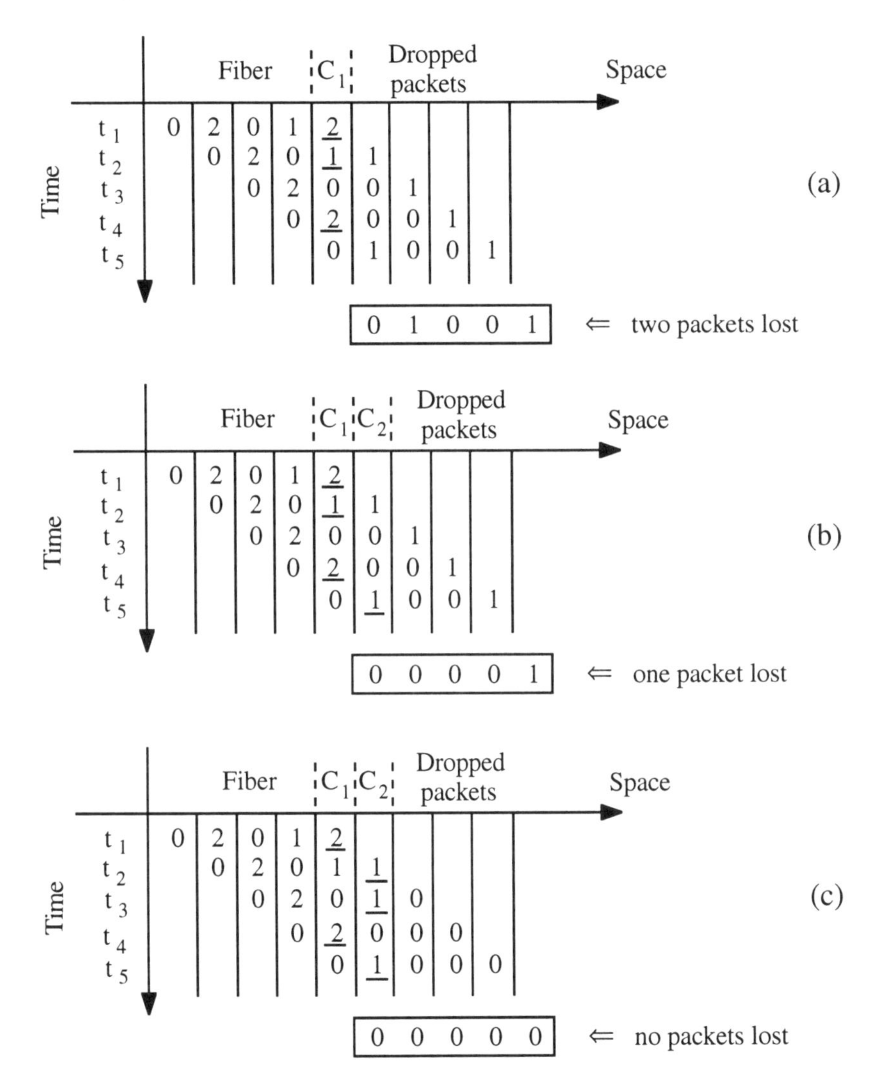


**Figure 2.14** Space-time diagrams for a particular sequence of packet arrivals: (a) no delay lines, (b) one delay line using LIFO reception strategy, and (c) one delay line using FIFO reception strategy. (*Source*: [40]. © 1994 IEEE.)

So, let us suppose that the distribution of packets destined to node $i$ in five contiguous slots, ordered form right to left, is (0, 2, 0, 1, 2), which means that there are two packets in the first slot, one packet in the second slot, zero packet in the third slot, etc. The average arrival rate is of one packet per slot, which implies that an optimum receiver should be able to detect all the packets without losses.

Figure 2.14 shows the position of arriving packets inside the receiver in different time slots: the horizontal axis represents the space, that is, fiber on the left, receiver in the middle, dropped packets on the right, while the vertical axis represents time, with $t_1$, $t_2$, $t_3$, $t_4$, and $t_5$ denoting five different contiguous time slots. The slot selected for the reception in each time slot is underlined; clearly, in the slot following a reception, the number of packets in the selected slot is decremented by one.

In Figure 2.14(a) the case of a system where QUADRO mechanism is not used is reported for comparison; in this case the receiver has only one slot in which to receive each packet. As a consequence of the specific packets distribution, two packets are lost. On the other hand, we observe that there are two slots, precisely $t_3$ and $t_5$, where the receiver remains completely inactive since no packets are available for reception. This is the inefficiency that QUADRO attempts to eliminate. For the considered example, using one delay line ($d$ = 1), one packet loss remains when the LIFO reception strategy is applied (see Figure 2.14(b)), whereas we have no packet loss when applying the FIFO strategy (see Figure 2.14(c)). In both cases the advantage over the system without QUADRO is evident.

It is interesting to observe that the performance of the QUADRO receiver is also affected by the way packets are selected for transmission. The transmission strategy, in fact, directly reflects on the probability of conflicts at the receiver. This is a general consideration, which was discussed in [40] by considering two different procedures, that is, random transmission and FIFO transmission. In random transmission, a packet is chosen randomly from the buffer queue at the source node, while FIFO strategy chooses from the buffers the packet that has spent the longest time in the system. The random strategy offers a higher capacity but the FIFO strategy minimizes packet delay variance. A lost packet needs to be retransmitted and, in this sense, the random strategy, which does not favor the transmission of any packet, is less sensitive to the effect of packet retransmission.

***Some keypoints.*** The RCA protocol is able to detect and avoid receiver collisions. The price to pay is in terms of increased processing time, with respect to many of the other schemes. In QUADRO the receiver conflicts are solved using delay lines as optical buffers. In practice, when two or more packets are received in the same time slot, the receiver tunes to one of the wavelengths for reception, while storing the other packets to detect them in subsequent time slots.

#### 2.2.3.4 *Dynamic Time-Wavelength Division Multiple Access Protocols*

Some of the protocols described in the previous sections [32,33,40,41] were conceived as extensions (sometimes with reduced hardware complexity) to the dynamic time-wavelength division multiple access (DT-WDMA) proposed in [34]. The original DT-WDMA requires two transmitters and two receivers per each node: one transmitter and one receiver are always tuned to the control channel, each node has exclusive transmission rights on a data channel to which its other transmitter is always tuned, and the second receiver is tunable over the entire wavelength range. Thus, the system requires $N + 1$ channels, $N$ for data transmission and the $(N + 1)$th for control.

The system is slotted with slots synchronized over all channels at the hub (star coupler). Slots on the data channels are called, as usual, data slots, since they contain the actual data packets. Slots on the common control channel, instead, are called status slots, since they carry status information about the packet and the transmitter. A status slot is divided into $N$ minislots, one per each node. Thus access to the control channel is TDM based. Each minislot contains the address of the destination, the delay experienced by the packet since its arrival at the source, and the mode of the transmission (which can be packet-switched or circuit-switched). In particular, the delay for a packet is computed as the sum of its total delay until the transmission instant (i.e., the difference between the time of transmission and the time of arrival at the transmitter) and the time taken by the packet to reach the hub from the considered station. It is used by the source node as a priority information.

The data and the status slots are of equal length, given by the largest number between the length of the $N$ status minislots and the maximum packet length plus the time necessary for the receivers to tune from one channel to another (here taken into account). Each transmitter maintains a queue of packets; each packet is tagged with its arrival time and a status flag, which can be outstanding or waiting. A packet is outstanding if it has been transmitted but the result of the transmission is not known yet; a packet is waiting if it has to be transmitted. Obviously, once a packet has been transmitted its status flag changes for waiting to outstanding.

The adopted policy for transmission scheduling is FIFO. When station $i$ has a packet to send to station $j$, it transmits the destination address and the delay of the packet in the $i$th minislot of a status slot, say slot number $t$. Then it transmits the data packet, at wavelength $\lambda_i$, in the next data slot, that is, slot number $t + 1$. The sending station receives the information about the result of the transmission after a time equal to the round-trip propagation time between the station and the hub (and not that between the transmitter and the receiver). This reduced latency can be explained simply observing that each station receives all the minislots, so that every transmitter can determine the outcome of its transmissions without explicit acknowledgment. If the transmission is successful station $i$ removes the packet from its queue; otherwise, the status flag of the packet is changed from outstanding to waiting.

As mentioned above, each receiver continuously monitors any minislot at the common wavelength. After examining all minislots in slot number $t$, station $j$ knows if some data packets in the subsequent data slot are intended for it. When more than one packet is sent to station $j$ in the same data slot, station $j$ selects one of them using a common deterministic arbitration algorithm. Then, the station adjusts its tunable filter to the wavelength of the chosen data packet before the next data slot starts.

As regards the adopted arbitration algorithm, basically it consists in checking the priority fields of the corresponding minislots, and in selecting the one with highest priority. A simple but reasonable possibility is as follows [34]: station $j$ cannot be interrupted by other stations if it is in the circuit-switched mode (in practice, it can select only a packet from a specific source) while, in the packet-switched mode, the packet that has the largest delay so far succeeds, and the others fail. In any case, exactly one of the contending transmissions will always be successfully received.

An example of application of the DT-WDMA protocol is shown in Figure 2.15.

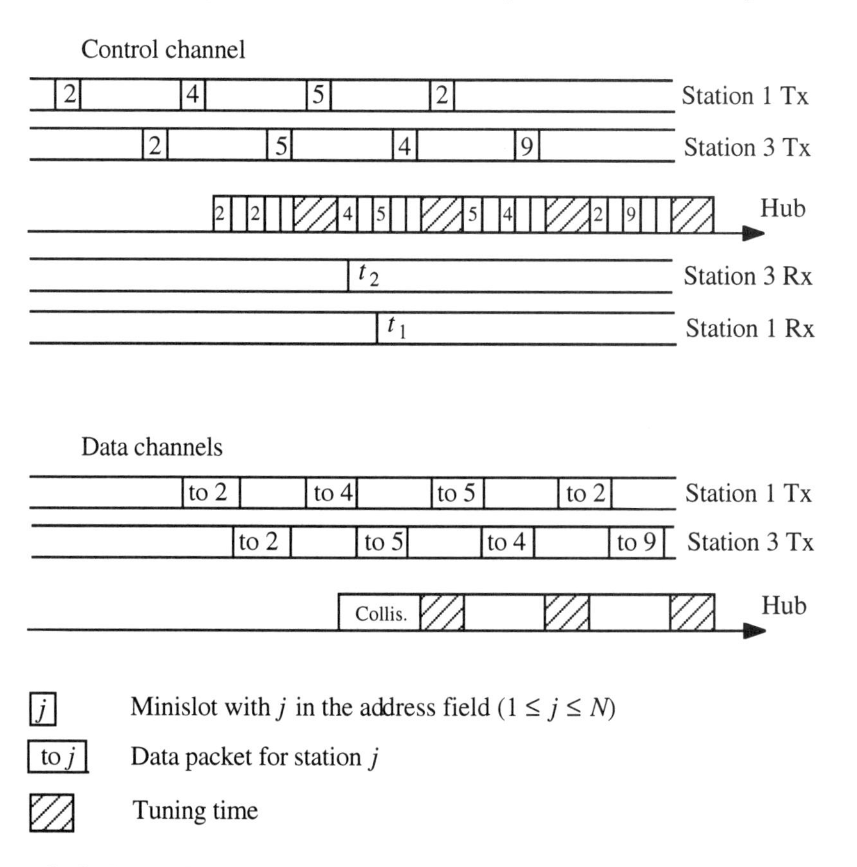


**Figure 2.15** The DT-WDMA protocol: time-space diagram for two stations, illustrating the method of operation (Rx = receiver, Tx = transmitter). (*Source*: [34]. © 1990 IEEE.)

Two stations are considered: number 1 and number 3, the latter located closer to the hub than the former. In their appropriate minislots in the first status slot, the two stations express their intention to send a packet to station 2. So, in the next data slot, both stations transmit the packet on their proper wavelength. On the basis of the arbitration algorithm described above, the receiver (station 2) decides to accept the packet from station 3 and, hence, station 1 will be unsuccessful. This information is detected by station 3 at time $t_2$ and by station 1 at time $t_1$ (greater than $t_2$, since station 1 is farther from the hub). After $t_1$, station 1 decides to retransmit the packet to station 2 as soon as possible. In the considered example, this can occur in the fourth slot, where station 1 can send again its intention to transmit to station 2, and this time the transmission is successful. In the meanwhile, however, two data slots have passed by, during which station 1 has successfully transmitted to stations 4 and 5, and station 3 has successfully transmitted to stations 5 and 4 (without collisions).

Looking at the performance, an interesting feature of the DT-WDMA mechanism is the capacity to support arbitrary propagation delays between the various nodes and the hub. Furthermore, the average packet latency is small: as an example, for a network where all stations are five slots away from the hub, the average delay is less than 17 slots. As a counterpart, we have poor scalability, because of the TDM-based control channel and the necessity to have as many data channels as the number of nodes.

Throughput, in turn, might be penalized by the inclusion of the receiver tuning time in the slot duration. Simulation indicates that the maximum normalized network throughput for systems of up to a few hundred stations, all equidistant from the hub, is 0.6 packets per station per slot. This value is comparable with that achieved using some fixed-bandwidth allocation schemes [13].

Among the recent proposals to reduce the number of collisions in DT-WDMA, the idea of using learning automata seems interesting [42]: each node is provided with a learning automaton that decides, on the basis of a network feedback information (common for all the stations), which of the packets waiting for transmission will be sent at the beginning of the next time slot. It was shown that such a new contention solving algorithm allows to have a reduction up to 20% in the number of packet losses.

Besides the above-mentioned improvements of [32,33,40,41] further evolutions of the basic DT-WDMA protocol were proposed in [43], where two reservation-based protocols with varying signaling complexity were presented:

1. The dynamic allocation scheme (DAS), which dynamically assigns slots on a packet-by-packet basis;
2. The hybrid time division multiplexing (HTDM) scheme, which combines the TDM and the DAS scheme, and allows both preassigned and dynamic slot assignment.

The network architecture is the same as in [34], with a fixed-wavelength transmitter and a tunable receiver at each node for data packets, and a separate transceiver for the control channel $\lambda_0$, which is shared by all stations. At each transmitter, packets destined for different receivers are stored in separate buffers. Assuming, for simplicity, that each station may send packets also to itself, each transmitter maintains $N$ receiver queues.

All packets, of fixed length, consist of $b$ bits and, like in the other DT-WDMA mechanisms, the system is time slotted-synchronized at the passive star. The duration of a slot is simply equal to a packet transmission time, and is denoted by $\Delta$.

In DAS, all the nodes execute an identical random scheduling algorithm, which selects packets for transmission from each station on a slot-by-slot basis. This algorithm, based on a common random number generator with the same seed at all nodes (a similar procedure was described in Section 2.2.2, relative to the protocols of [22]), is executed synchronously, at the beginning of each slot, by any node. Therefore, at the termination of the algorithm, each transmitter/receiver knows from which queue a packet will be transmitted/received in the upcoming slot.

A brief description of the random scheduling algorithm is as follows. First a transmitter, say $i$, is randomly selected. Next, among all its nonempty receiver queues, one queue, say $r$, is randomly selected in turn. The upcoming slot is reserved for transmitter $i$ to send a packet (the one at the head of the queue) to receiver $r$, that will tune to the transmission frequency of the chosen transmitter. If all the receiver queues at $i$ are empty, no packet is selected. The process is repeated to select the other transmitters and receivers, but with the expedient of excluding, at any step, the transmitters and the receivers that have been already scheduled. For example, after selection of transmitter $i$ and its nonempty receiver queue $r$, a second transmitter $j$ is randomly chosen so that $j \neq i$, and a nonempty queue is selected in the set $\{1, 2, \ldots, r-1, r+1, \ldots, N\}$.

We note that the random selection of transmitters ensures fairness, while higher priorities to the queues with larger arrival rates are implicitly provided through the action of selecting receiver queues only among the nonempty ones.

In order to implement DAS using the random scheduling algorithm, every station must know the state of any receiver queue in the network. This information is distributed using wavelength $\lambda_0$ (signaling channel), according with the usual TDM scheme, with dedicated minislots (one per each station), described above for the basic DT-WDMA mechanism. The signaling protocol proposed in [43] for DAS is very simple: at the beginning of every slot, each transmitter, prior to executing the random scheduling algorithm, broadcasts the state of all its receiver queues. Because the information to transfer, for each queue, simply consists in one of two possible alternatives (empty or nonempty), 1 bit is sufficient for this purpose and, globally, each station transmits $N$ signaling bits per each slot. Multiplying by the number of stations and dividing by the slot duration, we obtain a required signaling channel capacity of $N^2/\Delta$. Since the link capacity is $C = b/\Delta$, we are forced to have $N^2 = b$ at most, and this may lead to severe limitations on the size of the network; for example, assuming packet sizes of length $b = 1000$ bits, we have $N \leq 31$. In order to relax this constraint, the HTDM was also presented in [43].

In HTDM, time on the data channels is divided into frames consisting of $N + x$ slots, where $x$ is a positive integer number. As in a fixed TDM scheme, the $N$ slots are preassigned, one for each destination station. A transmitter sends a packet to a receiver in its preassigned slot if the corresponding receiver queue is nonempty; otherwise the slot remains unused. After every $n = \lceil N/x \rceil$ slots, however, one slot is left open, which means that, into it, the transmitter can send a packet to any receiver. An example of time slot assignment for HTDM, under uniform load and assuming $n = 2$, is shown in Figure 2.16.

The contention and scheduling problems, which appear during an open slot, are solved using the random scheduling algorithm discussed for DAS. Here, however, only $x$ ($< N$) decisions about which packet to transmit need to be taken, by each station, every $N + x$ slots. The signaling scheme proposed above can be also used, with a minor variation to take advantage of the fact that the transmitters have knowledge of the packets that will be transmitted in the preassigned slots. Therefore, each transmitter sends $N$ signaling bits, but only every $n\Delta$ seconds. As a consequence, the required signaling channel capacity is $N^2/(n\Delta) = Nx/\Delta$, which is smaller than for DAS. The constraint on the network size is reduced accordingly. For example, assuming $b = 1{,}000$ bits and $x = 5$ we have $N \leq 200$.

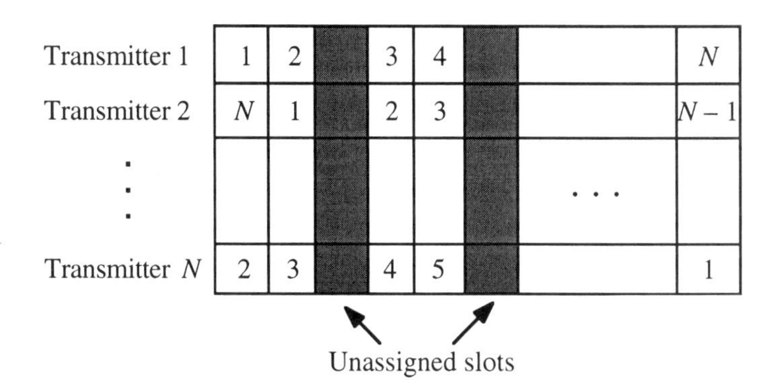


**Figure 2.16** Example of time slot assignment for the HTDM protocol. Every $n = 2$ slots, one slot is left open and can be used by the transmitter to send a packet to any receiver, in a nonpreassigned fashion.

The performance of DAS and HTDM was compared, through simulation and analytical studies, with that of TDM in [43]. It was shown that DAS provides the best result under ideal conditions, that is, zero propagation delay. On the other hand, as previously stressed, HTDM has lower signaling costs and can support a larger network. In comparison with TDM, when uniform traffic conditions are considered, HTDM behaves better under low loads, while it exhibits comparable performance for high loads. The advantage of the hybrid scheme, in terms of throughput and delay, becomes evident in the case of nonuniform traffic conditions, when the system can benefit much more by the dynamic allocation of slots. Moreover, the protocol can handle also bursty traffic.

We have previously stressed that one important drawback of the DT-WDMA protocol is in the fact that the system size is limited to the number of available channels. To overcome this constraint, the TDMA-W protocol was considered in [44]; this protocol is able to make the number of interconnected nodes and data channels independent, thus greatly simplifying system expansion through increases in nodes and/or channels. In the TDMA-W protocol each node is equipped with only one tunable transmitter for both control and data packet transmission. As in DT-WDMA, however, each node has two receivers, one fixed on the control channel $\lambda_0$ and one tunable over the entire wavelength range $\lambda_1, \lambda_2, \ldots, \lambda_W$ of the data channels. The control packet contains the source node address, the destination node address, the selected data channel $\lambda_i$ $(1 \leq i \leq W)$ and the data packet length $L$. This is an interesting feature of the protocol, which supports variable sized packets, without loss of utilization when the packets are small. Time is normalized to the control slot, so that the time required for the transmission of a generic data packet is $L$ times that required for the transmission of a control packet.

Access to the control channel is based on a static cyclic slot allocation scheme: each node is assigned one control slot per cycle, which it uses to reserve access on a data channel if backlogged. All nodes can transmit a control and data packet during each cycle; thus, a control cycle consists of $N$ control slots. The situation is shown in Figure 2.17; in the example considered, node 1 transmits a control packet in control slot $T_1$. Then it waits for $\alpha$ time slots ($\alpha = 1$ in the example considered) before transmitting the data packet on the selected data channel ($\lambda_2$). The delay $\alpha$ is defined as the maximum between the time required by the transmitter of the source to switch to the selected wavelength and the time required by the destination node to receive and decode the control packet and switch its tunable receiver to the selected data channel.

The difference with respect to the DT-WDMA protocol is evident; in the latter case, in fact, we have already observed that the data packet is transmitted, on the home channel of the source node, synchronized to the beginning of the next control cycle. The other substantial difference (not shown in the figure) has to do with the mechanism for solving contention. Instead of an arbitration algorithm executed at the end of a control cycle, the TDMA-W protocol eliminates the problem through the use of status tables. Each node maintains two tables: the first one (called the channel status table) keeps track of the status of the data channels, in order to eliminate data channel collisions, while the second (called node status table) keeps track of the status of the tunable receiver at each node, in order to avoid receiver collisions on the data channels. The two tables are updated, at each node, by the receiver tuned to the control channel, at the end of every control slot, after receiving and decoding a control packet. If node $j$ transmits a control packet preannouncing a data packet to node $i$ on the wavelength $\lambda_k$ ($1 \leq k \leq W$), then all nodes add $L + \alpha$ against entry $k$ in the channel status table and against entry $i$ in the node status table. The entries indicate the number of slots that the resources (data channel and receiver) will be busy. All positive entries of each table are decremented at the end of each control slot to update the remaining busy slots.

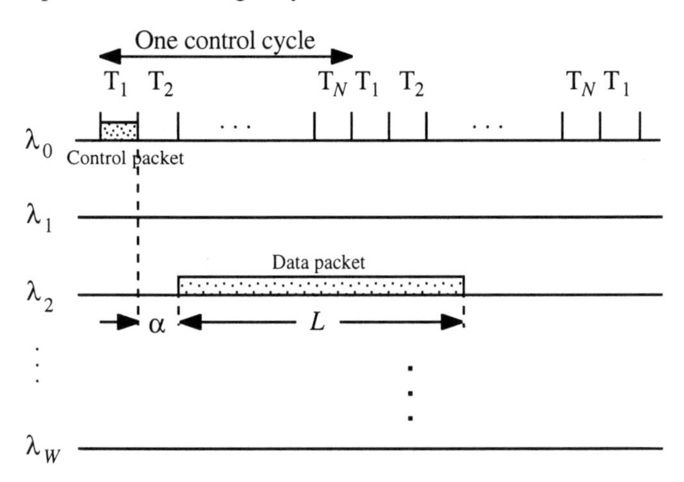


**Figure 2.17** The TDMA-W protocol. An example: node 1 transmits a control packet in the slot $T_1$, then waits for $\alpha = 1$ time slot before transmitting the data packet on $\lambda_2$.

The transmission procedure is as follows. Each backlogged node checks its status table at the beginning of the time slot preallocated to it. If the intended station has a status table entry equal to zero it is considered idle; in this case, the transmitter of the source node checks for any available data channel. A data channel is considered idle if its status table entry is less than or equal to $\alpha$. The reason for this is evident: because of the tuning functions, the considered data channel will not be used before the next $\alpha$ time slots. As a consequence, the switching latency is overlapped, and such property allows to decrease the impact of optical device switching characteristics. When the status of both the tables is favorable to transmission, the control packet is formed and transmitted. Otherwise (the intended node is busy or an idle data channel is not available) the transmitter waits for the next cycle, when it will attempt transmission again. Obviously, if a node is not backlogged, its control slot remains unused during the considered cycle.

The performance of the TDMA-W protocol was examined in [44], in terms of network throughput, packet delay, and control and channel utilization, looking at the sensitivity on the number of nodes and data channels, packet generation rate, data packet length, and optical device switching latency. In some typical cases, it was shown that the TDMA-W allows to achieve results better than (or, at least, equal to) those of the DT-WDMA protocol, with 50% fewer data channels.

***Some keypoints.*** DT-WDMA is a distinctive example of the $FT^2$-FR-TR system. It employs a deterministic arbitration algorithm, common to all the users, to select one out of a set of contending transmissions. There is no need for explicit acknowledgments, and this permits to reduce the packet latency. Since access to the control channel is strictly TDM-based, the main limitation of the protocol is in its scalability. Evolutions are found in DAS and HTDM. The former uses a random scheduling algorithm, which dynamically selects packets for transmission on a slot-by-slot basis. The latter, in turn, allows either preassigned or dynamic slot assignment. Both they are particularly suitable for rather small (e.g., local area networks) environments. Finally, in TDMA-W the number of interconnected nodes and data channels are independent, and this greatly simplifies system expansion through increases in nodes and/or channels.

### 2.2.4 Outline of Performance Comparison and Final Remarks on the Transmission Protocols

The main features of many of the protocols described in the previous sections are summarized, in comparative terms, in Table 2.5. The various schemes are recalled through their reference in the first column of the table. While, in fact, a more explicit notation would be advisable in principle, to help memory and comprehension, it is not possible in practice, since many of the considered protocols do not have their own names, since they are instead variants or improvements of the same basic scheme. Furthermore, in some of the listed references, more than one solution is proposed.

The second and the third columns give an idea of the processing requirements and of the throughput performance: L stands for low, M for medium, and H for high. Particularly, the rating on processing is approximate (and subjective, in some respects) since it is not so obvious to decide which system is more complex than others. Basically, the criterion we have followed is to assume low processing for those protocols that need to monitor only their control channel, and high processing when distributed algorithms are required.

**Table 2.5**
Summary of Access Protocols for Passive Star Networks

| *Reference* | *Processing* | *Throughput* | *Sync. Needed* | *Control Channels* | *Data Channels* | *Equipment* |
|---|---|---|---|---|---|---|
| Chlamtac and Ganz [13] | M | M to H | Yes | 0 | ≥ 1 | TT-TR |
| Dowd [19] | M | L | Yes | 0 | ≥ 1 | TT-FR |
| Habbab et al. [20] | M | L | Some | 1 | ≥ 1 | TT-TR |
| Mehravari [21] | M | L | Some | 1 | ≥ 1 | TT-TR |
| Ganz and Koren [22] | M | L, H | Yes | 0 | ≥ 1 | TT-$FR^m$ |
| Karol and Glance [24] | M | M to H | No | 0 | $N$ | TT-FR |
| Lee [26] | M | M to H | Some | 1 | ≥ 1 | TT-TR |
| Shi and Kavehrad [27] | M | M to H | Some | 1 | ≥ 1 | TT-TR |
| Sudhakar et al. [29] | M | L | Yes | 1 | ≥ 1 | TT-TR |
| Sudhakar et al. [30] | M | L | Yes | $n$ | ≥ 1 | TT-TR |
| Sudhakar et al. [31] | M | M to H | Yes | 1 | $N$ | TT-FR |
| Chen and Yum [32] | M | H | Yes | 1 | $N$ | $FT^2$-FR-TR |
| Humblet et al. [33] | L | H | Yes | $N$ | $N$ | FT-TT-FR-TR |
| Chen et al. [34] | M | M | Yes | 1 | $N$ | $FT^2$-FR-TR |
| Jeon and Un [36] | H | M to H | Yes | 1 | ≥ 1 | FT-TT-FR-TR |
| Senior et al. [37] | M | M to H | Yes | 1 | ≥ 1 | $FT^W$-$FR^W$ |
| Sudhakar et al. [38] | M | M to H | Yes | 1 | ≥ 1 | TT-TR |
| Jia and Mukherjee [39] | H | L | Yes | 1 | ≥ 1 | TT-TR |
| Chlamtac and Fumagalli [40] | M | H | Yes | 1 | $N$ | $FT^2$-FR-TR |
| Chipalkatti et al. [43] | H | H | Yes | 1 | $N$ | $FT^2$-FR-TR |

Many protocols, however, must monitor the complete network traffic, or follow procedures with intermediate complexity, so that it is more correct for such protocols to speak of medium processing. The column reflects the average situation in the sense that it is sometimes possible to improve the scheme by means of specific realizations. For example, it is known that the processing time required per packet depends on the implementation, and can be reduced by performing most functions in special hardware.

A similar interpretation of "average result" holds for the column of the normalized throughput. Here, we have assumed low values when the throughput is smaller than 0.4, medium values for throughputs between 0.4 and 0.7, and high values when the throughput is higher than 0.7. A double result appears for the protocols of [22], because the first of them (the multichannel slotted ALOHA) exhibits a low throughput while the second (the random TDMA) is characterized by high values of the same parameter (though at the expense of increased processing times and reduced scalability).

The fourth column of the table indicates whether the protocol needs any slot synchronization. Finally, the fifth, sixth, and seventh columns specify the number of control channels, data channels, and the NIU structure; $m$ and $n$ are variable quantities.

At the end of this section, we wish to underline that the review presented gives only a taste of the transmission protocols proposed till now for single-hop optical networks, focusing attention on the most consolidated and popular solutions. Actually, a large number of further alternatives and detailed analyses can be found in recent literature [45–51].

In many cases, the target is to optimize performance through the adoption of transmission schedulings that take into account aspects omitted or not sufficiently valued in previous treatments, such as variable-length messages [52,53], nonuniform [54,55] or variable [56] load conditions, multipriority traffic [57], wavelength reusing [58], or buffer sharing [59]. These ideas are not discussed in detail here to save space, but the interested reader can look at such references where further information can be found. Moreover, while most of the present protocols have been conceived thinking of the star physical topology, in the near future increasing interest will be probably devoted to the ring physical topology, which is becoming competitive thanks to the progress achieved in optical amplifiers to compensate the insertion losses at intermediate nodes. Specific protocols are therefore under consideration for these networks [60].

## 2.3 EXPERIMENTAL BROADCAST-AND-SELECT SINGLE-HOP NETWORKS

A number of practical demonstrations have been proposed for single-hop networks. They are generally very simple and do not employ any control channel for pretransmission coordination. The main features of some of them will be summarized below, together with examples of possible applications. The discussion does not pretend to be complete and must be seen as a sample of the many possible solutions and of the problems they involve. Other interesting information to integrate the historical review can be found in [61].

### 2.3.1 LAMBDANET

LAMBDANET, which is a typical FT-FR$^N$ system, was proposed by Bell Communication Research in 1986 [3]. It is formed by $N$ nodes connected, through single-mode optical fibers, to a hub location housing a passive $N \times N$ star coupler. Each node transmits its information on a unique fixed wavelength using, for example, a DFB laser in the range 1,527 nm $\leq \lambda \leq$ 1,561 nm, with a distance between adjacent channels of about 2 nm. Any node requires at least two optical fibers: one for transmission and one for reception. The fibers are coupled at the hub and, according with the broadcast-and-select function, any output fiber from the coupler collects a fraction of all the active channels carrying it to its corresponding node. A schematic representation of the network is shown in Figure 2.18.

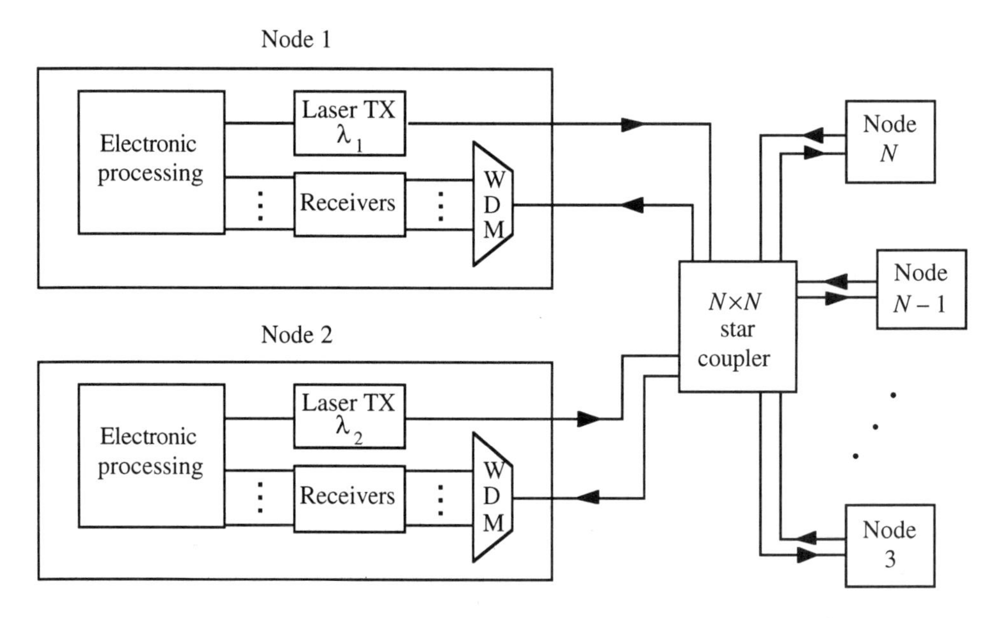


**Figure 2.18** Block diagram of a LAMBDANET network.

In order to have simultaneous transmissions from each node to any other, WDM is combined with TDM in a fixed bandwidth assignment protocol. The situation is similar to that presented in Figure 2.4(c) for the simple case of $N = 3$; the number of channels used, in fact, here necessarily equals the number of nodes. On the other hand, in the LAMBDANET, the transmission cycle is divided into $N$ time slots, since a self-loop from node $i$ to node $i$ ($i = 1, 2, \ldots, N$) is also available, though it is rarely used. For example, in a classical implementation [62], the network consists of $N = 18$ nodes; therefore, the TDM frame contains 18 slots (besides a synchronization tag) and the $i$th slot of the frame, transmitted on $\lambda_j$, contains the packet from node $j$ destined to node $i$. In practice, in the $i$th time slot, the $i$th node receives all the packets addressed to it; this is another difference with respect to the description given in Section 2.2.1, where the adoption of a TR allowed the detection of only one channel at a time. An additional control channel, common to all the nodes of the network, may also be present, which is adopted for call setup/takedown and signaling functions (not for pretransmission coordination). It is accessed by every node through a distinct transmitter and, for avoiding contention, it has a fixed time-slotted structure.

At each node receiver, a grating demultiplexer (noted by WDM in Figure 2.18) [63] separates the different channels, which are independently converted to electrical form by a dedicated photodiode, and then processed in an asynchronous and parallel way. Also the receiver array technology is now well consolidated [64] so that no particular difficulty exists in using $N$ distinct receivers at each node.

Among the advantages of the LAMBDANET structure, it is worth mentioning its transparency to the particular form and format of data, and its ability to manage digital as well as analog data streams. Furthermore, the system is intrinsically robust under optical fiber link failure. The aggregate transmission capacity $C = B \cdot N$ is limited by the individual transmitter rate $B$, and values as high as 36 Gbps have been experimentally demonstrated in the case of $N = 18$ channels. On the other hand it must be noted that only time multiplexing and demultiplexing are performed at rate $B$, since no other electronic equipment in the network needs to process data at higher rates. On the contrary, because of the power splitting realized to distribute the information, the performance of LAMBDANET could in principle be seriously limited by the available power budget.

Let us consider Figure 2.19 which shows a calculated power budget for the LAMBDANET, as a function of the number of nodes for some different node-node distances $z$ (assumed to be the same for all the connections). The horizontal dashed lines give the power available for losses, at the specified bit rate $S$, for a launch power $P_t$ of 0 dBm. In practice, their numerical value coincides with the opposite of the receiver sensitivity $P_r$, the latter evaluated for a bit error rate (BER) of $10^{-9}$ and, according with [62], considering a high-impedance receiver. Continuous lines, instead, represent the overall losses at each node $A_t$. Assuming that the star coupler is realized by cascading 3-dB couplers [65], so requiring $\log_2 N$ stages, $A_t$ is given by [66]

$$A_t = 10\log_{10} N + A_c \cdot \log_2 N + A_s \cdot (\log_2 N - 1) + A_d + A \cdot z + A_M \tag{2.3}$$

where the term $10 \cdot \log_{10} N$ takes into account the star-coupler splitting loss, $A_c$ represents the excess loss of each 3-dB coupler, $A_s$ the splice loss at the interconnection of each stage, $A_d$ the grating demultiplexer loss, $A$ is the fiber attenuation per unit length and $A_M$ a system margin. In Figure 2.19, in particular, we have assumed: $A_c = 0.2$ dB, $A_s = 0.1$ dB, $A_d = 4$ dB, $A = 0.3$ dB/km, and $A_M = 3$ dB. In general, $P_t \neq 0$ dBm, and, for the system to work, the following condition must be satisfied:

$$P_t - A_t \geq P_r \tag{2.4}$$

which, applied in Figure 2.19, allows to know the maximum number of connectable users as a function of the length of the links, the bit rate, and the power budget itself. For example, in case of $S = 2$ Gbps and $z = 40$ km, we have $N_{max} = 16$. Conversely, extending the analysis of Figure 2.19, once assigned $N$ and the bit rate, a maximum node-node distance can be determined that satisfies the power budget constraint; for example, assuming $N = 16$ and $S = 2.5$ Gbps, it is possible to show that the maximum distance equals approximately 30 km.

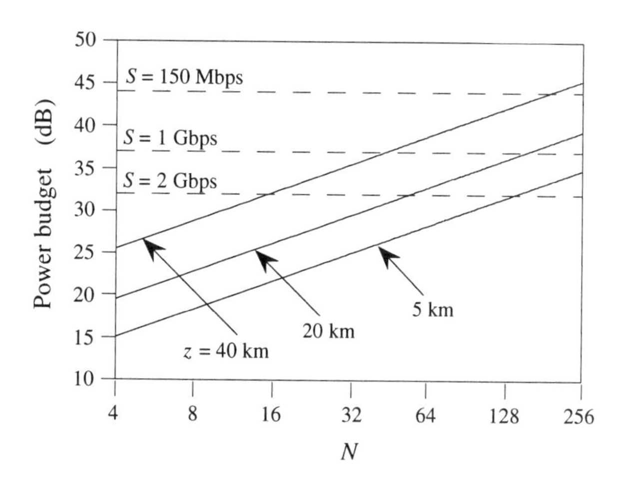


**Figure 2.19** Power budget of a LAMBDANET network.

It should be noted that, in principle, the adoption of a cascade of 3-dB couplers, to realize the star, forces the number of nodes $N$ to be a multiple of 2. In practice, however, this limitation is sometimes removed by exploiting the nonuniformity of loss to add further couplers in an asymmetric fashion. For example, adding 2×2 directional couplers at two of the low loss inputs of a 16×16 star coupler, the number of channels is increased to 18.

The above theoretical limits were experimentally demonstrated in a number of different cases [62,67,68]. Three examples are summarized in Table 2.6, where the achieved network performance is expressed in terms of the bandwidth-distance product, which can be seen as a figure of merit. The demonstration system had characteristics similar to those shown in Figure 2.19. More specifically, the star coupler was constructed from wavelength-flattened 3 dB directional couplers, and had a 14.4 dB average loss (12 dB theoretical splitting and 2.4 dB average excess loss) at $\lambda = 1.5$ μm, with up to 6 dB nonuniformity. The high-impedance receiver had a GaAs FET front-end, and it was based on a commercial InGaAs APD detector with a multimode fiber pigtail [65].

In the first experiment, the star coupler had 18 inputs and 16 outputs, the bit rate was 1.5 Gbps, and the transmission distance was equal to 57.8 km. In the second experiment, instead, the bit rate was increased up to 2 Gbps but, because of the laser chirping, the fiber dispersion and the limited power, only 16 wavelengths could be used and the transmission distance was reduced to 40 km. Finally, in the third experiment, labeled as WDM in Table 2.6, the star was replaced by a grating multiplexer, thus increasing the available power at the receiver; again 18 wavelengths were then successfully transmitted at 2 Gbps over 57.5 km. This last experiment cannot properly be called LAMBDANET, and has been described here for the sake of comparison.

**Table 2.6**
LAMBDANET Experimental Results

| | *Point-to-Point* [*Tbps·km*] | *Point-to-Multipoint* [*(Tbps·km)·Node*] |
|---|---|---|
| LAMBDANET 1.5 Gbps, 18 λs, 57.8 km | 1.56 | 21.5 |
| LAMBDANET 2.0 Gbps, 16 λs, 40.0 km | 1.28 | 18.0 |
| WDM 2.0 Gbps, 18 λs, 57.5 km | 2.07 | — |

(*Source*: [62]. © 1990 IEEE.)

It is interesting to note that, while the measured bandwidth-distance product for point-to-point communications can be easily obtained multiplying the aggregate throughput (which practically equals the transmission capacity) by the corresponding transmission distance, the same parameter for point-to-multipoint communications does not increase as the number of reached nodes. This behavior is due to some nonidealities encountered in using practical components; in particular, as stressed above, different output ports had slightly different insertion losses, and this resulted in the need of variable maximum link lengths. Accounting for these variations, the network capacity for broadcasting was smaller than the theoretically predictable value.

The power budget constraints can be somehow alleviated, either theoretically or practically, by resorting to optical amplification of the WDM signals. It was shown that the simultaneous amplification of 20 channels in an experiment using the LAMBDANET hardware and a semiconductor laser amplifier, allows one to achieve a 6 to 8 dB net gain [69]. Keeping in mind Figure 2.19, this allows to connect a larger number of nodes, on greater distances.

Among the suggested system application of LAMBDANET we can mention its potential use for voice traffic transport, broadcast video, or as a packet switch fabric [62].

At the end of this section it is worth observing that, in place of employing a multiple fixed receiver at each node, one could use, as mentioned above, one tunable filter. In this case, however, simultaneous utilization of information from several transmitting nodes is no longer possible, and a contention problem appears when two or more packets are addressed, in a time slot, to the same destination. An innovative solution to resolve this problem has been recently proposed [70] that is based on the adoption of a neural network. By knowing the destination addresses of the input packets, suitably stored in FIFO buffers, the neural network computes a nonblocking switching configuration and determines the appropriate tuning currents for the wavelength tunable filters or the acoustic frequencies for integrated acousto-optic filters. This variant of LAMBDANET has been called neuro-star packet switch; it can be further modified by grouping of optical channels of different wavelength allocation [71]. This way the number of channels becomes greater than the number of nodes, so that more than a single channel per port can be active at any time slot and the demand of traffic loads can sometimes be better met.

### 2.3.2 RAINBOW

RAINBOW, which has been developed by IBM Research for Metropolitan Area Networks (MANs) [5], is an FT-TR system designed to cover a network diameter up to 50 km connecting, in its first prototype, 32 IBM PS/2 personal computers (PCs) through a 32×32 star coupler. A scheme of principle, for this specific application, is shown in Figure 2.20.

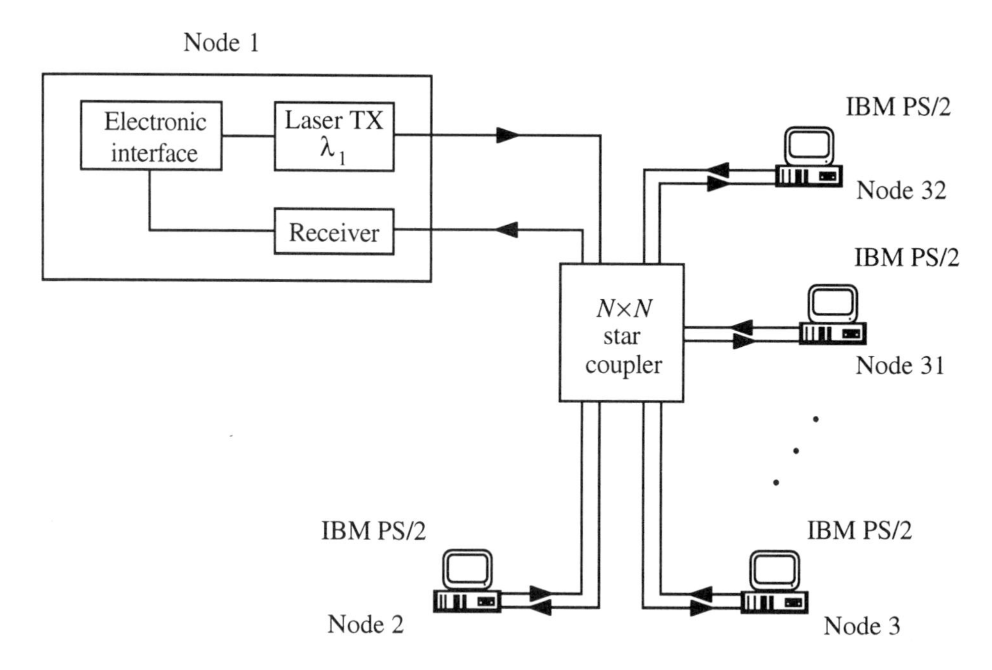


**Figure 2.20** Block diagram of a RAINBOW network (TX = transmitter).

Each PS/2 station transmits on a different channel, using a DFB laser at fixed wavelength, with a bit rate of 200 Mbps (which is the maximum bit rate of the PC bus). Intensity modulation is used for transmission, together with direct detection. As usual in broadcast-and-select networks, the transmitted data are combined in the passive central star coupler and sent to all the receivers. The wavelengths span a 50 nm range, and are separated one from another by approximately 1.6 nm; their number is equal to the number of stations, so that channel collisions cannot occur. A Fabry-Perot interferometer is used as piezoelectric-tunable filter at each receiver, and is characterized by a full-width half-maximum (FWHM) bandwidth of 0.35 nm. The fixed tuned laser and its support circuitry, the tunable filter and its control electronics, and the photodetector and the receiver electronics are all mounted on the same RAINBOW card (one for each station).

Another card houses the 200 Mbps electronic circuitry (clocking, for example); these functions have been obviously simplified in Figure 2.20. All the procedures of signaling, control and network management are realized through the central processing unit (CPU) of the PS/2; the time required for reconfiguration is on the order of a few tens of milliseconds.

The main problem to be solved by the transmission protocol is the coordination of the tuning of the receivers. A simple "in-band receiver polling" option was chosen for the RAINBOW prototype because of its intrinsic simplicity [5]. This means that, similarly to the other protocols with no pretransmission coordination, the coordination of the retuning of the receiver filters is managed within the band accessible to each user, and a separate wavelength channel (with an additional laser and receiver per station) is not required to carry the signaling information. More precisely, a node that has a message to send continuously transmits a connection request, specifying the address of the destination node. At the same time, it tunes its receiver to the intended destination's transmitting channel, waiting for an acknowledgment to establish the connection. Each idle receiver continuously sweeps its optical filter until it detects a connection request addressed to it. Once the request is found, the node sends the acknowledgment on its own wavelength, and the transmission session begins. In the case of point-to-multipoint connections, the transmitter must wait for acknowledgment from all the intended destinations. Therefore, the delay due to the setup acknowledgment, which is rather long on average, makes this simple protocol not very suitable for packet-switched traffic. On the contrary, it seems to work well for circuit-switched applications with long holding times.

A timeout mechanism was also introduced in the protocol to solve possible deadlocks. If two stations send connection requests to each other nearly simultaneously, they await for an acknowledgement that will never be sent. So, the connection attempt is aborted if the acknowledgment is not received within a certain timeout period.

The performance achievable with the described RAINBOW protocol has been analyzed for the first time in [72], using the equilibrium point analysis (EPA) technique. Once having defined a suitable state diagram [72] for describing the stations' evolution, EPA assumes that the system (as a whole) always operates at an equilibrium point. This implies that the expected number of stations entering each state, in the model, equals the number of stations leaving it, in each time slot. The effect of such an approximation is more pronounced in the case of low arrival rates, while it is generally acceptable for high arrival rates. The throughput performance was analytically derived in [72] (and verified by simulation) looking at its dependence on various system parameters, that is, message arrival rate, message length, and timeout duration. It was shown that, for a given set of parameters, an optimal timeout duration exists, which yields to maximize the throughput.

An important role, in fixing the network size, is played by the finesse $F$ of the Fabry-Perot interferometer, defined as [73]

$$F = \frac{\mathrm{FSR}}{\Delta\nu_{\mathrm{FWHM}}} \tag{2.5}$$

where FSR is the free spectral range and $\Delta\nu_{\mathrm{FWHM}}$ the full-width half-maximum bandwidth. It is in fact possible to show that the crosstalk penalty introduced by the filter is strictly related to the value of $F$: for a given finesse, if the wavelengths are equally spaced within FSR, the crosstalk depends only on the number of channels [74] and the penalty is maintained below 1 dB assuming $N < F/2$ [75], or below 0.5 dB assuming $N < F/3$ [76]. These constraints were largely satisfied in the prototype, where the measured finesse was between 200 and 250; therefore, the crosstalk was not a relevant problem in the considered application.

For a given number of users, a link budget allows us to determine the maximum distance between two connected nodes. As an example, assuming a loss through the star coupler $A_c = 16.5$ dB, a loss due to the filter $A_f = 3$ dB, an emitted power (output of the laser pigtail) $P_t = 0$ dBm, a receiver sensitivity (at 200 Mbps and with a BER of $10^{-9}$) $P_r = -39$ dBm, and a margin $A_M = 3$ dB, we have an available budget

$$P_x = P_t - A_c - A_f - A_M - P_r = 16.5 \text{ dB} \tag{2.6}$$

which allows to design a single point-to-point link 55 km long using standard fibers with an attenuation of 0.3 dB/km in the neighborhood of the wavelength $\lambda = 1.55$ μm. Such a distance could be obviously increased using optical amplifiers. Alternatively, keeping the distance constant, the number of connected users can be increased; the amplifier noise should be taken into account, together with an additional penalty due to saturation and crosstalk effects when more channels are present, but the advantage generally overcomes the disadvantage. In [77], for example, it was shown that, with typical parameters, the number of stations in the RAINBOW can be increased by a factor of more than 5. The amplifiers should not be placed just before the star, where the high power levels (on the order of –10 dBm) can cause gain saturation. The best arrangement, instead, is based on positioning amplifiers just before or after the tunable filter at the receiver.

Further evolution of the first RAINBOW prototype has been, more recently, proposed, potentially able to guarantee better performance. Therefore, in [78] the bit rate at each node was increased up to 300 Mbps, while in [79] the so called RAINBOW-II was proposed, which supports the usual 32 nodes but each at 1 Gbps over a distance of 10 to 20 km.

The main goal of RAINBOW-II is to study real systems that require Gbps capacities, gaining experience with higher-layer protocols and end-user supercomputer applications. To this purpose, the problem of providing suitable attachment to hosts was addressed, identifying a suitable and widely implemented solution in the standard high-performance parallel interface (HIPPI) [80].

A very simple optical transport protocol was also developed, based on a go-back-$N$ error recovery strategy [81] with fixed-length packet headers, and able to use large packet sizes. In the go-back-$N$ protocol, the receiver discards all the packets following an error until it is retransmitted and correctly received. $N$ is the number of packets that can be sent without waiting for the next packet to be requested. Therefore, the protocol is intrinsically not very efficient since, in the case of high error rate, a lot of bandwidth may be wasted. On the other hand, such protocol is very easy to implement, and this is the reason for its choice in RAINBOW-II, where the error rate was estimated to be low.

### 2.3.3 FOX

Fast optical crossconnect (FOX) is a $TT^2$-$FR^2$ system, proposed in 1988 [4] to provide an interconnection network for parallel processor computers. Its structure is shown in Figure 2.21.

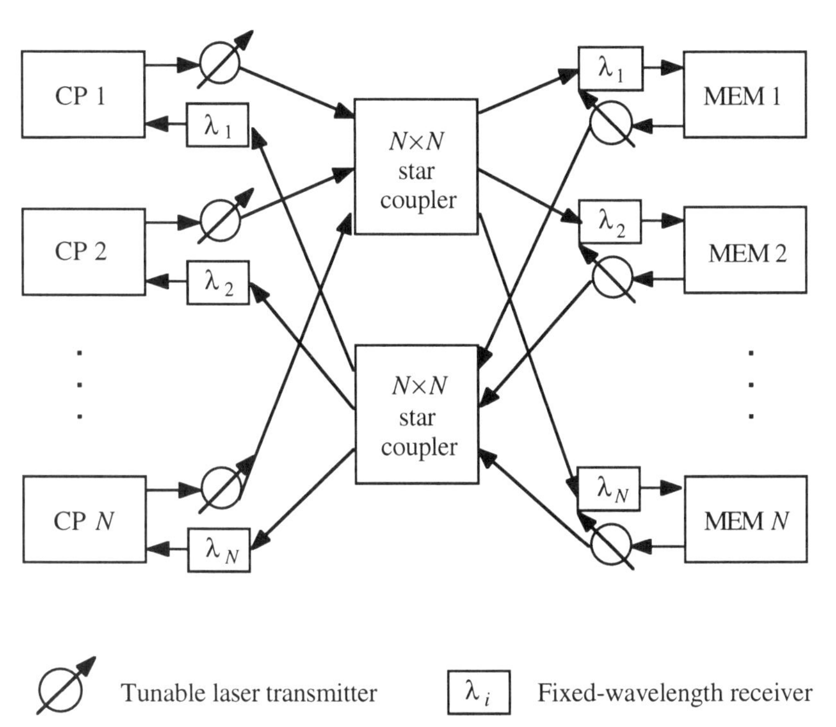


**Figure 2.21** Block diagram of a FOX network.

The $N$ processor nodes (noted by CP), each provided with a local cache memory, and the $N$ external memory nodes (MEM) are connected using two separate interconnection networks: an $N \times N$ passive star network for transmission to the output ports and a second network, of the same type, for receipt of responses from the output ports. As shown in the figure, the CP and MEM nodes have separate optical transmission and receiver modules, with a unique fixed-wavelength filter at each receiver and a tunable-wavelength source at each transmitter. So, the number of wavelengths is equal to $N$, whereas the number of channels equals $2N$. The number of nodes is limited by the available power budget, and a maximum $N$ of 256 is generally recommended [82].

Data packets arrive at the input ports, on the CP's side. To address a particular memory, each processor tunes its wavelength to that memory's filter, and vice versa in the opposite direction. FOX is internally nonblocking, which means that simultaneous access to different memories is possible without contention, but if a memory node is simultaneously addressed by more than one CP, then a channel collision occurs. In this case, when an input port does not receive a successful response indicator from the desired output in a prefixed time period, it can retransmit the packet, following a suitable algorithm, until it is correctly received. The need of retransmission obviously affects the throughput as well as the network response time. On the other hand, the FOX network's performance relies on the fact that the probability that more than one packet is simultaneously destined to the same output port is small. For example, we can assume that packet arrivals on the $N$ input trunks are independent and identical Bernoulli processes, and denote by $\rho$ the probability that a packet will arrive on a particular input (mean utilization). If the traffic is uniformly distributed, each packet has equal probability $1/N$ of being addressed to any given output, while successive packets are independent. Under such hypotheses, the probability $p_k$ that $k$ packets are simultaneously destined to the same memory is binomially distributed [83], according to

$$p_k = \binom{N}{k}\left(\frac{\rho}{N}\right)^k\left(1-\frac{\rho}{N}\right)^{N-k} \tag{2.7}$$

With $N = 32$, $k = 4$, and $\rho = 0.9$, we have $p_k \approx 10^{-2}$, and this value decreases for increasing $k$.

When a collision is detected, the transmitting node may execute a binary exponential backoff [17]: transmission is aborted and it will be repeated after $m$ time slots, the length of each slot being equal to the worst-case round trip propagation time $\Delta T$. The feature of the algorithm is that the value of $m$ depends on the number of collisions the packet has already suffered: after $i$ collisions, $m$ is randomly chosen in the range $[0, 2^i - 1]$. Hence, the number of slots to wait is 0 or 1 after the first collision, 0, 1, 2, or 3 after the second collision and so on. The effect of such an algorithm in terms of reduction in the normalized throughput per each node is shown in Figure 2.22, as a function of the mean utilization at the input trunks. $\rho$ has also the meaning of fraction of memory references that are external; therefore, $\rho = 1$ corresponds to the situation in which any processor always makes a new external memory request in the memory cycle immediately following its own satisfied request. Curve b refers to the application of the binary exponential backoff algorithm, whereas curve a, reported for comparison, corresponds to the ideal situation in which, in the event of a conflict, one of the requests is satisfied. Throughput degradation when $\rho = 1$ is rather relevant, passing from about 0.58, in the ideal case, to about 0.25, in the real one. Nevertheless, if the cache miss-rate is maintained relatively low, the adopted conflict resolution algorithm ensures good performance as well. For example, assuming a typical value of $\rho \approx 0.1$, both the curves give a normalized throughput on the order of 0.98.

In any case, everything considered, FOX is suitable to realize the transmission of short ($\approx$100–200) bit packets, in applications where access to the output ports is sufficiently rare. So, consistent with the original proposal, its appropriateness for many parallel processors provided with internal cache memory is confirmed.

The key component in the FOX design is the tunable laser, which must be fast enough not to slow down the network, waiting for the transmitter to tune its desired wavelength. At the same time, it must address as many channels as possible. For data packets durations ranging from 100 ns to 1 μs, tuning times less than a few tens of nanoseconds should ensure acceptable efficiency. A very good target should be a switching time of about 5 ns over a tuning range of 25 nm [4].

Actually, in [4], packets of 100 bits at a bit rate of 1 Gbps were transmitted using a 1.5-μm DFB laser, switched by current injection. The laser was switched between two wavelengths separated by 0.06 nm in less than 20 ns. Even better results should be obtained using sources specifically designed for wavelength tuning, such as two- and three-section distributed Bragg reflector (DBR) or DFB lasers [84]. In this way, the desired target of a 5-ns switching time may be approached.

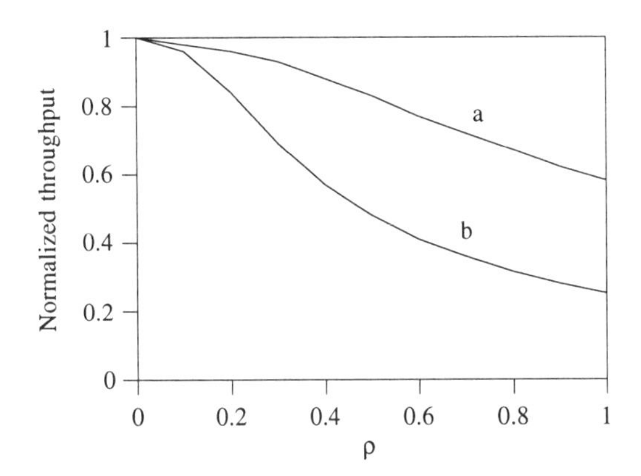


**Figure 2.22** Normalized throughput per each node: (a) in the ideal case, and (b) using binary exponential backoff retransmission. (*Source*: [4]. © 1988 Institution of Electrical Engineers (IEE).)

## 2.3.4 HYPASS

HYPASS is another Bellcore proposal [6], which realizes an $N \times N$ opto-electronic packet switching network. The electronic components are used, inside the system, for memory and logic operations, whereas optical components are used for routing and transport. Its overall function, schematically summarized in Figure 2.23, consists in receiving packets on optical fiber input trunks, for switching them to the appropriate optical fiber output trunks.

In the prototype solution, the data were assumed to arrive at the input trunks with a rate of 2 Gbps, assembled in packets of 1,000 bits. After switching, the packets were transmitted on the output trunks at the same bit rate. As shown in Figure 2.24, the system is formed by two separate networks, one for transporting data packets from the inputs to the outputs and one for transporting control information, in the opposite direction. The data packet transport network uses tunable transmitters and fixed tuned receivers, while the control information transport network uses fixed tuned transmitters and tunable receivers. Globally, the system requires either high-speed tunable lasers or tunable receivers, so it is correct to classify it as a TT-TR network [9].

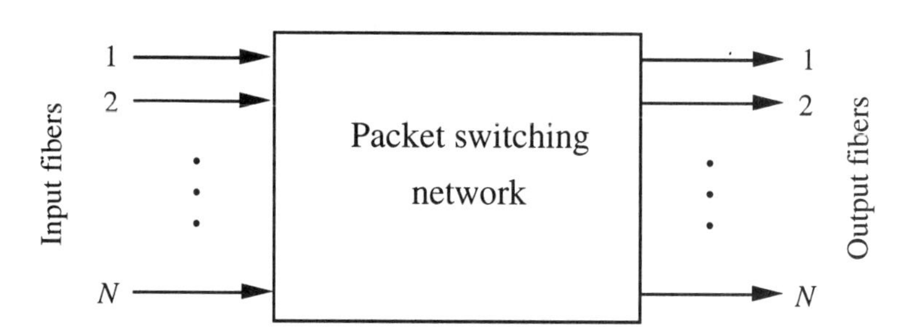


**Figure 2.23** Block diagram of the packet switching system.

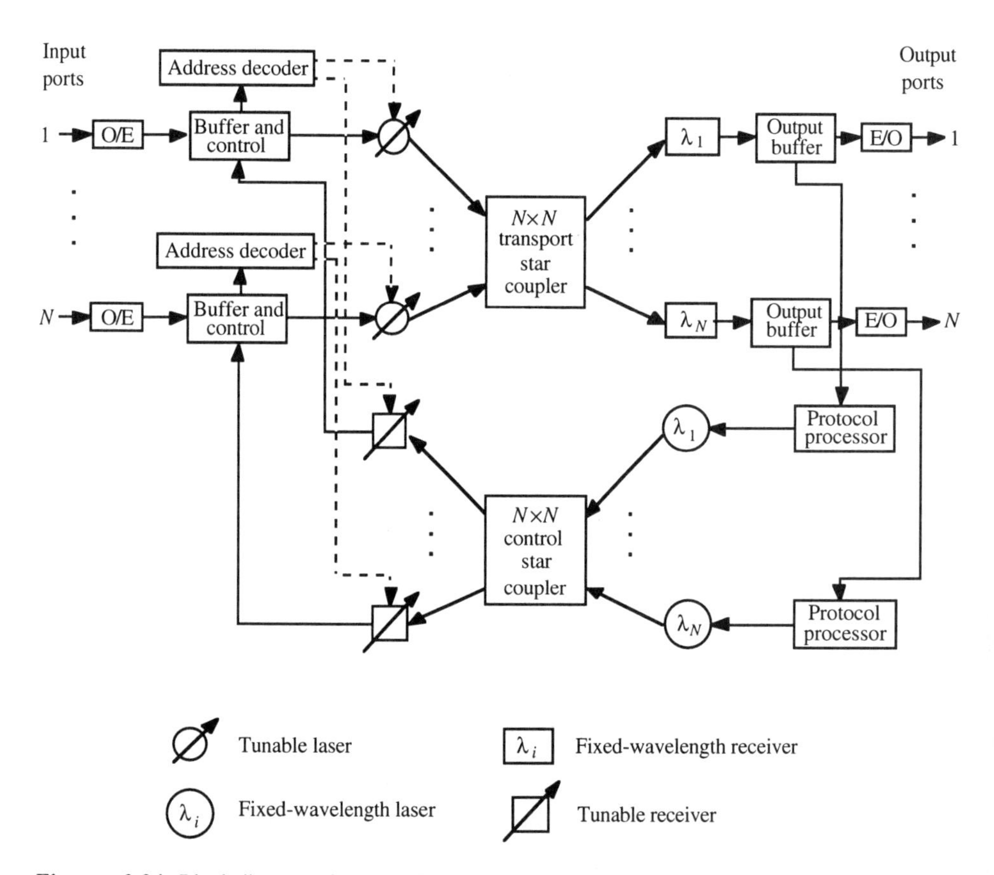


**Figure 2.24** Block diagram of a HYPASS network (E/O = electro-optic converter; O/E = opto-electric converter).

The data packet transport network is based on a passive $N \times N$ star coupler; the packets, coming from the transmitters at the input nodes are coupled and sent to all the output receivers. Only one wavelength can be detected by each receiver, so that addressing is made by tuning the laser to the destination wavelength. More specifically, the optical signal incoming an input port is converted into an electric signal by an opto-electronic (O/E) receiver and sent to a series-parallel (S/P) shift register. The sequence of bits is then transformed into a sequence of word-parallel electronic data, of width $b$, temporarily stored in a buffer. The principle is shown in Figure 2.25. The address of the destination node, contained in the packet header, is decoded and used by a digital-analog (D/A) converter. The electric current so produced allows to tune the transmitter as well as the receiver of the input node, both at the wavelength of the destination node, the first being connected to the transport network and the second to the control network. Transmission begins when the input, through the control network, receives a request-to-send signal from the desired output port. At this point, the stored data pass through a fast parallel-series (P/S) shift register and the signal obtained modulates the laser, previously tuned.

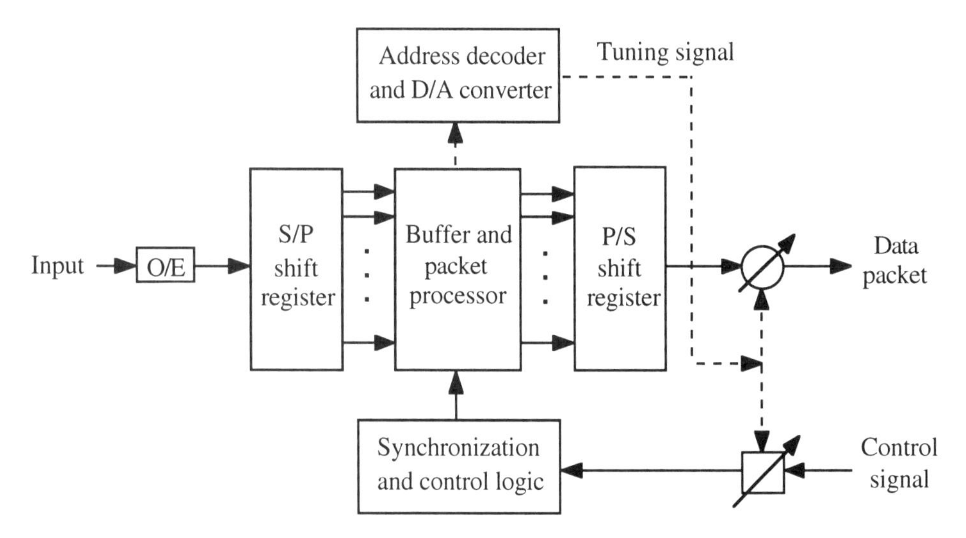


**Figure 2.25** Schematic representation of the input subsystem in a HYPASS network (D/A = digital-analog; P/S = parallel-series; S/P = series-parallel).

The conversion into parallel data permits to use electronic devices working at speeds $b$ times slower than the incoming bit rate. This way, the critical components, which must operate at high bit rates, are limited to the O/E receiver and the S/P and P/S shift registers. The buffer and the address decoder manage the sequence of words outgoing the S/P. The inner frequency of the buffer depends on the bit rate and the word width; for example, assuming a bit rate of 2 Gbps and $b = 16$, the buffer must store $125 \cdot 10^6$ words per second, that is, the required internal clock rate is 125 MHz. This speed can be obviously reduced by increasing $b$.

At the output of the fixed wavelength receiver, the packet, after an S/P conversion, is stored into a buffer. Once again, operation is made in a word-parallel format, which relaxes constraints on the buffer's clock rate. Before being transmitted on the output optical fiber trunk, the signal is accurately retimed (i.e., synchronized), P/S converted and electro-optic (E/O) regenerated. These functions are summarized in Figure 2.26, which also shows the control functions used to arbitrate the HYPASS operation. As stressed above, each output node probes its own status and, when idle, sends a request-to-send signal to the input ports, transmitting on its identifying wavelength. The signal is distributed, through the $N \times N$ control star coupler of Figure 2.24, to all the input nodes, but it is detected by the only nodes that, having some packet to transmit toward the considered output port, have tuned their receivers to its wavelength. Packets coming from different inputs can be simultaneously transmitted to different outputs thanks to the adoption of a distinct channel for each connection. Thus, the network is internally nonblocking. On the other hand, collisions occur when more packets are simultaneously addressed to the same output node. Collisions are detected by the electronic control at each port, and then eliminated by the specific protocol adopted.

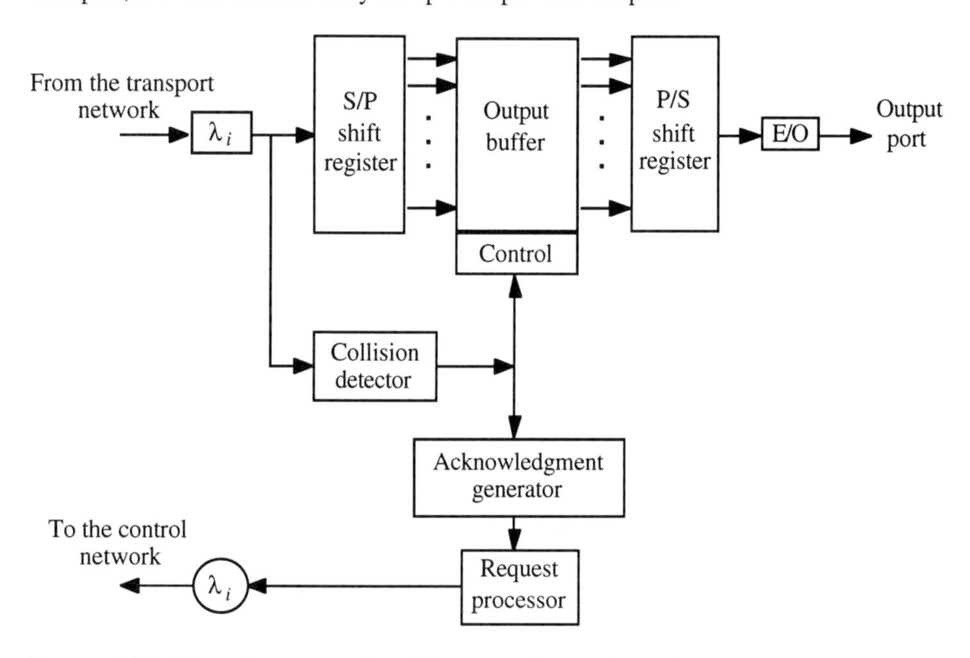


**Figure 2.26** Schematic representation of the output subsystem in a HYPASS network.

It is clear from the description developed so far that the control protocol for the HYPASS network is based on an input-buffered/output-controlled arbitration. Several algorithms can be used to implement such a procedure, and the choice of a specific solution depends mainly on the type and extent of the traffic demand. For random (uniform) traffic, which means transmission of uncorrelated packets, uniformly addressed, with probability $1/N$, toward any output port, a protocol based on the tree-polling algorithm has been proposed [85–87]. The algorithm manages transmission, on the basis of the collisions detected, by means of suitable control signals emitted by the outputs synchronously. The time quantile is $\Delta t$, which represents the duration of the sequence of steps which are necessary to transmit a packet, including processing of the input data, packet transmission and reception of an acknowledgment from the tagged output.

It is assumed that the number of inputs equals a power of 2, say $N = 2^k$, and that the input ports are the leaves of a full binary tree, with $k$ levels. Starting from the root, a cycle begins with a start signal that enables all the inputs having one packet stored in the buffer to transmit. If an input port receives a packet after the start signal, it must wait for the subsequent cycle. The output port, once having received the data, checks if collisions have occurred or not. In the latter case, the cycle terminates in just one interval $\Delta t$. Otherwise, the collided packets are retransmitted, first enabling the leaves of the left subtree and then those of the right subtree. The procedure is recursively repeated for all the subtrees of the upper levels, until it resolves all collisions. An example is shown in Figure 2.27, for a network with $N = 8$, where the input ports 1, 4, 6, and 7, at the beginning of a cycle, have to transmit a packet to the same output.

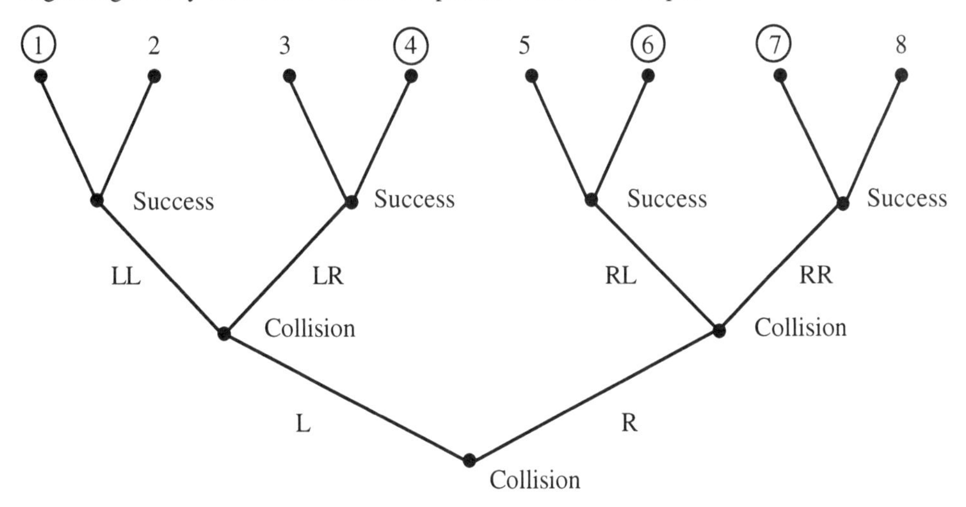


**Figure 2.27** Example of tree-polling for a 8×8 HYPASS network. At the beginning of a transmission cycle, the circled ports have to send a packet to the same destination.

Because of the group polling strategy, after the first collision (at the root of the tree) the branches are visited in the following order: L, LL, LR, R, RL, RR. Correspondingly, the subsets of polled inputs and collisions or transmissions occurred are those reported in Table 2.7. In this particular example, seven intervals $\Delta t$ are necessary for transmitting four packets. In general, the network throughput is directly related to the offered load and to the number of ports. In some cases, it may be preferable, for shortening waiting times, to begin the cycle from the subtrees of a higher level. More specifically, a dynamic tree protocol can be conceived, which adapts the starting point to the load changes [87]: when the traffic is light, polling begins at the lower levels (for very light loads polling begins at the root of the tree), while when the traffic becomes heavy, polling begins at the upper levels (for very heavy loads polling begins at the leaves of the tree). The choice on the depth of the tree from where to start is made on the basis of the duration of the previous cycle.

Table 2.8 shows the average number of probes (intervals of duration $\Delta t$), required by the tree-polling, for different values of the probability $p$ that an input port has a packet destined to a given output port, for various load conditions and number of nodes. Noting by $q$ the probability that a port has a packet to send and by $\gamma$ a load balance parameter, we have in fact

$$p = \frac{q}{2^k}\gamma \tag{2.8}$$

When the system is uniformly loaded $\gamma = 1$, whereas $\gamma = 2$ implies that there are twice as many packets at different input ports destined for the same output; the system is therefore temporarily unbalanced.

**Table 2.7**
Detail of Polling and Transmission for the Example of Figure 2.27

| *Cycle* | *Polled Inputs* | *Transmission from Port* |
|---|---|---|
| 1 | 1–8 | Collision |
| 2 | 1–4 | Collision |
| 3 | 1–2 | 1 |
| 4 | 3–4 | 4 |
| 5 | 5–8 | Collision |
| 6 | 5–6 | 6 |
| 7 | 7–8 | 7 |

**Table 2.8**
Average Number of Probes in the Tree-Polling Algorithm

| $N$ | $(q = 0.5)$ $(\gamma = 1)$ | $(q = 1)$ $(\gamma = 1)$ | $(q = 0.5)$ $(\gamma = 2)$ |
|---|---|---|---|
| 16 | 1.33 | 2.10 | 4.36 |
| 32 | 1.36 | 2.19 | 4.41 |
| 64 | 1.38 | 2.25 | 4.58 |
| 128 | 1.39 | 2.29 | 4.70 |

(*Source*: [6]. © 1988 IEEE.)

The values in Table 2.8 demonstrate that even in presence of heavy loads ($q = 1$) and unbalanced conditions ($\gamma = 2$), the average number of intervals required is very small, while the dependence on $N$ is rather weak. The network throughput, in turn, can be expressed as

$$T_r = N \cdot S \cdot \frac{\tau_p}{\tau_o} \tag{2.9}$$

and depends on the algorithm and the performance of the devices adopted. In (2.9), which is a general definition, $S$ represents the data rate per port, $\tau_p$ is the packet transmission time excluding control overhead, and $\tau_o$ is the total packet transmission time. In accordance with the description above and assuming the availability of standard components, a possible time sequence of events for HYPASS is as follows ($S$ = 2 Gbps) [6]:

- Less than 10 ns to tune the laser and the receiver;
- About 50 ns to process the poll;
- About 20 ns to add the preamble (≈ 64 bits);
- 512 ns to transmit the packet (1,024 bits);
- About 20 ns for control and acknowledgment.

The sum of the listed contributions defines $\Delta t$, which equals $\tau_o$ in the case of an ideal protocol without any need of resolving collisions. Setting $N$ = 128, as an example, we have $T_r \approx 214$ Gbps, which represents a sort of peak throughput goal for this number of users and external bit rate. In reality, the throughput evaluation must take into account the penalty due to the tree protocol with collision resolution, as well as the queueing strategy adopted. In the HYPASS, in fact, every input has a buffer for incoming packets, which are served on a FIFO basis at the beginning of each slot.

Therefore, the system suffers the so-called head-of-line (HOL) blocking phenomenon: the packets buffered at the head of the FIFO queue and that cannot be transferred to the desired output (since the latter has been already selected by another input) block the network access to other packets, even if they are destined to idle nodes. The extent of the delay $D$ (and then of $\tau_o = D \cdot \Delta t$ ) has been determined in [6], for the case of uniform traffic distribution, assuming a dynamic tree polling and taking into account the HOL effect.

The result is shown in Figure 2.28, as a function of the input link load ρ, which gives the probability that a packet will arrive at each port.

Analytically, $D$ can be expressed as

$$D = \frac{1-\rho}{f(\rho)-\rho} \tag{2.10}$$

where $f$ represents the probability of winning a switch arbitration in a slot, and is a decreasing function of ρ (exact expressions can be found in [6]). $D$ becomes higher and higher when $f$ approaches ρ (the limit value is $\rho = 0.318$) but, in practice, loads greater than about 0.25 cannot be accepted because of their intolerable delays. Taking this into account, the maximum network throughput, for $S$ = 2 Gbps and $N$ = 128, is on the order of 80 Gbps.

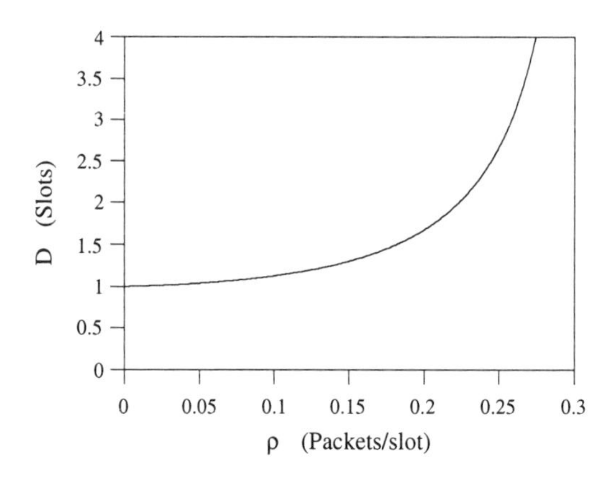


**Figure 2.28** Mean delay as a function of the link load.

## 2.3.5 BHYPASS

BHYPASS is a modification of HYPASS, conceived with the goal of reducing the hardware complexity [82]; contrary to the HYPASS, in fact, it only requires tunable transmitters and fixed-wavelength receivers. At the same time, control, which occurs at the outputs in the HYPASS, is here transferred to the inputs.

The BHYPASS switch is shown in Figure 2.29: as usual in a TT-FR system, the transmission of the data packets is made by tuning the laser to the wavelength of the desired receiver.

The main characteristic of the BHYPASS, however, is the usage of an electronic Batcher-Banyan network for contention resolution [88], which drives an ensemble of buffers for control and storing of the input data. The packets coming from the various sources are temporarily stored in the buffers, while the pairs source address-destination address, which specify the desired connections, are sent to the Batcher-Banyan network. The latter processes the packet header only, and not the data (this is an important difference compared to many entirely electronic switch designs that are based on the Batcher-Banyan approach [89]). The required operation speed at electronic level is therefore relatively low.

The destination addresses are processed by the Batcher sorting subnetwork [90] (Figure 2.30): if the packets are addressed to distinct receivers, the corresponding inputs are permitted to transmit, through suitable "activation" signals (triggers); on the contrary, if more packets are simultaneously addressed to the same output node, the selection subnetwork activates the transmission by one input only, maintaining the other packets in their buffers, for another attempt in the subsequent slot.

The activation signals are sent through the Banyan connection subnetwork (Figure 2.30) which links the output of the Batcher with the control devices of the input nodes. The selection of the inputs is managed, inside the connection subnetwork, on the basis of the source addresses. If the packets are characterized by priorities, in case of contention the packet with higher priority is transmitted first. Otherwise (no priorities or equal priorities) the choice is made randomly. Priorities can be inserted by the contention resolution algorithm itself. In fact, in order to avoid the possibility that, because of repeated contentions, a packet suffers too long a delay, the packets which remain in the queue are marked by a binary code that represents their age, that is the number of unsuccessful attempts they have suffered. The older the packet is, the higher the priority assigned to it, and then the possibility that it will be transmitted in the subsequent slot.

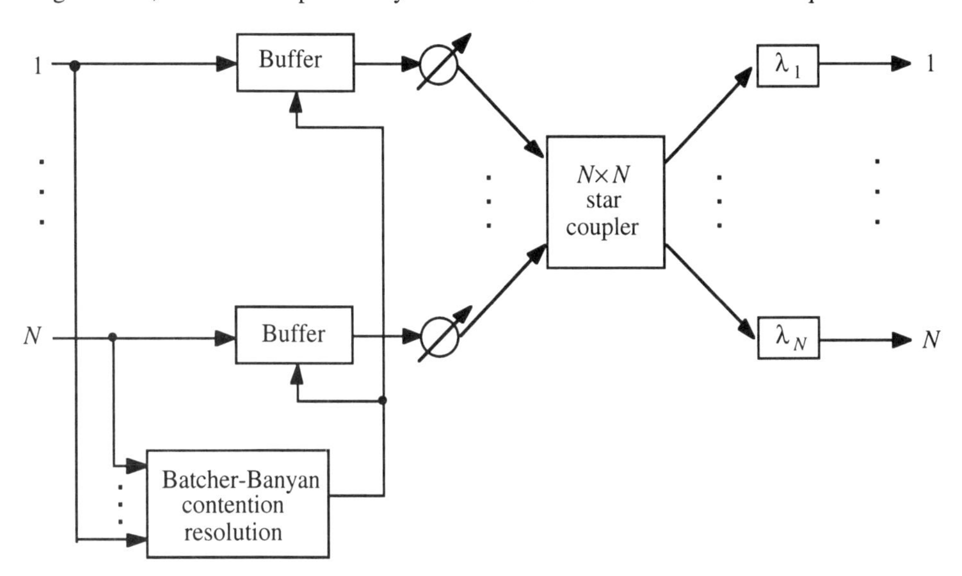


**Figure 2.29** Block diagram of a BHYPASS network. Tunable transmitters and fixed wavelength receivers are shown as in Figure 2.21.

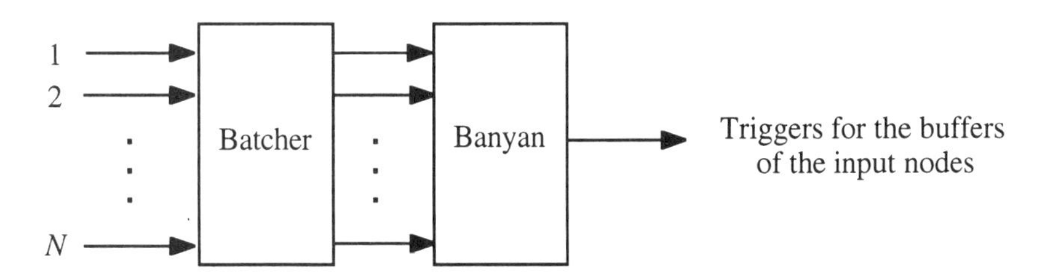


**Figure 2.30** Schematic representation of the Batcher-Banyan function.

## 2.3.6 Photonic Knockout Switch

The photonic knockout switch, whose general structure is shown in Figure 2.31, is a FT-TR$^L$ packet switching system, proposed in 1988 [91]. The $N \times NL$ optical space-division fabric consists of a passive $N \times N$ star coupler, whose generic output termination is split, at the corresponding output node, into $L$ branches.

The situation is shown in Figure 2.32, where, coherently with the previous system definition, every input is provided with a fixed wavelength laser, and every output has $L << N$ tunable receivers. This way, up to $L$ packets can be simultaneously detected, coming from different inputs, and temporarily stored in the buffer for subsequent retransmission, according with a FIFO strategy, by the output interface. Since the architecture uses output buffering, no HOL effect appears, and good delay-throughput performance is theoretically achievable.

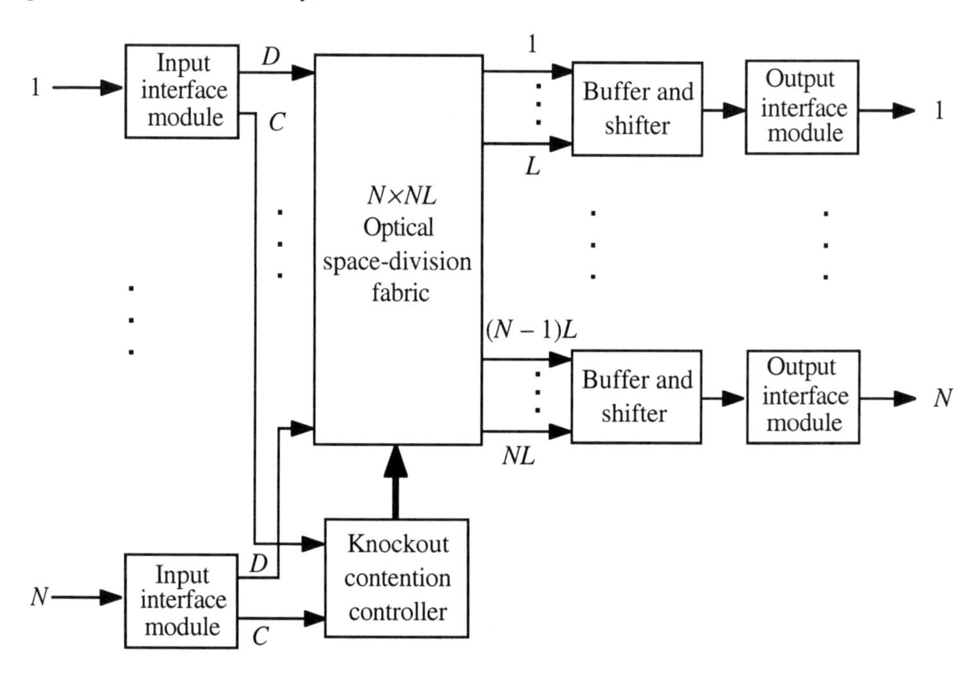


**Figure 2.31** Block diagram of a photonic knockout switch. (*Source*: [91]. © 1988 IEEE.)

Tuning of the optical receivers is governed by the knockout electronic device, which resolves the contentions at the output ports using, for this purpose, a dedicated channel.

The system can be adapted to treat fixed length packets or variable length packets. For the two cases, the internal optical structure remains the same, whereas some differences appear as regards the control electronic devices as well as the input and output interface modules [92]. For the sake of simplicity, we will limit discussion to the case of fixed-length packets.

The bit rate of the input data, conveyed by fiber links or any other transmission medium (e.g., microwave links, open air optics, etc.) to the input interface modules, will be denoted by $S$ (e.g., $S$ = 2 or 4 Gbps). Each packet contains an "address field" (i.e., a sequence of bits that specify the address of the source) and a "data field."

Each input interface has to process the arriving packets, generating the electrical signals noted by $C$ and $D$ in Figures 2.31 and 2.32. $D$ is the data signal, which is sent to the optical connection network, while $C$ is the control signal, which contains information about the presence of a packet to transmit and, if this is the case, its own destination. $C$ is obviously sent to the knockout controller.

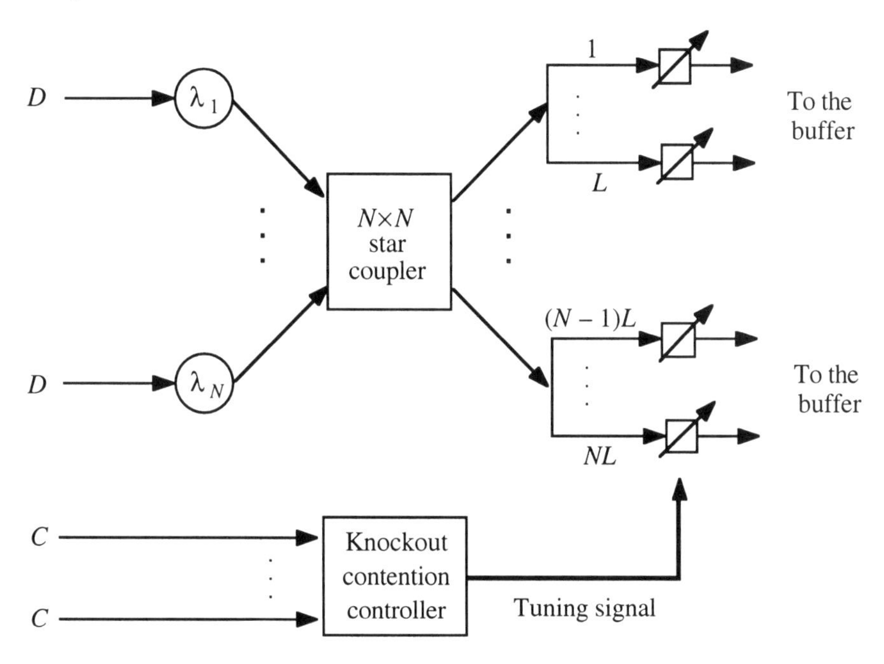


**Figure 2.32** The $N \times NL$ optical space-division fabric used in the photonic knockout switch.

More specifically, the structure of the input interface module is that shown in Figure 2.33, where the case of an input at optical frequency has been considered (this justifies the presence of the photodiode). The incoming data packets are transformed, through the O/E conversion realized by the photodiode, into baseband electrical signals; they are reclocked, and individual packets are inserted into synchronous time slots entering the switch as signal $D$. During retiming of the data packets, the address information in the header of each packet is analyzed, thus determining the output port to which it is destined. The address of the output port constitutes part of a word (the "destination word") which, in turn, is included in a control packet, containing the information necessary for data routing. For example, in a destination word of 8 bits, the first bit represents the activity bit, denoting the presence or absence of a packet at the considered input, while the other 7 bits designate the output to which the packet, when present, is destined. Obviously, with 7 bits for addressing, the maximum size of the switch will be 128×128. To form the entire control signal, another 8-bit word is added (the "source word"), whose first 7 bits represents the address of the input node while the eighth is unused. The signal $C$ is sent to the knockout contention controller synchronously with the data (i.e., at the same time instant) but with different rate. More precisely, the knockout device, which processes the control sequence only, can work at rather low speed (on the order of 200 Mbps) easily achievable by current very large scale integrated (VLSI) circuits. The data signal, instead, can be characterized by higher bit-rate, on the order of some Gbps, since it is processed by fully optical devices. In practice, the connection network can operate at the same bit rate as the input and output links. At the same time, the difference between the bit rates of *C* and *D* obliges to limit the length of the control packet; to maintain synchronization, in fact, the control packet cannot be longer than the data packet. Continuing the example above, a 16-bit control packet at 200 Mbps spans 80 ns, and the data packet will be larger than that on condition of having a length exceeding 160 bits at 2 Gbps or 320 bits at 4 Gbps.

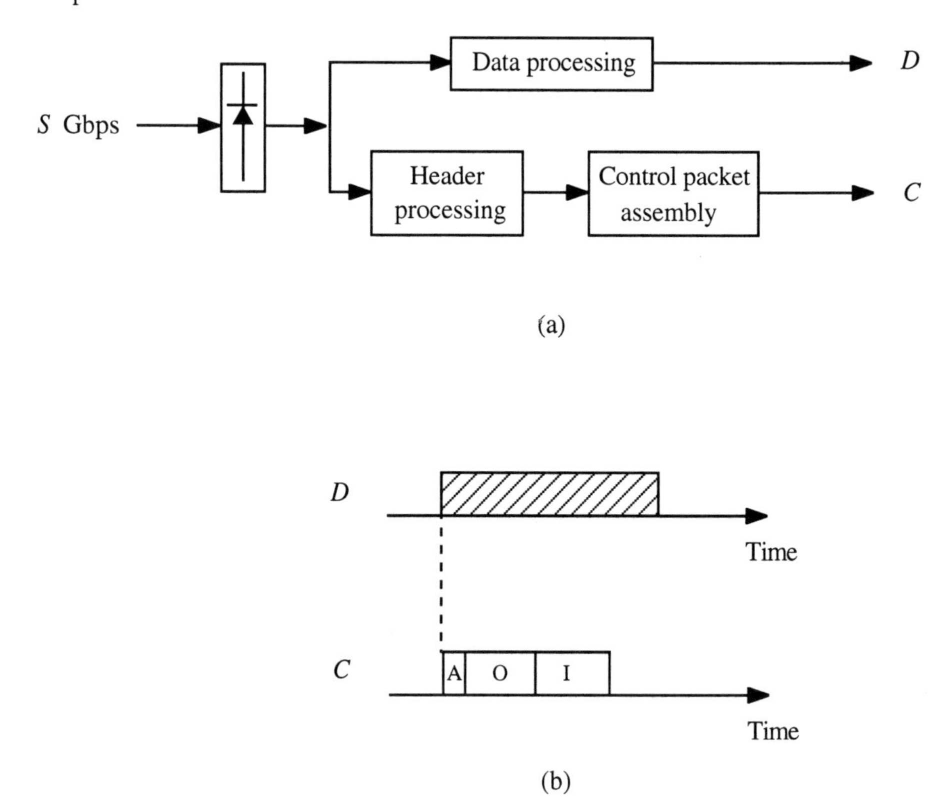


**Figure 2.33** (a) Structure of the input interface module in the photonic knockout switch and (b) data (*D*) and control (*C*) packets (A = activity bit(s); I = input address; O = destination address).

Once entered the switch, any *D* signal can be used, as shown in Figure 2.32, to modulate a fixed wavelength laser. The data packet is therefore converted to optical form and, after a predetermined delay Γ, it is sent through the star coupler to all the outputs. Γ takes into account the time required by the knockout circuit to generate the tuning signals as well as to reconfigure the receivers. As stressed before, any output node has $L$ parallel tunable receivers, and anyone of them can lock the frequency of one input node having a packet destined to the considered output node, up to a maximum of $L$ simultaneous connections per each slot. Detection can be coherent (heterodyning), if frequency tunable lasers are used, or direct, based on tunable filters [91].

The channel selection is controlled by the knockout device, which is formed by $N$ blocks of size $N \times L$, one for each group of $L$ receivers [93]. As shown in Figure 2.34, in each block, the control packets $C$ generated by all the input interface modules are processed in the packet filters, which examine the destination words of the active packets (i.e., those having the activity bit at logic level 1). The signals destined to the output port linked with the block maintain their activity bit unchanged; the others (active but not destined to the considered output) are disactivated, by setting their activity bit at logic level 0. In practice, a distinction is made between valid and invalid packets for each output. In so doing, any incoming control packet is halved in size since, at the output of each packet filter, a word is produced that contains only the activity bit (modified or not) and the source address. These shortened sequences are sent to the concentrator, which has the duty to manage the active words, selecting up to $L$ of the incoming valid packets and discarding the others, together with the words with activity bit 0 (absence of data packets or different destinations).

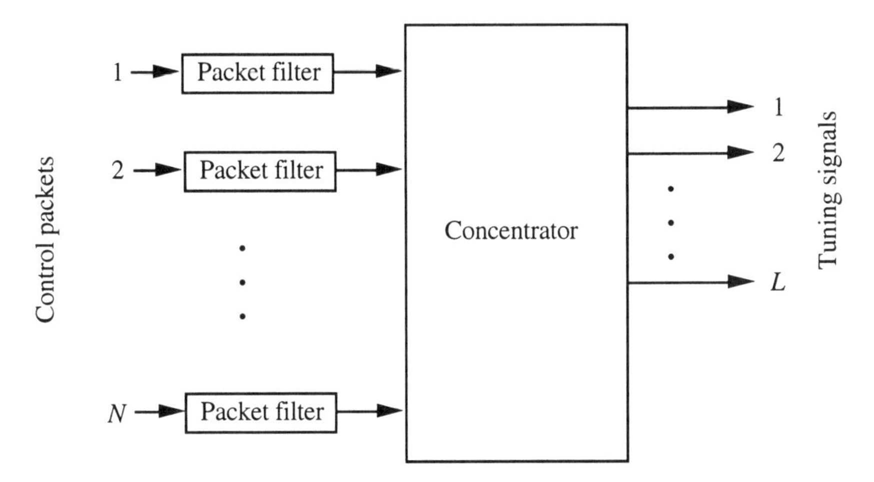


**Figure 2.34** Knockout device for the $i$th output node ($i = 1, \ldots, N$).

The signals generated by the knockout device are used to tune the receivers (Figure 2.32) to the wavelengths of the input nodes that had data addressed to the considered output. The received packets, converted to electronic form, are sent to the buffers for storing, slot by slot. To reduce the latency times and then to increase the speed of the switch, $L$ parallel buffers are used at each output (shared memory). The principle is shown in Figure 2.35: the electronic signals outgoing the $L$ receivers are sent to the buffers through an $L \times L$ shifter. The shifter is a switch with $L$ states, which provides a circular shift of the inputs to the outputs in such a way that the $L$ separate buffers are filled in a cyclic fashion. For example, let us suppose to have $L = 8$; if five packets enter the shifter, in the first time slot, they will be simply transferred to buffers 1 through 5. Now, if in the subsequent time slot four packets arrive, thc shifter will circularly shift the inputs five outputs to the right so that the arriving packets will enter buffers 6, 7, 8, and 1. In the third time slot, storing will recommence from buffer 2, and so on. This way, at any time the number of packets stored in the buffers can differ by one at most. Retransmission is managed, cyclically again, according with a FIFO strategy; the buffers with a greater number of packets are then selected first. The order of the packets in the received sequences is preserved.

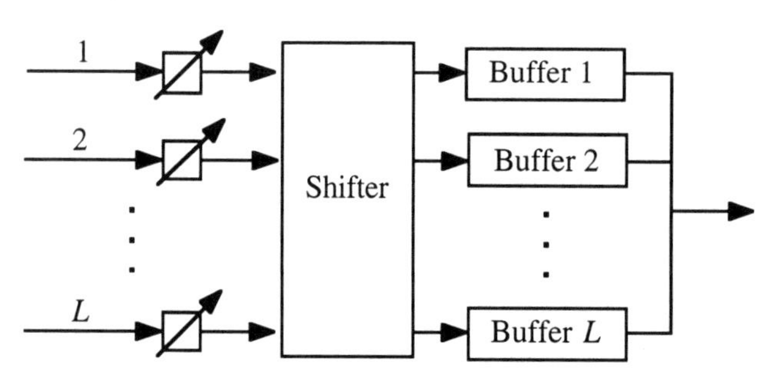


**Figure 2.35** Shared memory and shifter at any output.

Finally, the function of the output interface modules of Figure 2.31 is to reconvert the electronic signals in the original format (incoming the device), which is typically but not necessarily optical. As stressed above, in fact, the switch can also be inserted in a nonoptical transmission system simply varying the interfaces that have to interpret the format and protocol of the incoming data. Input conversion into electronic form and final conversion to the original format allow to maintain the internal switch structure unchanged.

Although the architecture of the photonic knockout switch could seem rather involved at a first sight, it is possible to verify that the adoption of $L$ receivers, with $L << N$, allows to reach very good performance with a relatively simple hardware. The knockout principle relies on the observation that the probability $p_k$ that $k$ packets are simultaneously (i.e., in the same time slot) destined to the same output, given by an expression like (2.7), is generally very small for not negligible $k$ and, most of all, rapidly decreases for increasing $k$. Since a dropping event only occurs if $k$ exceeds $L$, it is always possible to dimension the number of receivers at each output node in such a way as to have sufficiently small loss probability. In particular, the object could be to have a loss probability comparable with all the other causes of performance degradation (transmission errors, buffer's overflow, and so on) which affect any practical system.

On the basis of (2.7), where $\rho$ is the probability that a packet arrives at each input of the switch, independently of the others and equally likely destined to each output, the loss probability is given by

$$P_l = \frac{1}{\rho} \sum_{k=L+1}^{N} (k-L) \binom{N}{k} \left(\frac{\rho}{N}\right)^k \left(1-\frac{\rho}{N}\right)^{N-k} \tag{2.11}$$

$P_l$ depends mainly on $L$, while the dependence on $N$ is rather weak. So, a $P_l < 10^{-6}$ can be achieved with $L$ not smaller than 8 when $\rho = 0.9$, independently of the number of nodes [93]. An increase of $L$ by one unit, at a parity of the other parameters, allows to reduce $P_l$ by one order of magnitude, but the improvement is paid in terms of increased complexity, since the network requires $N$ receivers more, one for each output.

On the opposite side, a further reduction in the number of receivers can be achieved by considering that the probability that more than eight packets (to continue the example) are simultaneously addressed to the same destination in two consecutive slots is really very small for a fixed $\rho$ (surely lower than the probability of having eight packets equally destined in a single slot).

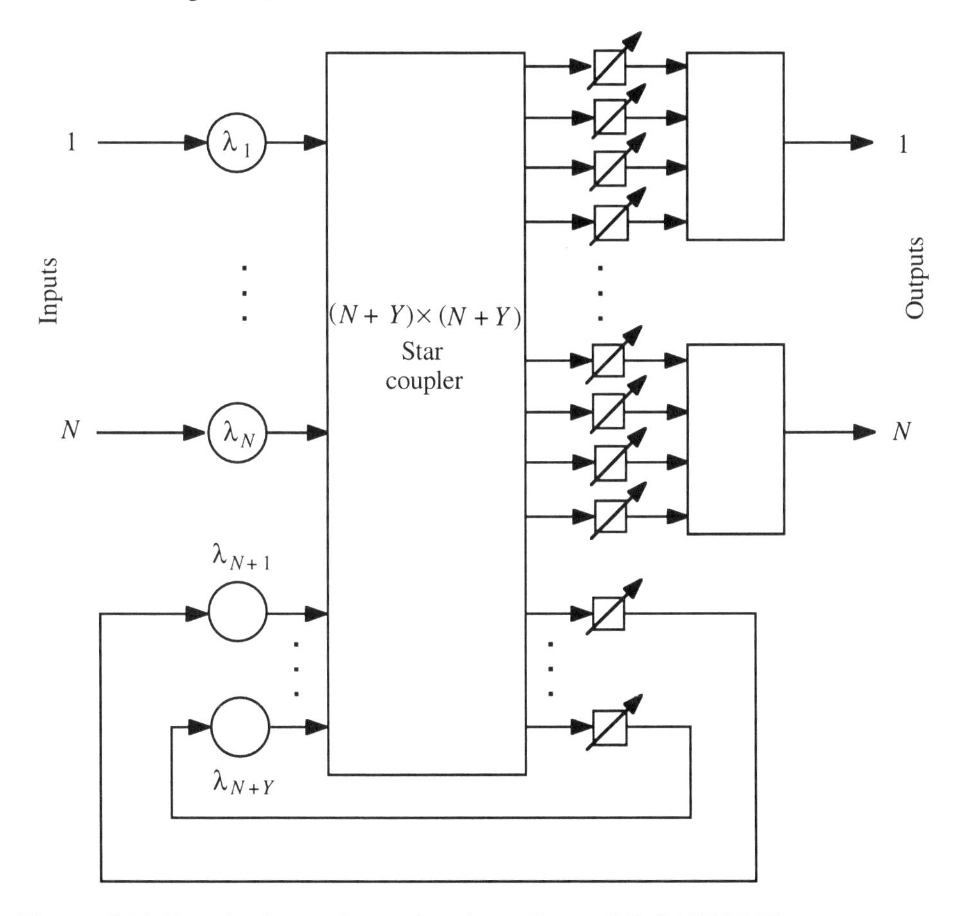


**Figure 2.36** Example of repeated contention scheme. (*Source*: [91]. © 1988 IEEE.)

Therefore, if the $k - L$ packets that are "losers" in a time slot are retained and reinserted in the system in one or more subsequent time slots, they can become "winners" later on, and the packet loss probability should decrease significantly (or, equivalently, $L$ could be reduced to attain the same $P_l$).

In particular, assuming that the reinserted packets are marked with a higher priority (compared to those contemporaneously arriving for the first time), but that a packet that has lost the second chance is definitively dropped, a packet loss probability of $10^{-6}$ can be reached using only $L = 4$ receivers per node [91]. Increasing $L$ by one unit, in this repeated contention scheme, allows us to reduce $P_l$ by two orders of magnitude. It is evident that such an advantage must be paid in terms of system complexity. An example of possible implementation is shown in Figure 2.36, where the losing packets are sent to an additional set of star coupler outputs, temporarily stored, and then brought back into the coupler inputs for retransmission.

The system requires a coupler with $Y$ additional ports for the rerouting of the losing packets: those at the output must each be equipped with a tunable receiver and those at the input with a fixed-wavelength laser. To achieve a packet loss probability on the order of $10^{-6}$, with $L = 4$, $Y$ must be approximately $N/2$, so that the total number of lasers is $N + N/2$ and the total number of receivers is $4N + N/2$.

### 2.3.7 Passive Photonic Loop

The passive photonic loop (PPL), proposed by Bellcore in 1988 [94], is a fiber subscriber-loop architecture based on the so called double-star topology [95]. The solution is sketched in Figure 2.37: each user is connected, through a single-mode optical fiber, to a remote terminal (RT) which, in turn, is linked to the central office (CO) through another common fiber (feeder). Two wavelengths are assigned to each subscriber, one for transmission and one for reception. Hence, the network globally requires $2N$ wavelengths.

At the CO, the signals destined to the nodes are modulated onto their assigned downstream wavelengths and multiplexed onto the feeder. Then, at the RT, they are demultiplexed on the appropriate distribution fiber. Since the channels are filtered by the WDM demultiplexer, each subscriber receives only on its unique wavelength; in this way, security is preserved and, at the same time, no need arises for wavelength filters at the subscriber premises.

Transmissions in the opposite direction are performed similarly: each subscriber transmits on its unique wavelength, the different signals are multiplexed at the RT and then demultiplexed at the CO. The network size, that is, the number of nodes that it is possible to connect ensuring satisfactory performance, mainly depends on the features of the devices used, particularly as regards the dimensions of the multiplexer/demultiplexer. Bellcore's model indicated a maximum of 40 to 50 nodes using the grating-based technology available at the time of the proposal [63,96]. More recent research, however, has demonstrated the possibility of realizing multiplexers with 64 inputs [97].

The PPL architecture uses only passive WDM devices to perform routing and multiplexing functions and does not need electronic components for control. This avoids back-to-back O/E and E/O conversion at the RT as well as the need for air-conditioned controlled-environmental vaults (CEVs) and remote power backup. The system requires minimum maintenance, while the relative compactness of the optical devices permits great flexibility in the placement of RTs. Moreover, the reduced number of subscribers connected to the RT allows one to install the remote terminal rather close to the nodes, thus minimizing the length of the distribution fibers.

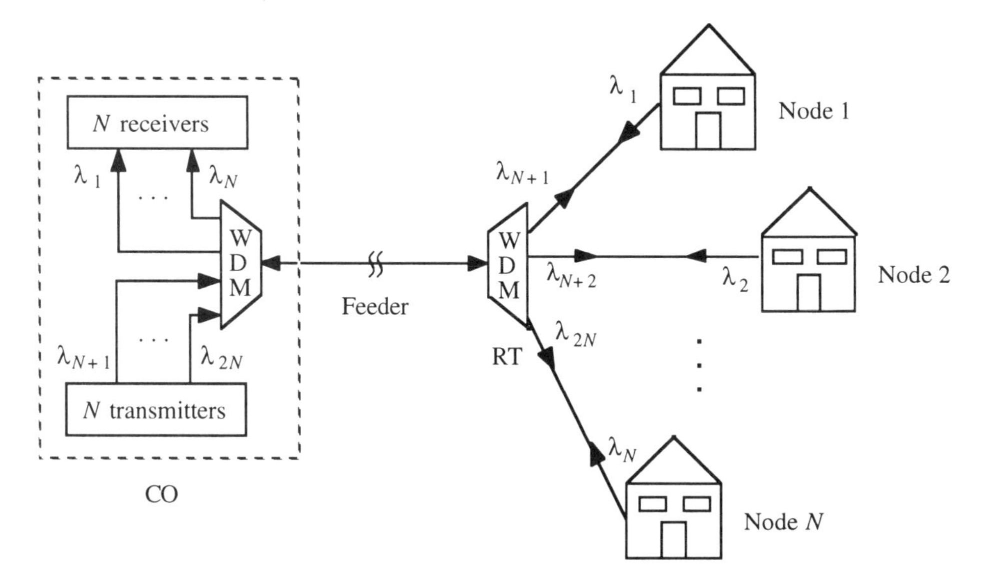


**Figure 2.37** Block diagram of the passive photonic loop. (*Source*: [94]. © 1988 IEEE.)

The main property of the network, however, is in the possibility of accommodating heterogeneous traffic; the WDM channels, in fact, are independent from each other and can accept, in principle, any format, protocol, or modulation technique, adapting themselves to the specific requirements of any user. Hence the network seems to offer full compatibility with any kind of standard and possible future changes or service developments that can be proposed, maintaining the optical connection structure unchanged. The PPL, providing a bidirectional link to each subscriber, can be used for point-to-point services, such as telephony, data communications, and switched video. On the other hand, if the RT is equipped with a passive power splitter, the network can also accommodate, in parallel, point-to-multipoint transmissions, such as video broadcasting. In the latter case, the array of signals could be multiplexed at the CO using either WDM [98] or electronic subcarrier techniques [99]. The hybrid solution, removing a significant portion of the video traffic from the switched portion of the network, allows one to reduce the complexity of the switch and to facilitate the mixture of analog and digital video services [94].

An experimental evaluation of the PPL has been made on the basis of the prototype shown in Figure 2.38 [100]: it uses a pair grating-based multiplexer/demultiplexer, with 20 channels in the 1.5 μm band. The WDM channels are spaced at intervals of 2 nm and the FWHM passband of the multiplexer, for each channel, is approximately 0.3 nm. The system consists of 10 nodes, placed at a variable distance in the range 2.2 to 3.3 km from the RT, which in turn is at a distance of 9.7 km from the CO. All the connections employ single-mode optical fibers. The wavelengths λ between 1,523 and 1,541 nm are used by the subscriber to transmit, while the wavelengths between 1,543 and 1,561 nm are used to receive. As an alternative, one could use two separate transmission windows (like those centered at 1.3 and 1.5 μm), one for each monodirectional link.

The system has been analyzed proposing, as optical sources, either DFB lasers or cheaper light-emitting diodes (LEDs). It is evident that performance is optimal in the case of lasers, not necessarily DFB, characterized by high spectral purity, since their bandwidth, on the order of 10 MHz, is much narrower than a WDM channel (0.3 nm corresponds approximately to 40 GHz in the 1.5 μm band). All power going out from the laser is therefore available for transmission and this acts favorably on the power budget. In particular, it can be supposed that DFB lasers will be used, with a minimum launched power $P_t = -3$ dBm, and that there will be a system attenuation $A_a = 7$ dB due to the fibers and the connectors, a loss $A_1 = 6$ to 7 dB at the RT, and a loss $A_2 = 2$ dB (when it acts as a demultiplexer) or 11 dB (when it acts as a multiplexer) at the CO. Then, computing a sensitivity $P_r = -36$ dBm at a bit rate per node of 1.2 Gbps and a BER of $10^{-9}$, we remain with a minimum link margin $A_M = 6$ dB. It must be noted that the loss asymmetry in the CO WDM device is due to its supposed multimode pigtails on ports 1 to 20, while the WDM device at the RT does not suffer the same effect since it is supposed to be fully single-mode.

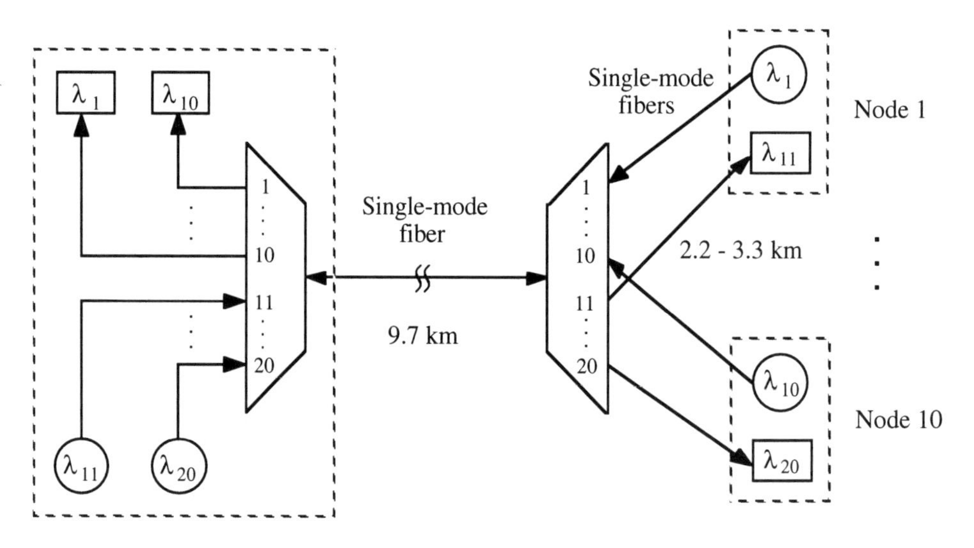


**Figure 2.38** Experimental demonstration of the passive photonic loop. (*Source*: [100]. © 1988 IEE.)

If LEDs are used, in place of lasers, at the subscribers, the system performance degrades because of the reduction in the power available for the signals. Transmission is based on the so-called "spectrum slicing" concept illustrated in Figure 2.39 [101]. LEDs are characterized by emission spectral widths of some 10 nm to about 100 nm; then, the passbands of the grating based multiplexers are, in comparative terms, much narrower and densely spaced. As a consequence, a lot of transmission channels fit into the LED spectrum.

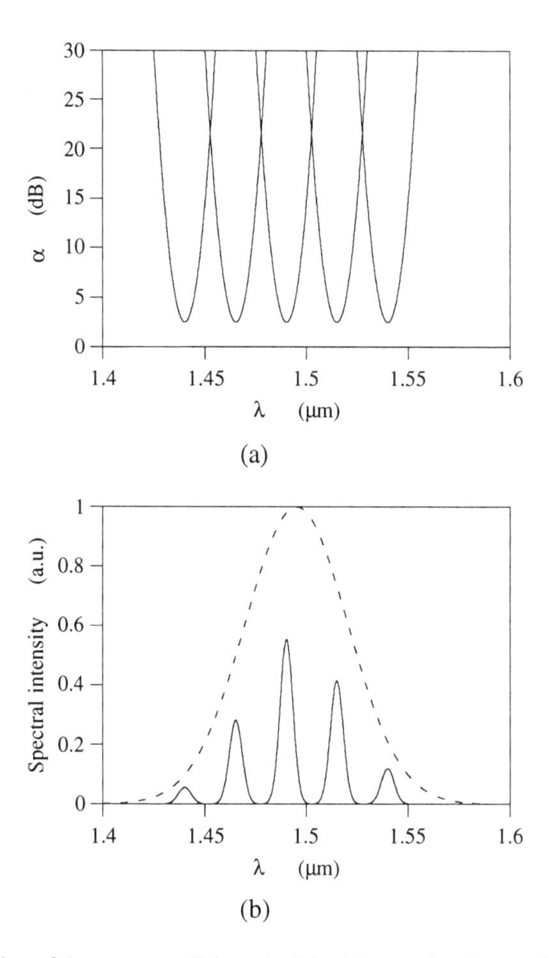


**Figure 2.39** Illustration of the spectrum slicing principle: (a) example of spectral attenuation α for a grating multiplexer with five input ports; (b) example of spectral intensity for an LED (dashed line) and spectrum sliced LED emission (continuous line) at the output of the multiplexer in (a) (a.u. = arbitrary units).

The emission spectra of the LEDs are filtered according to the passband characteristics of the multiplexers, and the multiplexed signals exhibit low crosstalk (less than –30 dB in the experiment of [100]). Another interesting feature is that the spectra of the LEDs may widely overlap or even be equal; in other words, the subscribers can use identical light sources. The small portion of spectrum selected by the multiplexer for each signal contains the original information and then represents the subscriber data channel. The penalty paid, however, is in terms of reduced power budget since only a fraction of the LED's spectral power is available, for each channel, at the output of the multiplexer (attenuation is on the order of 31 to 34 dB, also including the multiplexer's intrinsic excess loss). The reduction in the power budget limits the bit rate per channel up to a few megabits per second using standard LEDs. Otherwise, one could think of adopting superluminescent diodes (SLDs) which make it possible to increment the bit rate remarkably (150 Mbps in a system with 10 WDM channels or 50 Mbps with 16 channels) [102].

### 2.3.8 STAR-TRACK

STAR-TRACK is a packet switching FT-TR system proposed in 1990 and mainly conceived for the management of selective point-to-multipoint (i.e., multicast) traffic [103]. The switch fabric is formed from two internal networks: an optical star network for the data, and a one-ring electronic network for control (Figure 2.40). Each input port of the star network is equipped with a fixed-wavelength laser and each of its outputs is equipped with a tunable receiver.

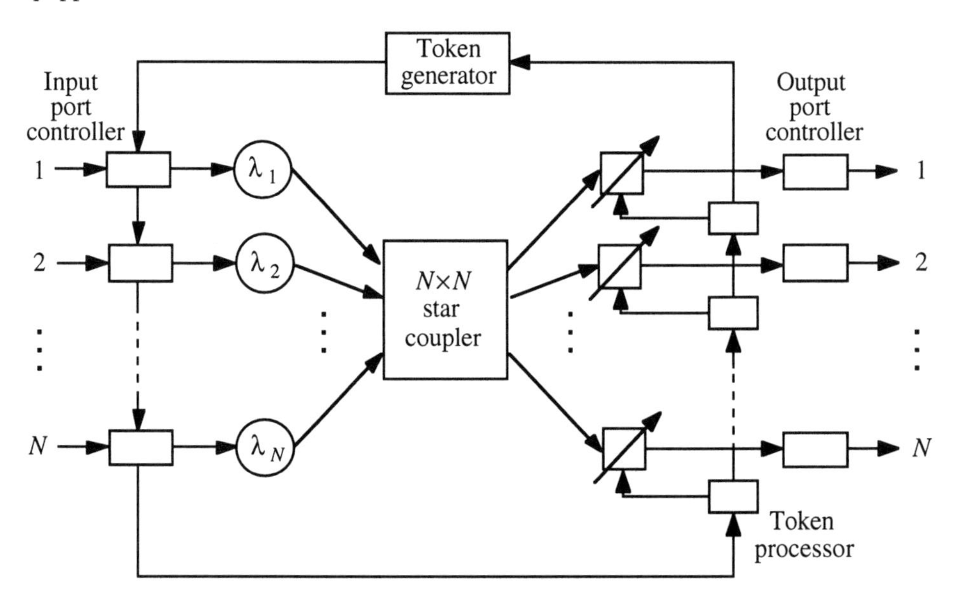


**Figure 2.40** Block diagram of a STAR-TRACK network.

The network control functions as well as the management of the transmissions (including collisions resolution) are fulfilled by the control system. In particular, the packets that arrive from the input optical fibers are stored in the buffers of the input ports, awaiting the permission to be transmitted. The control system generates a token, formed from as many subtokens as the number of nodes, which passes through all the input ports. The control cycle can be subdivided into two phases. In the first phase, that of writing, each input having some data to transmit examines the subtoken of the output port it intends to reach; if the subtoken is empty, which means that the desired output has not yet been reserved, the input writes its own address in the subtoken, thus booking the transmission for the next slot. On the other hand, if the subtoken is not empty, the input must wait for the subsequent cycle, when it will try again to acquire a reservation. Such a protocol is inherently favorable to multicast transmissions, since point-to-multipoint connections can be arranged simply writing the address of the same input port into the subtokens of many output ports.

At the end of the first control phase the token collects all the accepted reservations from the input ports. Then, the second phase, that of reading, begins: any output examines its subtoken for tuning the receiver to the wavelength of the appropriate input port. As the token is emitted from the token generator, packets transmission is commenced at the input ports. Thus, the transmission and control cycles overlap in time and the control phase is executed during the previous transmission cycle. Collisions are obviously avoided by virtue of the multiwavelength transport.

Among the factors that limit the performance of the STAR-TRACK system, besides those typical of all the networks based on a star topology, which has already been mentioned in the previous sections, a prominent role is played by the constraints imposed by the electronic components; they have an effect on the time needed to perform a control cycle and then limit the capacity of the control channel. Furthermore, the whole process of reservation for the input ports and tuning for the output ports is sequential, so that the control tag must be processed by all the ports, even if they have no packets to transmit or to receive. The situation is quite different, for example, from the HYPASS or BHYPASS systems, where the input ports are processed in parallel.

The need for the ring to complete its cycle within the transmission time of the previous packet limits the maximum number of nodes and makes this number dependent on the packets' length, the bit rate at the input ports, and the tuning times of the receivers [7]. The number of nodes reflects on the length of the addresses to be written in the subtokens: $N$ nodes require $\log_2 N$ bits to be represented. Also, increasing the network size increases the length of the addresses, with augmented difficulties for the electronic control which must process information at a bit rate generally lower than the optical transmission but within the same time. Therefore, the number of nodes cannot be arbitrarily large.

At the same time, the limited working speed of the electronic ring makes the network capacity dependent on the packet length and on the bit rate of the ring as well. Even though the time required to resolve contentions is negligible, the limited capacity of the control channel can seriously penalize the capacity of the whole network. To explain this concept better, let us suppose that synchronized sources transmit packets of equal length $P$. Each output port analyzes its own subtoken, containing the address of the reserved input ports, consisting of $\log_2 N$ bits. Thus, globally, the control track must transport $N \cdot \log_2 N$ bits, at most, for each cycle, that is to say, $N$ addresses. Noting by $T_c$ the duration of a control cycle, the capacity of the control channel is given by

$$C_c = \frac{N \cdot \log_2 N}{T_c} \tag{2.12}$$

But $T_c$ also equals the duration $T_p$ of a data transmission; so, assuming that each source transmits at a bit rate $S$, we have

$$T_c = T_p = \frac{P}{S} \tag{2.13}$$

Substituting (2.13) into (2.12) we obtain

$$C_c = \frac{\log_2 N}{P} \cdot C \tag{2.14}$$

where $C = S \cdot N$ represents the network capacity. If the control channel consists of a bus with $k$ parallel lines, each characterized by a bit rate $S_c$, since $C_c = k \cdot S_c$, (2.14) can be used to derive the network capacity as

$$C = k \cdot \frac{P}{\log_2 N} \cdot S_c \tag{2.15}$$

The maximum value of $k$ is obviously $\log_2 N$ and, correspondingly, the maximum capacity is $C = P \cdot S_c$. To obtain high values of $C$, then, implies having proportionally high values of $P$ or $S_c$. The packet length $P$ is normally prefixed by the specific service that must be ensured, by particular requirements on the transmission times, or by standards. Considering the case of packets for asynchronous transfer mode (ATM) transmission, the packet length must be assumed equal to 53 bytes (424 bits) [104]. With this choice, and setting $S_c$ = 1 Gbps and $S$ = 5 Gbps, the maximum number of nodes is 85 for $k = \log_2 N$, but drops to 20 for $k$ = 1 (serial bus). In any case, the limits imposed by the control channel are generally more restrictive than those due to the power budget. Taking into account that the efficiency of the protocol becomes worse and worse with increasing distances (which implies greater propagation times) we can conclude that STAR-TRACK can be considered interesting for multicast transmissions but principally in the case of centralized networks and given the availability of very fast electronics. Moreover, considering that, at optical level, the network requires fast tunable receivers, it is easy to foresee that STAR-TRACK will take advantage from the future technological developments.

### 2.3.9 Fiber Delay Line Switching Matrix

The fiber delay line switching matrix, also called OASIS [1], is an ATM centralized photonic switching fabric, prototyped by Alcatel Alsthom Recherches in 1993 [105,106]. The system, shown in Figure 2.41, is TT-FR but with the interesting characteristic that the tunable transmitters are replaced by wavelength converters. It is worth noting that, according to the ATM standard, the objects of the transmission are very short packets of fixed size, more properly called cells. This term will be therefore used in the course of this section.

The network, described in detail in [107], consists of three different functional units:

- The wavelength routing unit, which can be split into three parts: (1) a set of cell encoders, one for each input, including the wavelength converters; (2) one $Q \times N$ optical star coupler, which mixes all the wavelength-encoded cells; (3) a set of $N$ fixed-wavelength optical filters, which enables each output port to select only the cells addressed to it; (2) and (3) constitute the wavelength demultiplexer block of Figure 2.41;
- The cell buffer unit, which realizes the cell time switching on suitable delay lines; the unit is constituted by $N \times Q$ optical gates, arranged in a shuffle network and providing access to $Q$ delay lines operated in a wavelength-multiplexing mode;
- The routing and control logic unit, essentially based on an electronic command memory, that drives the wavelength converters (for routing) and the $N \times Q$ optical gates (for queueing) according to the cell routing tags, all processed simultaneously at each time slot.

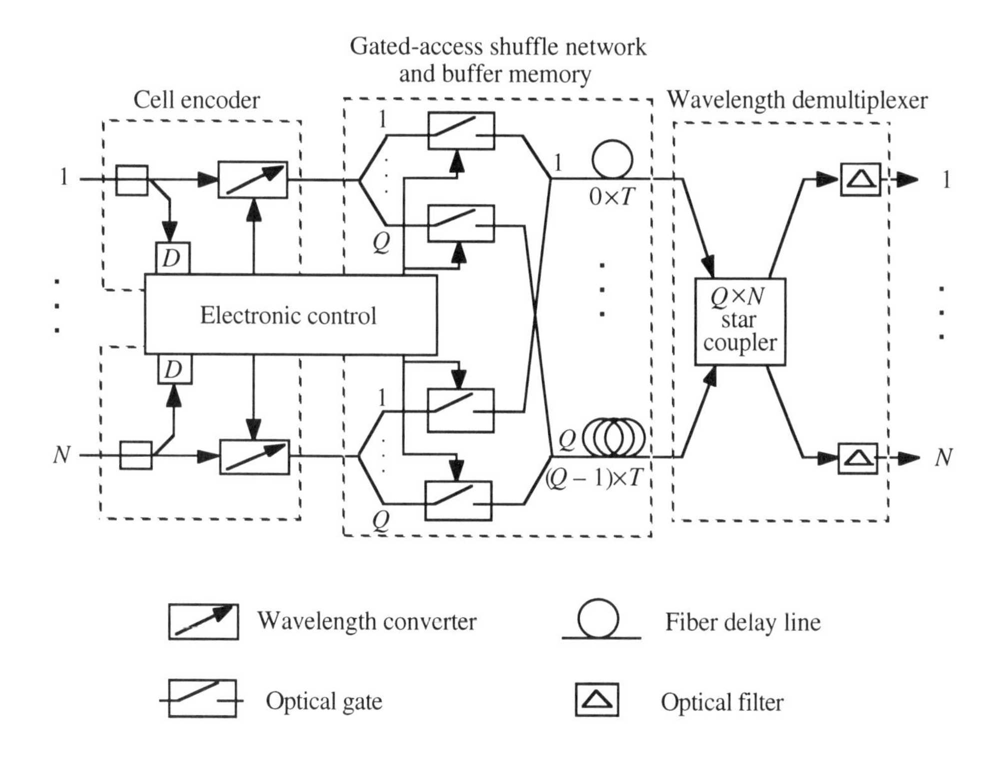


**Figure 2.41** Block diagram of a fiber delay line switching matrix.

To explain better the basic operation of the system, let us suppose that all the cells arrive in phase, that is, synchronously, at each of the $N$ input ports. A tap on each input link extracts part of the optical power, converting it into electronic form (through the detector $D$), which is sent to the control system; the remaining fraction of optical power, instead, is sent to the wavelength converter. The electronic control processes the ATM cell headers, containing the addresses of their destinations and, on the basis of such information, produces suitable control signals for the converters and the optical gates. Each converter, commanded by the control signals, transforms the wavelength of the optical signal at its input into the wavelength of the destination output port which will therefore be able to detect the cell using an optical bandpass filter (fixed tuned receiver). In practice, we can say that, at the output of each converter, the cells have been wavelength encoded.

Collisions between synchronous cells with identical wavelengths are avoided by the time switching block, which can be seen as an optical partially shared buffer. It consists of a set of $Q$ optical fiber delay lines, with increasing lengths corresponding to relative propagation times ranging from 0 to $(Q - 1) \cdot T$. $T$ is the duration of a cell; for example, with a bit rate per node $S = 2.488$ Gbps, recalling that the size of an ATM cell is 53 bytes, we have $T = 424/(2.488 \cdot 10^9) \approx 170.4$ ns.

The output of each wavelength converter is connected to all the $Q$ fiber delay lines through a $1 \times Q$ power splitter and a set of $Q$ E/O gates. Access to the delay lines is determined by the optical gates commanded, in turn, by the electronic logic: contending cells are sent to different delay lines, and then will undergo different delays. For example, if two cells, in the same time slot, have equal destination, one of them will be sent to the first line (null delay) and will be transmitted in the same time slot, while the second will be sent to the second line (delay $T$) and will be transmitted in the subsequent time slot. The buffer is then managed with a FIFO discipline, and we have as many logical queues as the number of outputs. The fiber delay lines are physically shared by the whole set of queues, thus minimizing the amount of optical hardware.

In evaluating the performance, the traffic and service parameters considered are $\rho$ and $P_l$, defined respectively as the input traffic load on each link and the cell loss probability. Losses occurs when the number of contending cells exceeds the capacity $Q$ of the corresponding logical queue. The output queueing model [108] can be used to dimension $Q$ and $N$, to satisfy assigned requirements on $P_l$ for different values of $\rho$. Considering uniform traffic distribution on all the outputs and assuming $\rho = 0.5$ and $P_l = 10^{-9}$ (which is typical for ATM services), it can be seen that $Q = 13$ is requested for $N = 8$, while $Q = 15$ is the minimum buffer size for $N = 16$. At a parity of $N$, the value of $Q$ increases as the load increases, slowly at first up to $\rho \approx 0.6$, and very quickly for higher values. For this reason, applications for loads that are not too great are advisable, to limit optical hardware dimensions and complexity.

A specific experimental demonstration of the fiber delay line switching matrix has been presented in [106] for a small network ($N = 4$) and $S = 2.488$ Gbps. The system uses two different types of wavelength converters: (1) polarization-insensitive semiconductor optical amplifiers, each arranged with a tunable DBR laser operated in continuous wave (CW) mode, to perform all-optical wavelength conversion; (2) tunable DBRs combined with O/E detectors and amplifying/leveling electronic chains to form hybrid wavelength converters. The devices can switch the emission wavelength, driven by the electronics, over four modes centered around 1,548 nm and spaced by 0.6 nm, with a maximum reconfiguration time of 9 ns. The buffer memory is accessed through 16 gates based on semiconductor optical amplifiers ($Q = 4$) exhibiting a response time of less than 200 ps and providing more than 12 dB fiber-to-fiber gain with less than 2 dB polarization and wavelength dependence.

Finally, the demultiplexing block is made with a 4×1 combiner followed by a 1×4 splitter and four Fabry-Perot etalons. The combiner and the splitter realize the star coupler and are connected by an amplifier board based on one erbium-doped amplifier, with 30 dB gain at 1,548 nm, operated in the nonsaturated regime. The optical filter bandwidth (FWHM) is on the order of 0.3 nm and the filtering crosstalk from adjacent wavelength encoded channels is less than –15 dB.

Actually, the main factors limiting the system performance are the crosstalk and the amplified spontaneous emission (ASE) noise [73] due to the amplifiers. In [107] it has been demonstrated that, working in the neighborhood of $\lambda = 1{,}564.5$ nm with channels spaced 0.28 nm apart, a bit error rate (BER) of $10^{-9}$ is compatible with a maximum number of nodes $N = 16$.

### 2.3.10 SYMFONET

SYMFONET is a synchronized, star-connected, FT-FR$^N$ system proposed in 1990 [109]. Each node is equipped with a fixed tuned laser (typically a DFB) and a multichannel grating demultiplexer (MGD) receiver [110]. The latter consists of a grating demultiplexer, projecting each wavelength channel onto a separate detector in a monolithic array. Thus, each node transmits on a different wavelength and is capable of receiving all the wavelengths, its own included. The transmitted signals are routed to all the nodes via a passive star coupler (e.g., such as in the LAMBDANET).

The main feature of SYMFONET, however, that makes it different from many other solutions, is the precise control of the timing of the transmitted information, so that all the channels arrive synchronously at each receiver. In this way it is possible to adopt a common clock for all the nodes to regenerate the signals (thus reducing complexity) and, even more important, the control protocol can make use of both wavelength and time dimensions (saving the cost of a separate control channel).

Synchronization is ensured by electing one of the nodes of the network as the reference node, which must provide the clock and the frame structure for all the other nodes. The choice of the reference node is arbitrary and can be made by any method, on the condition that all nodes select the same reference. One possibility consists of choosing the lowest frequency present when the process starts up. Once synchronization has been achieved at bit level, the reference clock is used to regenerate the incoming signals.

In Figure 2.42, the main components of a SYMFONET node are shown; the considered node is labeled as 3 (this is evident from the use of $\lambda_3$ as the transmission wavelength), while the reference node has been assumed to be node 1 (not shown in Figure 2.42, which refers to a single node). The network is supposed to have eight nodes. The information and control data form a regular structure, called a *frame*, of fixed length.

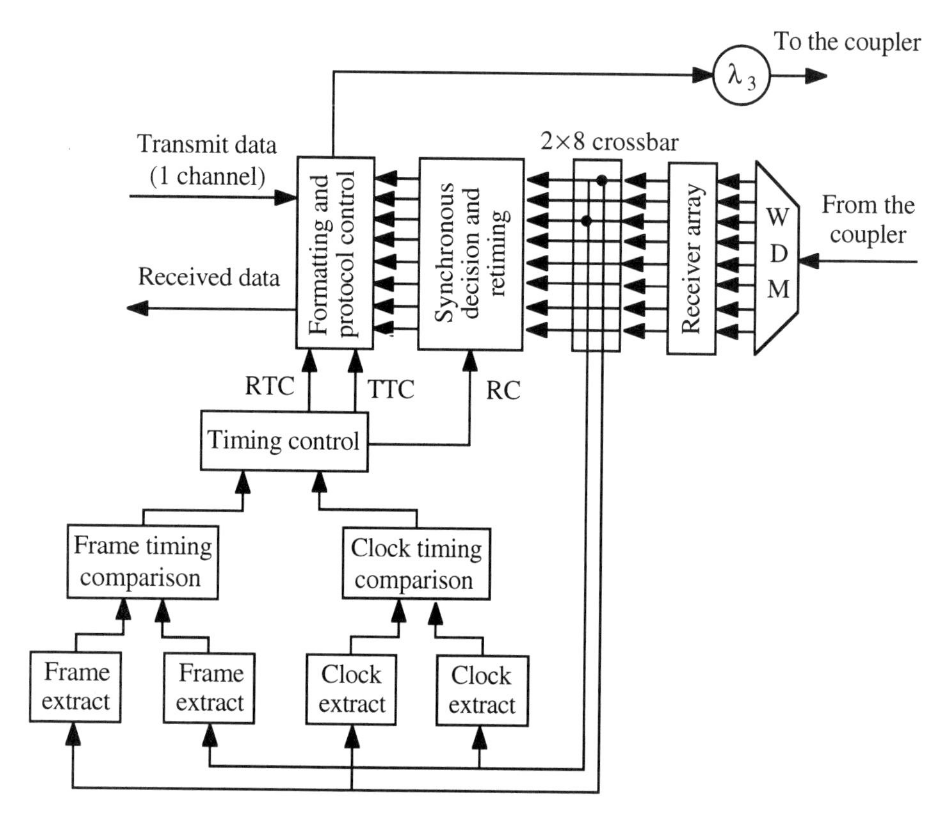


**Figure 2.42** Block diagram of a node in the SYMFONET network. (*Source*: [109]. © 1990 IEE.)

Each node controls the timing of its own transmitted clock and frame, continuously comparing them with those transmitted by the reference node. More precisely, with reference to the specific example in Figure 2.42, at the output of the receiver array an electronic crossbar switch extracts the signals emitted by the node itself (3 in Figure 2.42) and by the reference node (1 in Figure 2.42), sending them to the circuits for clock and frame timing extraction. The clock and frames of the two nodes are then compared to verify that the considered node is always synchronized with the reference. The timing control devices introduce the necessary adjustments, generating the timing signals for the receiving apparatus [receiver clock (RC) and receiver timing control (RTC)] and the transmitting apparatus [transmitting timing control (TTC)]. The procedure is repeated for any node of the network.

Neglecting the effects of wavelength dispersion, it is evident that separate frame synchronization between the reference node and each one of the other nodes also ensures global frame synchronization. To reinforce this concept, let us consider the frames emitted by nodes 1 (reference), 2, and 3; we would like to demonstrate that all they arrive synchronously, for example at node 2. To this purpose, let us denote by $c$ the common speed of propagation and by $L_{xy}$ the distance between the nodes $x$ and $y$; $L_{xx}/2$ represents the distance of node $x$ from the star coupler. Thus, owing to the structure of the network, the following relationships hold:

$$
\begin{aligned}
L_{12} &= \frac{L_{11} + L_{22}}{2} \\
L_{13} &= \frac{L_{11} + L_{33}}{2} \\
L_{23} &= \frac{L_{22} + L_{33}}{2}
\end{aligned}
\tag{2.16}
$$

Assuming the instant at which the frame is emitted by the reference node as the origin of the time axis, synchronization between nodes 1 and 2 implies that the latter emits its frame at the instant

$$t_2 = \frac{L_{12} - L_{22}}{c} \tag{2.17}$$

Obviously, $t_2 = 0$ when $L_{12} = L_{22}$ while, if the distance between node 2 and the star coupler is greater (smaller) than the distance between node 1 and the star coupler, node 2 must anticipate (delay) the transmission of its own frame.

With similar arguments, synchronization between nodes 1 and 3 implies that the latter emits its frame at the instant

$$t_3 = \frac{L_{13} - L_{33}}{c} \tag{2.18}$$

Now, the frames emitted by nodes 1 and 2 are received by node 2 at

$$t_4 = \frac{L_{12}}{c} \tag{2.19}$$

On the other hand, the frames emitted by node 3 are received by 2 at

$$
\begin{aligned}
t_5 &= t_3 + \frac{L_{32}}{c} = \frac{L_{13} - L_{33} + L_{23}}{c} \\
&= \frac{\frac{L_{11} + L_{33}}{2} - L_{33} + \frac{L_{22} + L_{33}}{2}}{c} = \frac{\frac{L_{11} + L_{22}}{2}}{c} = \frac{L_{12}}{c} = t_4
\end{aligned}
\tag{2.20}
$$

Thus, the frames coming from nodes 1, 2, and 3, and received by node 2, are all perfectly synchronized. Extending the reasoning to all the other nodes of the network, frame timing is therefore demonstrated.

Synchronization occurs at the receiver but it is evident that, in reality, the frames arrive simultaneously at the central star and maintain synchronism as they travel the common distance to the destination node. Like the previous demonstration, however, this conclusion is also valid, in principle, only on condition that the wavelength dispersion is negligible. If the speed of propagation is different for each wavelength, in fact, frame timing may be no longer ensured and some form of optical or electrical dispersion compensation may become necessary. Actually, this occurs when the wavelength dispersion, generally acting on different fiber lengths, causes more than about 1/10 period delay difference between the extreme wavelengths used in the system [109]. By employing standard fibers in the neighborhood of $\lambda = 1.5$ µm, with 32 channels at 2 nm spacing, each modulated at 1 Gbps, the maximum length difference, without the need for dispersion compensation, is about 100 m.

The problem disappears when the distance between the central star and the receiver is the same for each node; in this case, in fact, the system automatically compensates the effects of wavelength dispersion. This is another interesting feature of SYMFONET, which is easily demonstrable as follows. With reference to the three nodes example discussed above, we can denote by $c_i$ the speed of light at the $i$th wavelength. Moreover, putting $L_{11} = L_{22} = L_{33} = \overline{L}$, in this case we have

$$t_2 = \overline{L}\left(\frac{1}{c_1} - \frac{1}{c_2}\right) \qquad t_3 = \overline{L}\left(\frac{1}{c_1} - \frac{1}{c_3}\right) \tag{2.21}$$

so that

$$t_4 = \frac{\overline{L}}{c_1} \qquad t_5 = t_3 + \frac{\overline{L}}{c_3} = \frac{\overline{L}}{c_1} = t_4 \tag{2.22}$$

Here, however, synchronization is obviously at node level but not at star coupler level. As an example, denoting by $t_6$ and $t_7$, respectively, the instants at which the frames emitted by nodes 1 and 2 arrive at the central star, we have

$$t_6 = \frac{\overline{L}}{2c_1} \qquad t_7 = t_2 + \frac{\overline{L}}{2c_2} = \overline{L}\left(\frac{1}{c_1} - \frac{1}{2c_2}\right) \neq t_6 \tag{2.23}$$

By virtue of the type of receiver adopted [multichannel grating demultiplexer (MGD)], the SYMFONET can be implemented in such a way that each node receives one channel per slot (crossbar SYMFONET [109]) or all the wavelengths simultaneously (broadcast SYMFONET [111]).

In the crossbar SYMFONET, at each node, an electronic switch selects the desired output of the MGD receiver, detecting only one channel at a time. Anyway, the system allows us to monitor all the channels in parallel, and to switch from one channel to another with very low latency times. To avoid the collisions that occur when two or more transmitters simultaneously link with the same receiver, a suitable resolving algorithm must be designed. Actually, the control information can be encoded in the frame in many different ways, mainly depending on the system design constraints. A simple arrangement is the one shown in Figure 2.43, again relative to a SYMFONET with eight nodes.

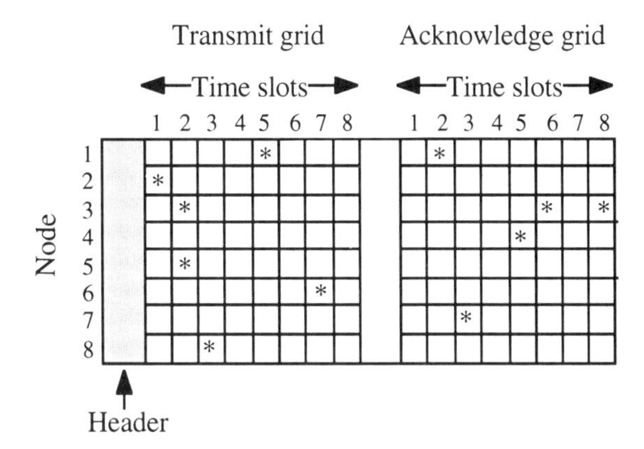


**Figure 2.43** Example of control protocol for a crossbar SYMFONET with eight nodes.

After a few bits of header, eight time slots (one for each node) are reserved to the connection requests (transmit grid). These are represented by asterisks in Figure 2.43: for example, transmission of a bit in the seventh time slot by node 6 means that node 6, in the considered frame, wishes to transmit to node 7. Eight further time slots are allocated later in the frame, reserved to the confirmation of previous correct message receipt (acknowledge grid): for example, the presence of a bit (marked by an asterisk) in the third slot of the acknowledge grid, emitted by node 7, means that the message transmitted by node 3 to node 7 two frames before has been correctly received. The delay of one frame in sending the acknowledgment takes into account the propagation delay; the duration of a frame, however, must be dimensioned in such a way that all messages reach their destination within the time of one frame.

In general, this particular structure of the control field of a frame allows an easy management of the multicast and broadcast transmissions and a fast resolution of the contentions. When two nodes (e.g., 3 and 5 in Figure 2.43) are in contention to transmit to the same node (node 2 in the figure), they emit a bit in the same time slot. Before the end of the frame, all nodes of the network (including 3 and 5, in the example) will have received this kind of information. Using a common algorithm, for example, the lowest node number always wins, contention is resolved; the loser (5) will be aware that its message cannot be correctly received by sampling all wavelengths in the tagged time slot (2) of the transmit grid returning from the star network, and will try to transmit the message again in the subsequent frame. Two frames later, instead, the winner (3) will receive the acknowledgment of the correct receipt of its message.

Figure 2.43 describes the structure of the control field; the remaining part of the frame is devoted to transmission of the data.

In the broadcast SYMFONET, the receiving apparatus of any node, at the output of the MGD, has $N$ electric channels, one for each wavelength. Thus, every node detects, in parallel, the data transmitted from all nodes, including itself, and no arbitration is required for access to the network or destination nodes. The main potential application of the broadcast SYMFONET is for shared memory multiprocessors, where it enables to meet the global time ordering and sequential access characteristics required [111]. More precisely, the inherent nonblocking access feature ensures that all packets in frame $n$ are, in time, before all packets in frame $n + 1$ and after all packets in frame $n - 1$, while time order within a frame is determined by an arbitrary fixed-number scheme, like that based on the wavelength order. What is relevant, compared to other competing solutions, is that each node is able to determine the global time order of all the packets in the network in an independent way.

### 2.3.11 Mesh with Broadcast-and-Select

The so-called mesh with broadcast-and-select (MBS) is an example of self-healing ring network that utilizes WDM to provide survivable transport without the need for high-speed electronic add-and-drop multiplexers [112]. Its general structure is shown in Figure 2.44; the approach is analogous to that of some star architectures (like the LAMBDANET, described in Section 2.3.1) but the ring topology greatly improves the survivability of the network compared to the star.

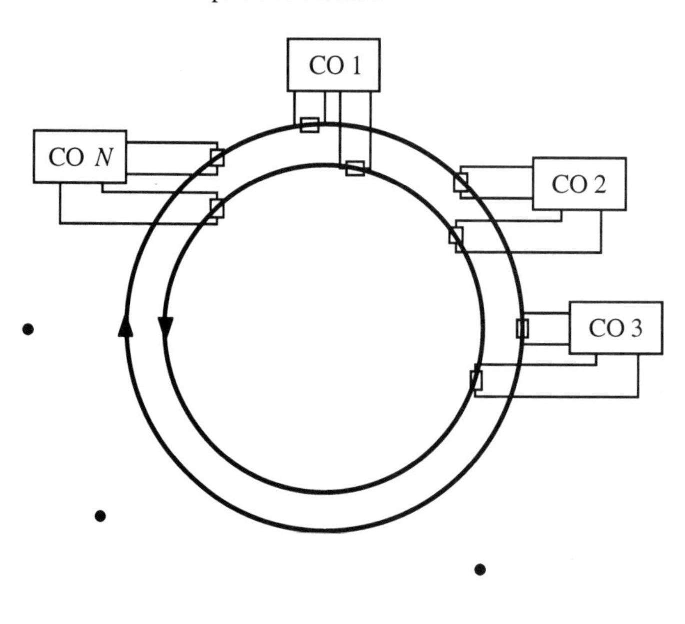


**Figure 2.44** Architecture of the MBS network (CO = central office).

The $N$ central offices (COs) are connected through two fibers (rings). If one fiber is damaged (typically because of a cable cut) the link is maintained on the other fiber. Also internally to each node, the probability of failure is relatively small since the transmitting and receiving apparatus are doubled: each CO is connected to a ring with a separate transceiver, as shown in detail in Figure 2.45. The two lasers are assigned the same wavelength (which is unique of the considered node) and the transmission on both rings differ only for the direction of propagation.

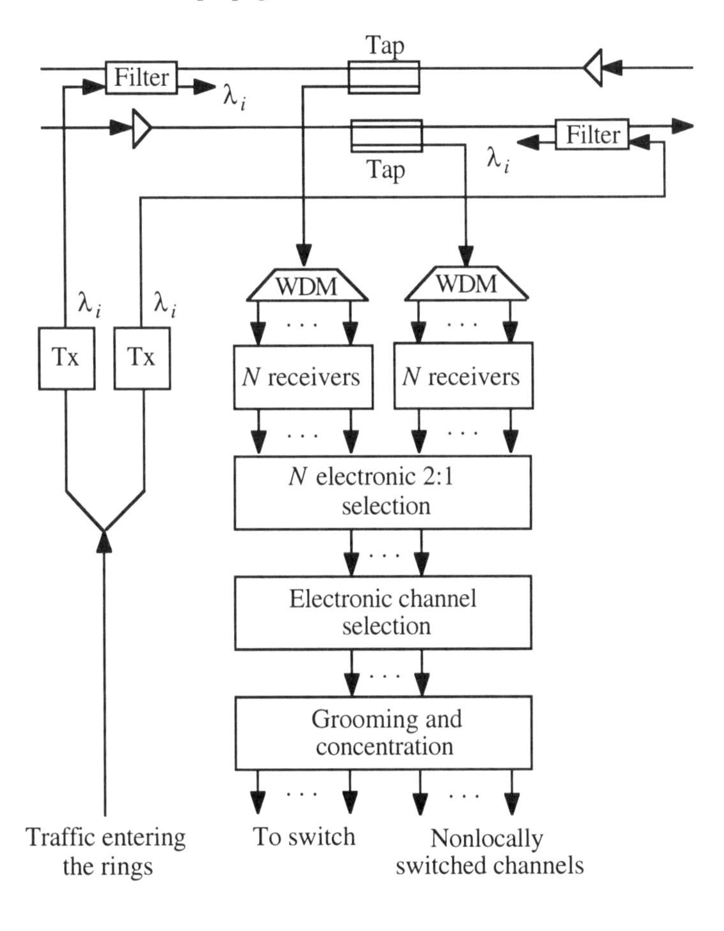


**Figure 2.45** Transmission equipment at each CO.

The various COs are independent from each other and can be also characterized by unequal transmission speeds. There is, in fact, no specific constraint to have the traffic symmetric (i.e., equal for all the nodes). Furthermore, one, several, or all COs can contain switches, so increasing the overall capability of the network. Interconnection, however, is completely symmetric (Figure 2.44) so that there is no need to distinguish between local and switching COs as required in other configurations, more suitable for switch-consolidation applications [112].

In the structure of Figure 2.45 each CO has access to all $N$ wavelengths. More precisely, a wavelength insensitive tap followed by a wavelength demultiplexer is used to monitor the transmissions of all other COs. The tap extracts part of the optical power circulating along the ring and, once demultiplexed, the corresponding signals are detected by the receiver array. One of the receivers in each array monitors its node's own transmission, which has made a complete circuit around the ring. Each of the other $N - 1$ receivers, instead, detects full transmission from one of the other COs in the ring. Only the signals intended for the node are then extracted, through an electronic processing based on high-speed sampling techniques [113,114]. The node's own transmission, remaining in the ring after the tap, is eliminated by the downstream optical filter that operates, as an add-and-drop multiplexer [115], at the wavelength assigned to the node. Actually, the "drop port" of the filter does not need to be used, since the tap-WDM combination is already capable of monitoring all signals entering the CO. Once having eliminated the old signal, the node can send a new signal, at its own wavelength, through the "add port" of the filter.

We have already stressed that one of the main features of the MBS is its self-healing capacity; the network is, in fact, able to detect cuts in one ring and to find a remedy for this instantaneously, thereby preserving the continuity of the transmissions. The principle used is very simple: if both the rings are not damaged, the signals coming from all COs will be received by both arrays; if one cable cut occurs, some signals will be available only from the left array, others from the right, and all they will be suitably chosen by the electronic 2:1 selectors.

In [112] it was proved that the MBS is able to interconnect about 10 COs, with throughputs up to 10 Gbps per CO. The major limits to the performance and the network size are those resulting from the WDM technology (add-and-drops filters and wavelength multi-demultiplexers) and from the adoption of optical amplifiers. The optical amplifiers (typically EDFAs) are placed periodically along each ring, every amplifier having a great enough gain to compensate the loss suffered by the signal between two subsequent amplification stages. The number of amplifiers cannot be increased indefinitely because of the ASE noise contribution which accumulates stage by stage, therefore degrading the signal-to-noise ratio (SNR). Moreover, the noise, adding to the useful signals, could cause the saturation of the gain [116,117]. In any case, even with negligible noise, saturation of the gain puts an upper bound to the number of nodes, as will be shown below with a numerical example.

Ignoring, for the moment, this limiting effect, one obtains the maximum number of COs in the network, reported in Table 2.9, where limitations due to the WDM technology and the amplifier spectral bandwidth have not been addressed as well. Thus, Table 2.9 refers to quite an optimistic evaluation. In the power budget, the following values have been considered for the parameters: $P_{in}$ = 0 dBm transmitter average launched power, spontaneous emission factor $n_{sp}$ = 1 [118], detector responsivity of 0.7, optical bandwidth for each receiver of 1 nm. Furthermore, 6 dB have been assumed for fiber attenuation and splices (per CO-to-CO span), 2 dB for an add-and-drop filter, 4 dB for a WDM demultiplexer, and 1 dB for a waveguide tap. The receiver sensitivities at the considered bit rates are –30 dBm (at 2.5 Gbps), –25 dBm (at 5 Gbps) and –18 dBm (at 10 Gbps), in the absence of ASE. Finally, 3 dB of margin have been maintained in the received signal power.

**Table 2.9**
Maximum Number of COs that Can Be Accommodated in the MBS Network, as a Function of Bit Rate and Amplifier Positioning, Neglecting Gain saturation of the Amplifiers

| *Bit Rate* | *Amplification at Every CO* | *Amplification at Every Other CO* | *Amplification at Every Third CO* |
|---|---|---|---|
| 2.5 Gbps | 190 | 150 | 54 |
| 5 Gbps | 60 | 54 | 27 |
| 10 Gbps | 11 | 12 | 12 |

(*Source*: [112]. © 1992 IEEE.)

When the gain saturation is taken into account, some of the numbers in Table 2.9 must be significantly reduced. As an example, we can consider the case of an amplifier at every CO; under such a hypothesis, we have as many amplifiers along each ring as the number of nodes. Noting by $P_{out}$ the power level at the output of the amplifiers (assumed to be all equal) and by $P_{sat}$ the saturation power, we must impose

$$P_{out} \le P_{sat} \quad (2.24)$$

The output power results from the sum of the regenerated useful power ($N$ channels in the worst case) and of the collective ASE due to all the amplifiers, each contributing with a power $P_{ASE}$. In formula

$$P_{out} = NP_{in} + NP_{ASE} \quad (2.25)$$

$P_{ASE}$, in turn, is defined as

$$P_{ASE} = 2n_{sp}(G-1)h\nu B_a \quad (2.26)$$

where $G$ is the gain of each amplifier and $h\nu$ is the photon energy ($h = 6.62 \cdot 10^{-34}$ J/Hz = Planck's constant and $\nu$ = frequency); finally $B_a$ is the optical-amplifier bandwidth.

Assuming $P_{sat}$ = 15 dBm, $B_a$ = 3.7 THz (corresponding to 30 nm, as typical for EDFAs), $h\nu = 1.3 \cdot 10^{-19}$ J and $G$ = 13 dB (which is a conservative value), while maintaining for $n_{sp}$ and $P_{in}$ the same values considered in Table 2.9, we have

$$N \le \frac{P_{sat}}{P_{in} + P_{ASE}} \approx \frac{P_{sat}}{P_{in}} \approx 31 \quad (2.27)$$

This value is significantly smaller than those reported in the second column of Table 2.9 for a bit rate of 2.5 or 5 Gbps, thus confirming that, in the considered conditions, the maximum number of nodes is not imposed by the requirements on the SNR but, instead, by the gain saturation effect. The value considered for $G$ is rather low since it is assumed that the distance between adjacent COs is relatively small. Using EDFAs with higher gain, greater distances can be covered in principle, but the constraint on $N$, fixed by the second member of (2.27), becomes more and more stringent. For example, assuming $G$ = 30 dB, while maintaining unchanged all the other parameters, $P_{ASE}$ becomes comparable with $P_{in}$ and the maximum number of nodes is practically halved.

## 2.4 AN EXAMPLE OF WAVELENGTH-ROUTING WDMA NETWORK: THE LINEAR LIGHTWAVE NETWORK

In spite of their topological simplicity, broadcast-and-select networks show a number of drawbacks which imply that they are not the most efficient solution for single-hop architectures. First of all, the power from each transmitter is broadcast to all receivers, so determining unnecessary power dissipation on receivers that do not use it. Second, the maximum number of nodes the network can have is related to (and often limited by) the number of resolvable wavelengths. Even though the spectrum available from an optical fiber is on the order of 25 THz, the wavelength resolving technology is today rather crude, and networks with more than 100 wavelengths cannot yet be built. Actually, this number may be further reduced by other factors, as discussed in Section 1.4. The resort to TDM, to be used in conjunction with WDM (thus relaxing the constraints on the number of wavelengths), may impose more and more stringent control requirements on transmitters and receivers, which combined with the problems of the tuning technology generally lead to increased costs. Furthermore, simultaneous use of a wavelength in many places of the network is not allowed, and no alternative paths are usually provided in case of failures.

An attractive way of overcoming most of the above limitations can be found in the adoption of wavelength-routing networks [119]. The basic idea consists in channeling the energy transmitted by each node along a restricted route to the receiver, instead of spreading it out over the whole network as with the broadcast-and-select structure. The key component in these networks is therefore a wavelength-selective device, which is able to route the light entering a node to one of its output ports. We could consider self-routing devices based on nonlinear operations [120], which are potentially addressable for this purpose. The technology of such devices, however, is at present still green, and the required power levels too high.

On the other hand, an important target in the network management is transparency, which should allow one to support a high degree of flexibility, including user-chosen modulation formats and user-chosen bit rates. The transported information should not be modified by the network nodes and by the protocols used; it is usual to say that the system should be as transparent as "a piece of glass."

The achievement of true transparency in the signal path requires linearity. For this reason linear lightwave networks (LLNs) have been proposed [121,122], where the routing functions are realized through linear components only. Such networks will be described in this section, with special emphasis on the routing problem and on the ability they show to be reconfigured in response to changing load conditions or component failures.

Although our attention will be focused on LLNs (here presented, in accord with the title of the section, as an example), it is worth mentioning that other very interesting approaches to the problem of routing and channel allocation in wavelength-routing networks are obviously possible [123]. The general concept used in most of these algorithms is that of (minimum cost of a) lightpath [124], to which that of (minimum cost of a) semilightpath [125] has been more recently added. A *lightpath* is a fully optical transmission path, while a *semilightpath* is a transmission path constructed by chaining together several lightpaths, using wavelength conversion at their junctions (differently from LLNs). In many respects, routing in LLNs can be seen as a special case of these more general approaches (which also include multihop networks). Nevertheless, because they are referred to a specific implementation, LLNs are simpler to describe, and thus more suitable for the purposes of the book.

Most of the results we will comment on below refer to theoretical analyses and simulations. This, however, must not lead one to think that the activity is still purely academic. A number of demonstrators and testbeds involving wavelength-routing networks have already been installed [126,127], and evaluation of their performance in the field is now under development and growing.

### 2.4.1 The LLN Architecture

An example of LLN is shown in Figure 2.46. It consists of a number of network nodes to which network stations are attached. Each link is bidirectional and formed by a pair of unidirectional optical fibers. Every network station must contain at least one tunable optical transmitter and receiver and an electronic interface to one or more user terminals. A wide flexibility for the user channels using the network is embraced; they may be digital or analog, and the users may be single workstations, video sources or displays, LANs, private branch exchanges (PBXs), or any other communication device.

Each network node must contain a linear combiner/divider (LCD) (see Section 1.5), with its control circuitry [121]. The LCD performs a controllable, waveband-selective optical signal routing, combining, and splitting. The combining function, in particular, allows multiplexing of channels from several input fibers onto the same output fiber, while the splitting function realizes the opposite action of multicasting optical power from a single input fiber to several output fibers. When the network is large, optical amplifiers may also be present in the LCDs, with the goal of restoring the appropriate power levels.

Routing is the result of subsequent combining and splitting functions. Through an external control, in fact, the coupling coefficient of each LCD can be varied almost continuously in the range from 0 to 1, thus allowing prescribed combinations of inbound signals at the node to be directed to each of its outbound fibers. Cascading this operation in more nodes, signals are forced to follow some specific paths instead of being broadcast over a common medium.

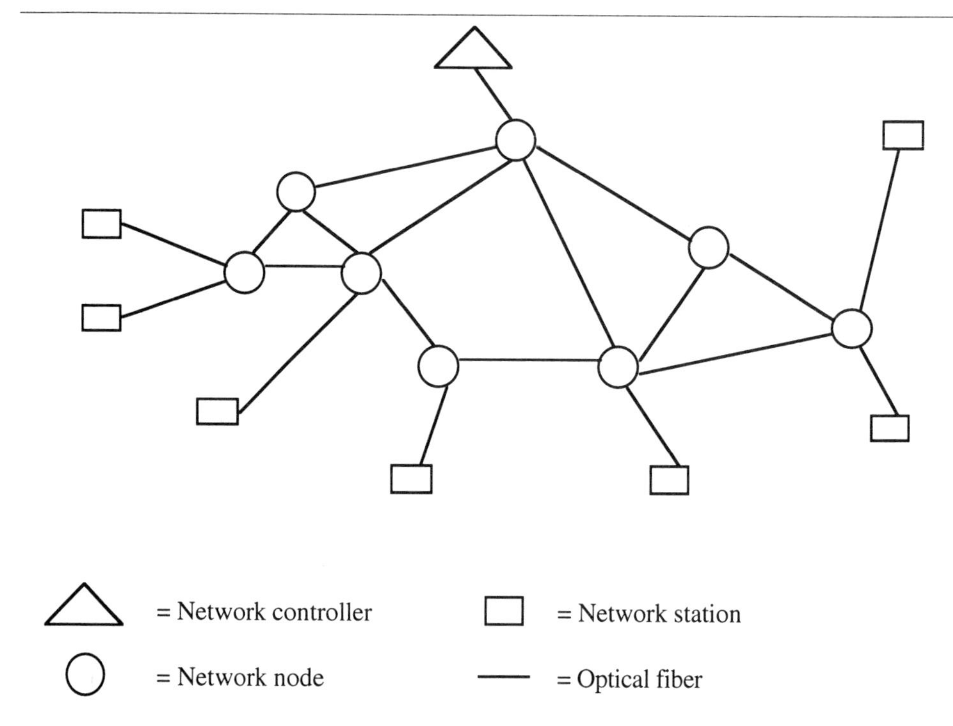


**Figure 2.46** Typical structure of an LLN.

An important point regards the need of coordinating the choice of the coupling coefficients in an LCD with the analogous choice in the other LCDs. Only in this way, in fact, is the appropriate signal routing established. Such a coordinating task is dispatched by the network controller (which can be centralized or distributed) with the exchange of preliminary information, subjected to propagation delays. For this reason, LLNs seem particularly suitable for circuit-switched systems, where the LCDs are required to switch on relatively slow time scales (on the order of milliseconds). Consequently, in the following we will assume that the LCDs change their setting only during optical connection establishment and termination. For the sake of completeness, however, we must say that packet-switched operation is also possible by adding suitable logical (electronic) overlays to the physical (optical) network [122].

A connection is established by creating a physical path for the optical signal from its source to its destination(s). Connections are set up and taken down through the supervision of the network controller, which employs a signaling channel. In response to a connection request, the controller determines a feasible path, assigns a channel, signals the stations interested in the transmission to tune to the chosen channel, and sends information to the LCDs to set up the routing path.

An important point concerns channel resolution at the network nodes and at the network stations. Although "optimum" performance could be achieved with maximum selectivity at the LCDs (which means capacity to discriminate up to one wavelength at a time) [128], we must realistically presume, on the basis of the currently available technology, that each LCD is able to resolve only wavebands with sufficient size and separation. Typically, we can consider wavebands 2 nm wide, placed at intervals of 6 nm. Any of these wavebands, which can be independently recognized and switched at each network node, has a bandwidth great enough to host a large number of individual channels. These must be distinguishable at the receivers but not, necessarily, at the LCDs. So we have two levels of partitioning for the optical spectrum: a "coarse" partition at the nodes, and a "fine" partition at the stations. Each transmitter is tuned to an assigned waveband and an assigned channel in that waveband. All the channels of the waveband are routed in the same way by the network nodes. Finally, the intended receiver is tuned to the waveband and channel of the sending node.

This specific organization naturally suggests the adoption of subcarrier frequency division multiple access (SFDMA) to mitigate the requirement of fine tuning in the optical domain [129]. In SFDMA, the transmitted data are first electronically modulated onto a microwave signal, and then that signal is used to modulate the optical carrier. There is the need to use lasers whose modulation bandwidth and linearity permit to handle many frequency-multiplexed microwave carriers, and these could be somewhat expensive. Anyway, such a drawback is compensated, on one hand, by the possibility of achieving tighter channel spacing (because of the increased stability of the sources) and, on the other hand, by the very low cost (in comparative terms) of the microwave technology. At the receiver, the optical signal is converted back to a microwave signal by the photodiode, and hence the data is extracted electronically. Each optical carrier supports many channels, each defined by its own microwave frequency. The advantage of such a scheme is clearly in the possibility of realizing direct detection with a wide optical bandwidth (without fine optical tuning), while accomplishing channel selection in the electronic domain (which is much simpler and reliable) with conventional microwave techniques. Some limitations due to optical beat interference could appear [130] but they can be overcome with a proper choice of the difference between the various optical frequencies and/or a suitable control of the laser's peak power.

### 2.4.2 Routing Constraints

The linearity of the operations performed in the LCDs (or, equivalently, the related network features described in Section 2.4.1) determines some conditions that must be satisfied to have a correct routing of signals within an LLN. We must distinguish between general rules, and specific rules, the latter being verified by signals placed in the same waveband. As general rules we mention:

a. *Channel and waveband continuity*: each connection must use the same channel and the same waveband, from source to destination, on all the traversed links;
b. *Distinct assignment of the channel and the waveband*: all the signals multiplexed on the same fiber must remain distinct, through the assignment of different wavebands or different channels within the same waveband.

The reason for condition (a) is evident: any change in the channel or waveband would require a frequency shift, which is not possible in a linear regime. Condition (b), in turn, does not preclude the possibility of using the same waveband-channel pair on disjoint paths within the network (wavelength reuse). Two or more paths are disjoint when their signals do not exhibit mutual interference, that is, they are independent one from the other. An obvious way to reach independence, also applicable to networks that are not wavelength routing, consists in using separate fibers, so realizing an SDMA scheme. With wavelength routing, however, the same result can be achieved, in principle, in some other cases. For example, we can have particular distributions of the input traffic, which make favorable to segment the network, permanently or temporarily, in two or more subnetworks. This would not be possible in a broadcast-and-select structure, but in a wavelength-routing network it happens, through a proper choice of the coupling coefficients in the LCDs. The same wavelengths can therefore be used within distinct subnetworks. Obviously, the segmentation of the network must be compatible with the physical topology and, in any case, it is necessary to satisfy the further constraints discussed below. More generally, one can take advantage of the attenuation suffered by a wavelength after passing through several LCDs and many kilometers of fiber. Remote transmitters can then use the same wavelength, with negligible interference, even when the network is fully broadcast.

Looking now at the signals in the same waveband, they must verify the following further constraints:

c. *Inseparability in a waveband*: the signals combined on a single fiber and within the same waveband cannot be separated within the LLN;
d. *Mutually independent source combining (MISC)*: a signal cannot be split among multiple paths and then recombined before reaching destination. Combination on the same fiber is allowed only for signals coming from mutually independent sources;
e. *Color clash*: it is necessary to avoid that a routing decision on a new call results in combining on the same fiber two or more calls that were previously assigned the same waveband-channel pair.

These additional conditions can be qualitatively discussed and explained with reference to simple practical examples.

In particular, condition (c) is obvious, since related to the ability of the LCDs to separate wavebands but not single channels. It is relevant, however, to emphasize the implications of this condition. To this purpose, let us consider Figure 2.47, where stations 1 and 2 are both connected to node A. If station 1 wishes to send some data to node $1^*$, its call can be routed via the path 1A-AB-BC-CF-FG-G$1^*$ (the shortest path, i.e., the path with the minimum number of LCDs). If, at the same time, station 2 wishes to send some data to node $2^*$, its call can be routed, with the same criterion, via the path 2A-AB-BD-DE-E$2^*$. The two paths are shown in the figure, where each arrow represents a unidirectional link. Signal $S_1$ (from station 1) and signal $S_2$ (from station 2) are combined at node A and then sent to the same outbound fiber. If $S_1$ and $S_2$ are assigned the same waveband (but different channels), they will be indistinguishable at node B, and cannot be separated downstream. Therefore, nodes $1^*$ and $2^*$ will receive either $S_1$ or $S_2$, and will select the desired connection tuning to the proper wavelength.

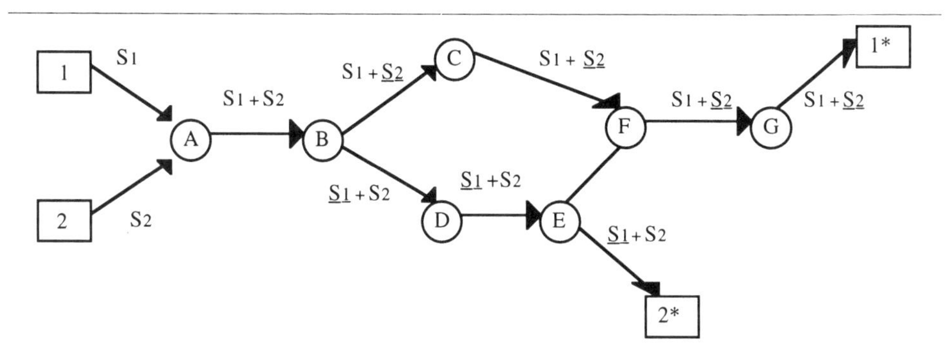


**Figure 2.47** Example showing the implications of constraint (c): inseparability in a waveband. Undesired signals are underlined.

In practice, we can say that inseparability tends to create fortuitous multicast connections. Undesired signals are underlined in Figure 2.47; besides the desired connections (1, 1*) and (2, 2*), the fortuitous connections (1, 2*) and (2, 1*) have also been established. In general, the number of fortuitous connections depends on the number of channels in each waveband, decreasing with it. Correspondingly, however, the required selectivity of the LCDs increases.

Condition (d) is necessary, to avoid more copies of the same signal reaching the destination with different power level and delay, therefore determining degradation in the quality of the received information. As an example, let us consider Figure 2.48.

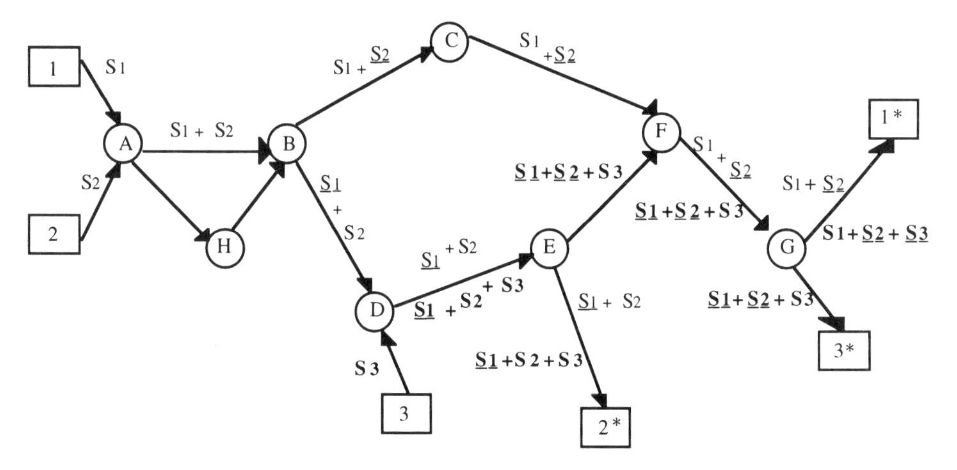


**Figure 2.48** Example of unrespected MISC condition. Connections (1, 1*) and (2, 2*) are in progress when a new call (3, 3*) arrives; its effect is boldface. (*Source*: [122]. © 1993 IEEE.)

With the connections (1, 1*) and (2, 2*) already in progress, a new call (3, 3*) is routed via the path 3D-DE-EF-FG-G3*. The effect of the new call is boldface in the figure. If all three connections use the same waveband, signal S3 carries with it signals S1 and S2 (see the link DE). Since, to reach 3*, S3 must be routed via the link EF, the same will occur for a fortuitous copy of signal S1, that will combine with the desired one, at node F. The resulting signal, addressed to station 1*, will be the sum of two signals differently delayed, and so unavoidably degraded. The MISC condition is violated and, to eliminate the problem, signal S3 must use a waveband different from that of S1 and S2. Incidentally, we observe that the latter assumption would permit to route via the EF link only signal S3, and to avoid the creation of any fortuitous connection involving nodes 3 and 3*.

As an example of color clash, we can consider the situation shown in Figure 2.49. Let us suppose that the connections (1, 1*) and (2, 2*) are in progress. If the LCDs at the nodes A and B operate in bar state (i.e., the power at their $i$th input port is completely transferred to their $i$th output port ($i = 1, 2$)) the signals S1 and S2 can propagate everywhere on different fibers, and their paths are disjoint even if they are assigned to the same waveband and channel. Reminding the comment on condition (b) above, the network in this starting situation is practically subdivided into two independent subnetworks. Color clash, however, occurs if a new call (3, 3*) is requested and allocated in the same waveband of S1 and S2 [though, obviously, on a different channel, to not violate condition (b) above]. The path for the new call is 3D-DE-EF-FG-G3*, and its effect is boldface in the figure. Once again because of the inseparability [condition (c) above], S2 and S3 cannot be divided at node E, and will combine at F with signal S1. Therefore, after the new call, the previous paths are no longer disjoint and the connection (1, 1*) is destroyed since S1 and S2 have the same waveband-channel pair. To avoid this catastrophic event, station 2 should be retuned to a channel that is different from those of 1 and 3. Alternatively, an even better result is obtained choosing (if possible) a different waveband for S3; in this case, in fact, the signals would be separated at nodes E and G, which does not affect the transmission at all.

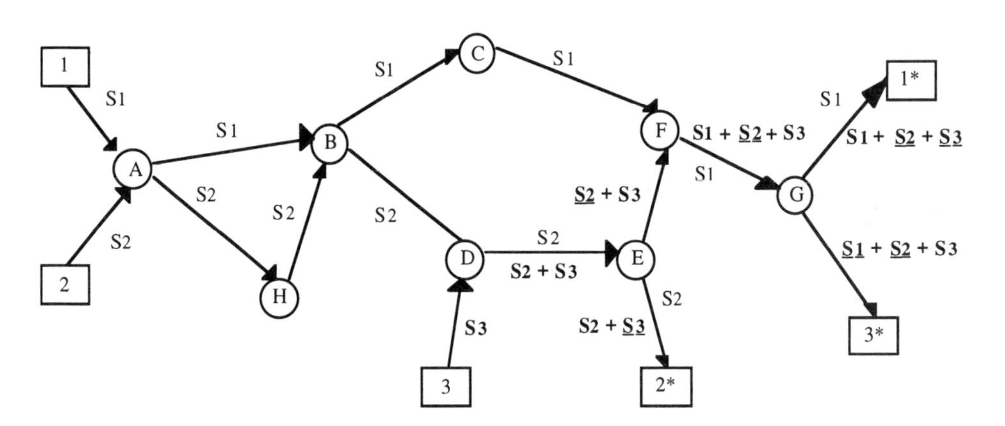


**Figure 2.49** Example of color clash. (*Source*: [122]. © 1993 IEEE.)

### 2.4.3 Performance with Different Routing Schemes

Keeping in mind the constraints discussed in Section 2.4.2, suitable routing strategies can be implemented in an LLN. These must take into account the criterion used for distributing the available resources (mainly the bandwidth) as well as the characteristics of the input traffic. Since it seems reasonable to assume that only a small fraction of the users needs to be connected without any interruption, an efficient scheme would allow to accept calls on demand, sharing the network resources only among the active users. This is the hypothesis we will make in the following. Each connection will require a rather large bandwidth (e.g., 1 Gbps) and, coherent with the preliminary considerations, it will be maintained for periods of time that are long compared with the propagation delays in the network. This assumption agrees with the circuit-switched techniques that are more frequently used in the original LLN proposal.

Two different approaches to the routing problem are possible in principle; routing, in fact, can be static or dynamic. We speak of static routing when the state of the LCDs within the network does not change with time, once having assigned a set of rules a priori. Really, occasional modifications of the routing patterns could occur, but only in response to faults [131] or heavy changes in demand.

Using static routing the values of the coupling coefficients are therefore set in advance, except for those relative to the LCDs interfacing the access stations, which must adapt themselves to the requests of call establishment and termination. In other words, the paths for any possible connection are fixed and stored in "routing tables," or something similar, which remain unchanged all the time (with rare exceptions, as stated above). When a new call arrives, if the corresponding path is, in part or completely, busy [i.e., no channels are available to establish the connection, safeguarding the satisfaction of constraints (a) through (e)], the call is temporarily blocked. Following this strategy, the connection between two nodes always happens on the same path, independently of the state of the network.

Static routing is not able, on average, to face changes in the distribution of the input traffic, with respect to the initial prevision. As a consequence, the network performance is not optimized, with some paths which may be near congestion and other almost empty, for some periods of time (as a function of the traffic fluctuations). Even the response to heavy changes in load patterns and network failures, which can lead to modify the routing tables, are generally slow.

To overcome these limitations, dynamic routing schemes have been conceived, where the path to assign to a new call is computed on the basis of the current state of the network, satisfying the constraints, and optimizing, according with a predefined criterion, the assignment of the waveband-channel pair.

The major flexibility and effectiveness of the dynamic routing schemes, with respect to the static ones, are out of question. Nevertheless, they are more difficult to implement, and their interest presently limited to very simple rules. Following [122] in Section 2.4.3.1 we will list a number of solutions to realize static routing, and in Section 2.4.3.2 we will mention three algorithms for dynamic routing, where the criterion for path selection consists in minimizing interference [128].

### 2.4.3.1 *Static Routing*

The static routing strategies presented in [122] are applied to a network organized as the so called Petersen graph shown in Figure 2.50. The figure refers only to the network nodes, while network stations are omitted for simplicity. The network has 10 nodes and 15 links; each node has degree 3 (i.e., it has three bidirectional links converging on it) and diameter equal to 2. Here the network diameter represents the maximum value of the minimum number of links which are necessary to connect two arbitrary nodes. Therefore, it should be noted the difference with the same term used for multihop networks (see Chapters 1 and 3) where it represents the maximum number of wavelength shifts that are necessary for establishing a connection.

In the figure, the links are oriented for the sake of notation. A lowercase letter, in fact, is used to denote the fiber carrying optical signals in the direction of the arrow (a: from A to B, b: from B to C, . . . , p: from H to J). A primed letter will be used instead, when necessary, to denote the fiber carrying signals in the opposite direction (a': from B to A, b': from C to B, . . . , p': from J to H). Two fibers are in fact used to connect a couple of nodes, one for each direction of propagation. Another interesting feature is that the graph is isotropic, which means that the topology looks the same as seen from each node. This property simplifies the analysis, permitting to prescind from the particular originating node.

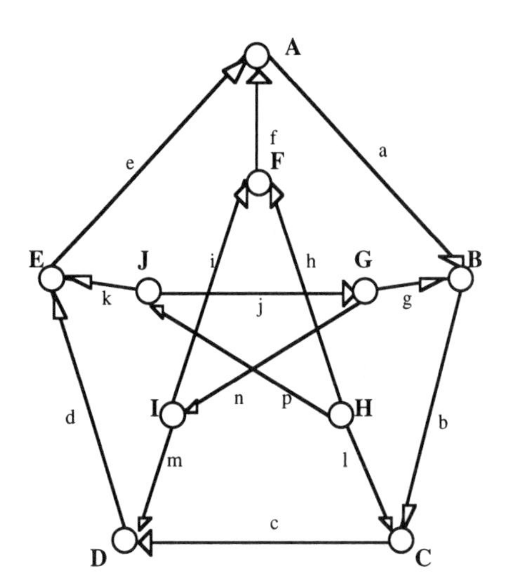


**Figure 2.50** Petersen graph.

In [122] attention was focused, for the sake of simplicity, on point-to-point connections. According to that model, the total usable spectrum is divided into $W$ channels grouped in $K_b$ wavebands, in such a way as to have $CH = W/K_b$ channels per waveband. As regards the characteristics of the input traffic, connections are requested by stations accessing the 10 network nodes. The traffic offered to each node is uniformly distributed among the other nine nodes; traffic among stations connected to the same node, however, uses only the local LCD, without loading the rest of the network.

Each station is active or idle for random intervals of time, governed by a two-state Markov chain [132]; when a call is accepted the chain passes from the idle to the active state, where the station remains for an exponentially distributed period of time. The hypothesis of large station population is also made, which allows us to assume that the block of a call is (practically) never due to the fact that the destination station is busy.

The assumptions above permit to use the so called Erlang B formula [81] to calculate the blocking probability $p_B$ for calls propagating along a given fiber and assigned to a given waveband. In detail, we have

$$p_B = \frac{ER^{CH}}{CH!}\left(\sum_{k=0}^{CH}\frac{ER^k}{k!}\right)^{-1} \tag{2.28}$$

where $ER$ represents the offered load, in Erlangs, on the given waveband and fiber.

Starting from the described model, which is not restrictive but only depicts a possible reference scenario, in [122] a number of static routing possibilities were considered and their performance compared. For saving space, we will not go into detail as regards the discussion of all those schemes, limiting ourselves to presenting some inspiring ideas, and finally summarizing the results.

First a distinction must be made between algorithms based on spanning trees and algorithms based on multiple nonspanning trees. A spanning tree is a connected subgraph of the network, that includes all the nodes but contains no loops. A possible spanning tree for the Petersen graph is that shown in Figure 2.51 (circles, denoting the nodes, have been omitted for simplicity).

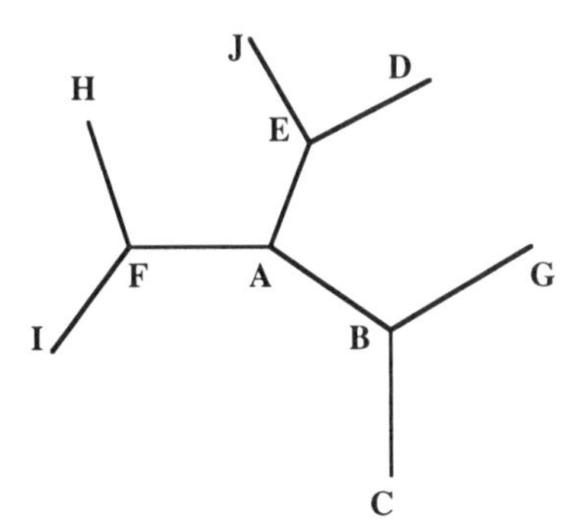


**Figure 2.51** Spanning tree for the Petersen graph.

By virtue of its definition, all the connections can be realized on it, while the MISC condition is automatically satisfied. The selected path for each connection is not, in general, the shortest one. For example, keeping in mind the notation of Figure 2.50, a connection (D, H) would be routed via the fibers d-e-f'-h' (remind that apices are used to denote transmissions in the opposite directions). Furthermore, many fortuitous connections can appear, because of the fiber sharing, when the subdivision of the spectrum in wavebands is not fine enough. Looking again at Figure 2.51, connection (J, B), for instance, would be routed via k-e-a; if it uses the same waveband of (D, H), the fortuitous connections (D, B) and (J, H) are created.

As an alternative to the spanning tree, we can use multiple nonspanning trees, whose union covers the network. A possibility consists in adopting the 10 subtrees $ST_i$, each with three links, shown in Figure 2.52. The subtrees are not disjoint, so that violations of the MISC condition are possible.

On the other hand, they can be made "waveband-disjoint" by assigning different wavebands to subtrees sharing the same fibers. Three wavebands are sufficient to achieve complete independence ($K_b = 3$): $ST_1$, $ST_7$, and $ST_8$ can operate on a waveband $WB_1$; $ST_2$, $ST_4$, $ST_6$, and $ST_{10}$ on $WB_2$; and $ST_3$, $ST_5$, and $ST_9$ on a waveband $WB_3$.

Every node can reach any other node through, at least, one of the subtrees to which it belongs. For example, node A can transmit to B, C, and G, via $ST_2$; to E, J and D, via $ST_5$; and to F, H, and I, via $ST_6$. Since each subtree is of diameter 2, all routing paths are shortest paths. Actually, connections between two adjacent nodes have a choice of two subtrees. In order to distribute the traffic as fairly as possible, one can choose the tree associated with the destination node.

Therefore, the routing table shown in Table 2.10 results, where *i/j* indicates that the connection uses the subtree $ST_i$ ($i = 1, 2, \ldots, 10$), with the corresponding waveband $WB_j$ ($j = 1, 2, 3$).

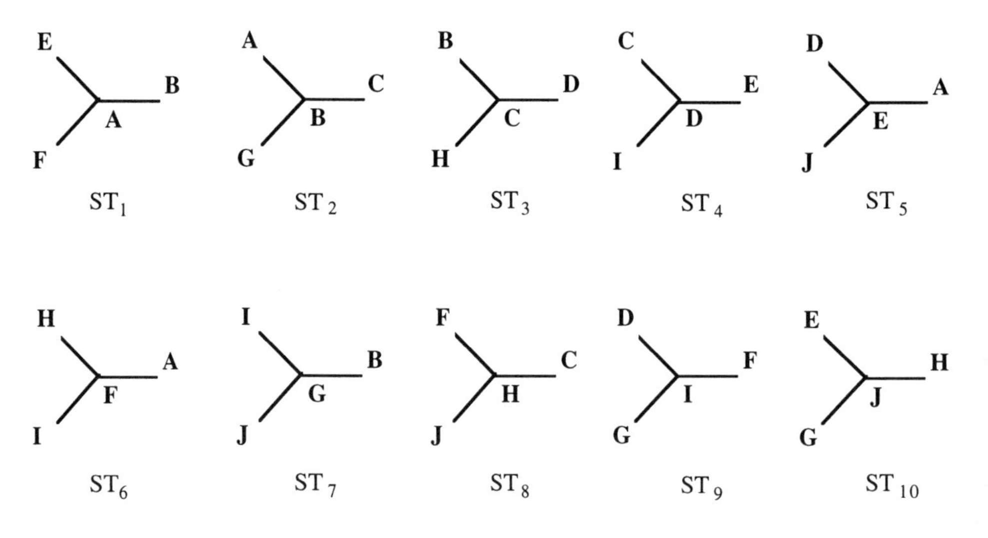


**Figure 2.52** Multiple nonspanning trees to cover the Petersen graph.

**Table 2.10**
Routing Table for the Multiple Nonspanning Trees of Figure 2.52, Assuming $K_b = 3$

| *Source* | *Destination* | | | | | | | | | |
|---|---|---|---|---|---|---|---|---|---|---|
| | A | B | C | D | E | F | G | H | I | J |
| A | | 2/2 | 2/2 | 5/3 | 5/3 | 6/2 | 2/2 | 6/2 | 6/2 | 5/3 |
| B | 1/1 | | 3/3 | 3/3 | 1/1 | 1/1 | 7/1 | 3/3 | 7/1 | 7/1 |
| C | 2/2 | 2/2 | | 4/2 | 4/2 | 8/1 | 2/2 | 8/1 | 4/2 | 8/1 |
| D | 5/3 | 3/3 | 3/3 | | 5/3 | 9/3 | 9/3 | 3/3 | 9/3 | 5/3 |
| E | 1/1 | 1/1 | 4/2 | 4/2 | | 1/1 | 10/2 | 10/2 | 4/2 | 10/2 |
| F | 1/1 | 1/1 | 8/1 | 9/3 | 1/1 | | 9/3 | 8/1 | 9/3 | 8/1 |
| G | 2/2 | 2/2 | 2/2 | 9/3 | 10/2 | 9/3 | | 10/2 | 9/3 | 10/2 |
| H | 6/2 | 3/3 | 3/3 | 3/3 | 10/2 | 6/2 | 10/2 | | 6/2 | 10/2 |
| I | 6/2 | 7/1 | 4/2 | 4/2 | 4/2 | 6/2 | 7/1 | 6/2 | | 7/1 |
| J | 5/3 | 7/1 | 8/1 | 5/3 | 5/3 | 8/1 | 7/1 | 8/1 | 7/1 | |

Instead of using more wavebands, the subtrees of Figure 2.52 can be made disjoint by running two fibers (instead of one) in each direction in every link. Hence, the number of fibers in the network is doubled and multiple access is now in wavelength and space [wavelength division multiple access/space division multiple access (WDMA/SDMA)]. From an economical point of view, this should be advantageous: on one hand, we may have a minor cost for using LCDs insensitive to the waveband while, on the other hand, four fibers (in each cable) instead of two do not imply major costs, since typical cables carry several fibers. Obviously, in the service of improving performance, it is possible to combine the idea of using more wavebands with that of using more fibers. Finally, to this more involved scheme, a further variation could be added, which consists in multiplexing the total traffic accessing each node, noted by $U$, on a set of $X$ access links, each carrying $U/X$ Erlangs.

Performance comparison can be based on the following method of reasoning: we can fix the value of $p_B$ and $W$, and for any routing scheme we can use the Erlang B formula to derive the value of $ER$. This, in turn, is related, through the structure of the network and the routing rules, to the offered traffic node $U$ or, equivalently, to the total network offered traffic $U_t = 10U$. The best routing scheme, in these conditions, will obviously be that characterized by the maximum value of $U$.

The solutions examined are listed below; between brackets we have evidenced the number of wavebands and the number of fibers per link used:

1. Spanning tree ($K_b = 1$, one fiber pair);
2. Multiple nonspanning tree ($K_b = 3$, one fiber pair);
3. Multiple nonspanning tree ($K_b = 1$, two fiber pairs);
4. Multiple nonspanning tree ($K_b = 4$, two fiber pairs);

5. Multiple nonspanning tree ($K_b$ = 12, two fiber pairs, $X$ = 3 access links per node).

Where not specified (schemes 1–4) the number of access links per node is equal to the number of access stations connected.

The value of $CH$ (as a fraction of the prefixed $W$) and the unknown $ER$ (in terms of the final result $U$) which must be inserted in (2.28) are summarized in Table 2.11.

**Table 2.11**
Number of Channels in the Waveband and Offered Load on Each Waveband and Fiber for the Considered Static Routing Examples

| *Routing Scheme* | *CH* | *ER* |
|---|---|---|
| 1 | $W$ | $9U$ |
| 2 | $W/3$ | $2U/3$ |
| 3 | $W$ | $2U/3$ |
| 4 | $W/4$ | $U/9$ |
| 5 | $W/12$ | $U/9$ |
| 6 | $W$ | $10U$ |

Besides the schemes listed above, another solution (6) has been inserted for further comparison. It refers to the extreme case of an LLN where all the stations access a single node, whose LCD behaves as an ideal star coupler. It has no practical interest in the present context (where, remember, the demand is random and each station is active only a portion of time) but constitutes a sort of "worst case," useful to help us appreciate the advantages of the various routing schemes.

Imposing $p_B = 0.04$ and $W = 96$ or 192, the maximum network offered load $U_t$ is reported in Table 2.12.

The best performance is offered by the scheme based on two fiber-pairs per link, with four wavebands; but the increase in the network load is relatively modest with respect to the much simpler solution which requires no waveband selectivity in the network (scheme 3). The latter, in turn, shows an improvement factor on the order of four with respect to the solution "waveband-disjoint" (scheme 2). On the contrary, to have preliminary multiplexing (scheme 5) does not entail any benefit: the number of wavebands must be increased to control the number of additional fortuitous connections ($K_b$ = 12 in the example) that are caused by the concentration of several desired connections on a reduced number of access links. This scheme has been proposed only because it represents the elementary block of a hierarchical structure that will be discussed, with others, in Chapter 4.

**Table 2.12**
Maximum Total Network Offered Traffic (in Erlangs) for Different Static Routing Schemes

| *Routing Scheme* | $W = 96$ | $W = 192$ |
|---|---|---|
| 1 | 100 | 207 |
| 2 | 375 | 840 |
| 3 | 1350 | 2790 |
| 4 | 1665 | 3735 |
| 5 | 414 | 990 |
| 6 | 90 | 186 |

(*Source*: [122]. © 1993 IEEE.)

### 2.4.3.2 *Dynamic Routing*

When dynamic routing is used, a path, a waveband, and a channel is assigned to a new call, not on the basis of routing tables fixed in advance and stored, but taking into account, in real time, the state of the network. This obviously involves an increased elaboration capacity, which must be provided by the network controller.

Because of its intrinsic complexity, the routing problem has to be decomposed into a number of smaller manageable subproblems. The overall approach, however, is essentially heuristic, and the performance is generally determined using simulation.

The first step concerns the choice of a waveband for a requested call. To this purpose, the controller maintains a sorted list $\Lambda$ of the wavebands. Sorting can be managed following two different criteria [128]: MAXBAND, in which the list is sorted in decreasing order of usage, and MINBAND, where it is sorted in increasing order of usage. When the usage level for two wavebands is the same, they are conventionally sorted in ascending numerical order. Since the usage of a waveband represents the number of connections using it at a given moment, it must be updated any time a connection is established or terminated.

When a new call request is perceived by the controller, an attempt is made to allocate it on the waveband at the top of the list $\Lambda$. This means that the allocation procedure goes on with the next steps (assignment of a path on the chosen waveband, checking for violation of the routing constraints, and assignment of an appropriate channel within the waveband), according with the rules that we will describe below. If one of these steps is unsuccessful, the call is blocked on the first waveband of the list, the controller passes to the second ordered waveband, and the procedures above are repeated. A call is definitely blocked if it is blocked on each waveband of the list.

The choice of the waveband follows rules which are clear and simple to implement. The problem, instead, becomes much more involved dealing with the choice of the path and the channel. Similarly to the static routing, in fact, but with the augmented difficulty of considering dynamic conditions, these choices must take into account the need to verify the routing constraints (particularly the MISC and the color clash conditions). As discussed above, many difficulties derive from the possible creation of unintended paths, associated with the desired ones. On these premises, a good strategy consists in searching for a path that is likely to satisfy the constraints (although one cannot be sure that it will do so), and verifying the implications of the given hypothesis. Likelihood of a path is measured by the amount of interference with other calls already in progress on the chosen waveband, with the goal of reducing the chances of violation of the routing constraints.

*Interference* is quantitatively defined as the number of independent signals that the call encounters on its intended path and on the chosen waveband. Three algorithms were presented in [128] for finding paths that tend to minimize this number; they are named:

1. K-SP;
2. BLOW-UP;
3. MIN-INT.

The K-SP algorithm finds $K$ shortest paths from source to destination. Thus, $K$ represents the number of paths to test before blocking the call on the given waveband. The shortness of a path is measured by the weight of its constituting links. Any meaningful weight definition can be used to this purpose; for example, we can refer to the attenuation suffered by the signal along each link or, more simply, we can assign weight 1 to every link, thus making the total weight proportional to the number of crossed LCDs.

As mentioned above, any of the $K$ paths is tested, checking its capacity to satisfy the MISC and color clash conditions for the intended connection as well as for all the unintended ones. This permits the definition of a subset of feasible paths. Finally, among the subset, a path is chosen that exhibits the least interference for the call.

Minimization of the interference with calls already in progress is therefore the dominant rule, which often leads to choose a path different from that with minimum total weight. An example is shown in Figure 2.53.

Let us suppose that the call (1, $1^*$) is in progress, and that a request arrives for connecting node 2 to $2^*$. Assuming $K = 2$ and link weights of 1, the shortest paths for S2 are 2A-AB-BD-DE-E$2^*$ and 2A-AH-HB-BD-DE-E$2^*$, respectively. The first would be the path with minimum weight but, as shown in the figure, the second one is selected, in such a way as to avoid interference with signal S1. Incidentally, we observe that this strategy tends to favor wavelength reuse: S1 and S2 are routed on disjoint paths, which permits them to be transmitted not only on the same waveband but also on the same channel.

The path chosen with the K-SP algorithm is that of minimum interference inside the subset of feasible paths, but we have no guarantee that it is, in absolute, the path of least interference. This path, in fact, is selected within a restricted number of possibilities ($\leq K$) which have been previously found on the basis of their weight. To overcome the risk that the path with absolutely minimum interference will be excluded because it is not sufficiently short (of course, in relative terms) it is necessary to modify the criterion, and to select as the shortest paths those with least interference. This also requires a new definition of path weight, which can be given after having introduced the concept of image network [128].

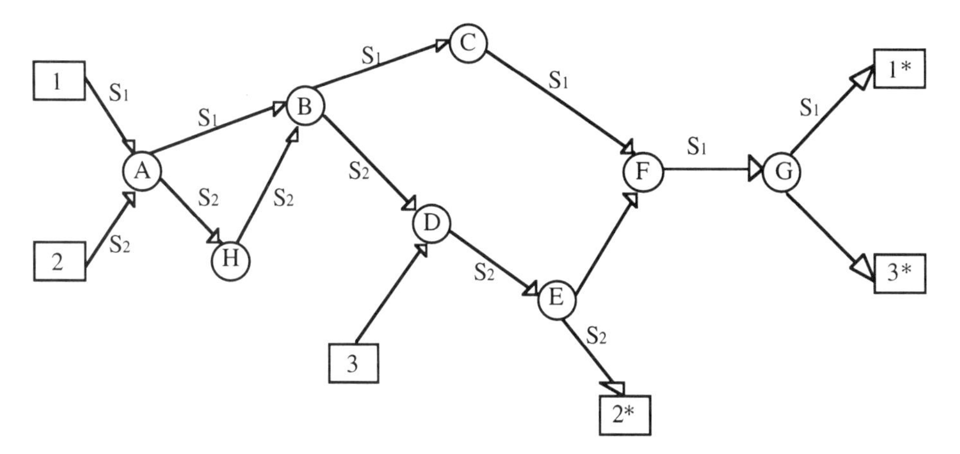


**Figure 2.53** Example of application of the K-SP algorithm; S1 and S2 are allocated on the same waveband. (*Source*: [128]. © 1995 IEEE.)

For an assigned waveband, the image network is obtained by creating, in each node, additional intranodal links between any input and output port pair. In practice, every node is blown up, showing the internal structure of its LCD. An example of image network is plotted in Figure 2.54(b), relative to the LLN of Figure 2.54(a).

Each node is split up into as many internal nodes as the number of its inbound and outbound links; a subscript is used to specify the number of the link, while superscript distinguishes between input (superscript i) or output (superscript o) fibers. This extended notation has been limited, in Figure 2.54, to node B, but only for the sake of simplicity. The image network provides a detailed view of the operation at each node, also useful, for instance, in isolating faults.

To each intranodal link *i-j*, a weight $w_{ij}$ is assigned as

$$w_{ij} = M \cdot \left| G_j - G_i \right| \tag{2.29}$$

In this expression $G_j$ represents the set of signals (on the chosen waveband) outbound from the internal node $j$, $G_i$ is the set of signals (on the same waveband) inbound to the internal node $i$, while $|G|$ is the number of signals in the set $G$. In Figure 2.54(b), for example, we have $G_{B_1^o} = \{S1, S2\}$, $G_{B_1^i} = \{S1\}$, $G_{B_2^i} = \{S2\}$, $G_{B_1^o} - G_{B_1^i} = \{S2\}$, $G_{B_1^o} - G_{B_2^i} = \{S1\}$ and, therefore, $\left|G_{B_1^o} - G_{B_1^i}\right| = \left|G_{B_1^o} - G_{B_2^i}\right| = 1$. Furthermore, a weight of 1 is assigned to each internodal link, and $M$ is an integer greater than the number of links in the original network.

It is easy to understand that the definition (2.29) is a direct way to count and weigh the signals interfering within a node. Apart from the multiplication by $M$, it results from the difference between the signals carried on a particular output link and those carried on a particular input link.

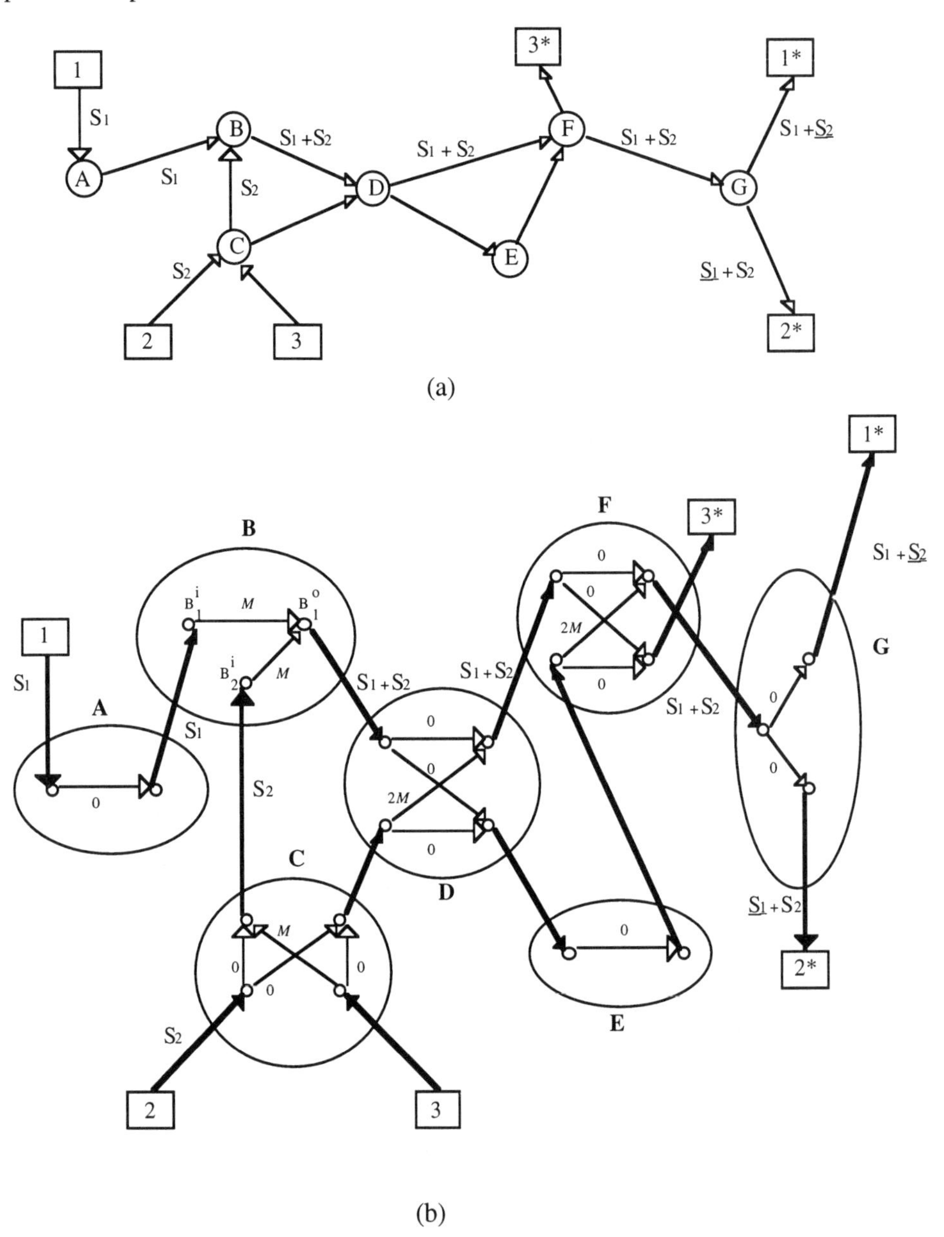


**Figure 2.54** Example of application of the BLOW-UP algorithm: (a) LLN and (b) its image network. S1 and S2 are on the same waveband. Internodal links are boldface, undesired signals are underlined. (*Source*: [128]. © 1995 IEEE.)

When referred to an intended path, this difference gives the signals that arrive to the considered outbound link but coming from an inbound link that is not of interest in the path. To assume $M$ greater than the sum of weights of all internodal links then corresponds to privilege the role of interference with respect to attenuation: just one interfering signal weighs more than all the links crossed.

The weights so defined are used in selecting paths through the network. Let us consider again Figure 2.54(a), and assume that the calls $(1, 1^*)$ and $(2, 2^*)$ are in progress, so that the controller maintains the image network, with the distribution of intranodal weights, that has been shown in Figure 2.54(b). When a new call $(3, 3^*)$ arrives, on the same waveband of $S_1$ and $S_2$, the controller can allocate it on one path chosen among the following four: (1) 3C-CB-BD-DF-F3*, (2) 3C-CB-BD-DE-EF-F3*, (3) 3C-CD-DE-EF-F3*, (4) 3C-CD-DF-F3*. In the image network, the total weight accumulated is, respectively: (1) $2M + 5$, (2) $2M + 6$, (3) 5, (4) $2M + 4$.

The BLOW-UP algorithm chooses the physical path that corresponds to the shortest path in the image network, according to this definition of weight, at the time the new call arrives. Therefore, in the example considered above, the path selected for the call $(3, 3^*)$ will be 3C-CD-DE-EF-F3*. The introduction of the new call in the network will usually cause an increase of the interference in some of the receivers. This additional contribution, for each receiver, is called incremental interference. It is possible to prove that, if the MISC condition is satisfied on the shortest path for a given source-destination pair in the image network, then this is the path along which the maximum incremental interference is minimized.

An important drawback of the BLOW-UP algorithm is its time complexity, especially when the nodes are linked through multiple fibers. The MIN-INT algorithm is equivalent to BLOW-UP as regards the physical path allocation (preserving moreover its property of minimizing the maximum incremental interference), but it is more efficient from the point of view of complexity [128]. The intranodal link weights used in BLOW-UP are replaced here by "additional weights" which allow us, as before, to estimate the interference, but with simpler calculations.

The path chosen with the BLOW-UP or the MIN-INT algorithms has to be verified as regards the satisfaction of the MISC and color clash conditions. If verification is successful, the controller can pass to the last step of the routing procedure, that is, the channel allocation. Otherwise, the call is blocked and a new path or waveband is considered. A fixed number of channels is available in the chosen waveband. The channel allocation of a new call should be done without requiring the calls already in progress to retune to new channels or, even more, to change their paths. On the other hand, a channel already allocated can be assigned to the new call, if the latter does not interfere with the calls that are using it. Then, assignment proceeds on the basis of usage, that is, the number of active connections on the given channel. Two simple criteria, both heuristic, can be adopted. In the first one, called MAX, the incoming call is allocated the most used channel (among those with which the call does not interfere); dually, in the second one, called MIN, the incoming call is allocated the least used channel (again, among those with which the call does not interfere). MAX tends to favor wavelength reuse, while MIN attempts to distribute the calls evenly among the available channels.

The performance, mainly in terms of blocking probability of the various algorithms presented, was evaluated in [128] by means of a simulator, written in the C language. It must be said that, because of the inherent complexity of the dynamic routing problem, simulation generally requires rather long time periods. At any rate, many results were presented in [128], either in the case of using a single waveband or in the case of multiple wavebands. Although it is rather difficult to draw absolute conclusions, it was shown that a scheme based on MAXBAND (as regards the choice of the waveband), MIN-INT (as regards the allocation of the path in the chosen waveband), and MIN (as regards the allocation of the channel) should result in the best performance.

The description given in this section obviously does not exhaust the ensemble of routing algorithms proposed. In [133], for example, with reference to multifiber linear lightwave networks (i.e., considering multifiber cables for node connections) a class of linear path (LP) schemes was presented, with the goal of simplifying the switch design and the path allocation mechanism, while maintaining the blocking probability sufficiently low. The LP schemes are based on shortest path allocations which minimize the number of fiber links occupied in the network. In case there is more than one path with the same shortest length, one of them is chosen randomly (LP-RS) or, deterministically, as the one having the least (LP-MIN) or the largest (LP-MAX) number of occupied wavelengths.

Instead of using multifiber cables, in [134] the adoption of TDMA in LLNs was discussed as the means to achieve better channel efficiency (similarly to what successfully made in broadcast-and-select networks). To relieve the synchronization problems, a rooted routing scheme was proposed, as an alternative to the shortest path criterion. In the rooted routing scheme, a node is selected to be the root, and all optical signals, instead of traveling on the shortest paths, must go through that root node before reaching their destinations. Propagation delays and power losses obviously increase, but the synchronization difficulties are completely overcome.

Finally, another attractive possibility, not necessarily restricted to the LLN scheme, consists in exploiting wavelength rerouting [135]. Let us consider Figure 2.55, where part of a wider network is shown, involving only three nodes. In Figure 2.55(a) connections (1, 2) and (2, 3) are in progress, the first on the wavelength $\lambda_2$ and the second on the wavelength $\lambda_1$. If a new call (1, 3) arrives, it should be allocated on a third wavelength (if available) thus reducing channel utilization. It is in fact evident, from Figure 2.55(b), that connection (1, 3) could be, instead, allocated on $\lambda_1$ (i.e., one of the wavelengths already in use) if connection (2, 3) is rerouted to $\lambda_2$.

This is the object of the rerouting algorithm. In a network where such a scheme is implemented, the controller first tries (an obvious solution) to route a new circuit without rerouting any existing circuits; however, if such an attempt fails, it tries to route the new circuit after rerouting some existing circuits. When also rerouting is unfeasible the new connection request is rejected.

Improvement in the channel utilization allows one to reduce the blocking probability. In [135] it was proved, with reference to a typical test network, that $p_B$ decreases, on average, by about 30%. As a counterpart, however, there is a throughput loss due to the disruption periods incurred during rerouting. To mitigate this penalty, a move-to-vacant wavelength-retuning (MTV-WR) mechanism was proposed, in which a circuit is moved to a vacant wavelength on the same path, without shutdown of the connected nodes. The main feature of the MTV-WR operation is the possibility of rerouting, in parallel, multiple circuits, on condition that they are on disjoint set of links. Combined with an optimal rerouting algorithm, which selects a minimum number of rerouted circuits to accommodate a new connection, this allows one to maintain the throughput loss sufficiently low, while improving the blocking probability performance.

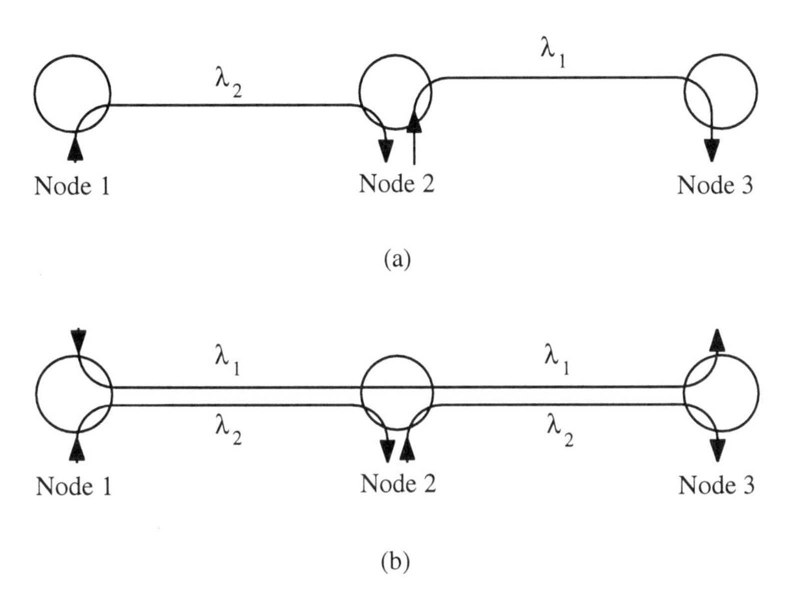


**Figure 2.55** A simple example where rerouting can improve wavelength utilization.

## REFERENCES

[1] Mestdagh, D. J. G., *Fundamentals of Multiaccess Optical Fiber Networks*, Norwood, MA, Artech House, 1995.

[2] Green, P. E., Jr., *Fiber Optics Networks*, Englewood Cliffs, NJ, Prentice Hall, 1993.

[3] Goodman, M. S., H. Kobrinski, and K. W. Lo, "Application of wavelength division multiplexing to communication network architectures," *Proc. ICC '86*, Toronto, Canada, 1986, Vol. 2, pp. 931–934.

[4] Arthurs, E., J. M. Cooper, M. S. Goodman, H. Kobrinski, M. Tur, and M. P. Vecchi, "Multiwavelength optical crossconnect for parallel-processing computers," *Electron. Lett.*, Vol. 24, No. 2, Jan. 1988, pp. 119–120.

[5] Dono, N. R., P. E. Green Jr., K. Liu, R. Ramaswami, and F. F.-K. Tong, "A wavelength division multiple access network for computer communication," *IEEE J. Select. Areas Commun.*, Vol. 8, No. 6, Aug. 1990, pp. 983–994.

[6] Arthurs, E., M. S. Goodman, H. Kobrinski, and M. P. Vecchi, "HYPASS: an optoelectronic hybrid packet switching system," *IEEE J. Select. Areas Commun.*, Vol. 6, No. 9, Dec. 1988, pp. 1500–1510.

[7] Brackett, C. A., "Dense wavelength division multiplexing networks: Principle and applications," *IEEE J. Select. Areas Commun.*, Vol. 8, No. 6, Aug. 1990, pp. 948–964.

[8] Ramaswami, R., "Multiwavelength lightwave networks for computer communication," *IEEE Commun. Mag.*, Vol. 31, No. 2, Feb. 1993, pp. 78–88.

[9] Mukherjee, B., "WDM-based local lightwave networks – Part I: single-hop systems," *IEEE Network*, Vol. 6, No. 3, May 1992, pp. 12–27.

[10] Sudhakar, G. N. M., M. Kavehrad, and N. D. Georganas, "Access protocols for passive optical star networks," *Computer Networks and ISDN Systems*, Vol. 26, March 1994, pp. 913–930.

[11] Chlamtac, I., and A. Ganz, "A multibus train communication architecture for high speed fiber optic networks," *IEEE J. Select. Areas Commun.*, Vol. 6, July 1988, pp. 903–912.

[12] Banerjee, S., and B. Mukherjee, "An efficient and fair protocol for a multichannel lightwave network," *Proc. OFC '92*, San José, CA, Feb. 1992, p. 183.

[13] Chlamtac, I., and A. Ganz, "Channel allocation protocols in frequency-time controlled high-speed networks," *IEEE Trans. Commun.*, Vol. 36, No. 4, April 1988, pp. 430–440.

[14] Ganz, A., and Y. Gao, "A time-wavelength assignment algorithm for a WDM star network," *Proc. INFOCOM '92*, Florence, Italy, May 1992, pp. 2144–2150.

[15] Ganz, A., and Y. Gao, "Time-wavelength assignment algorithms for high performance WDM star based systems," *IEEE Trans. Commun.*, Vol. 42, No. 2/3/4, Feb./March/April 1994, pp. 1827–1836.

[16] Ganz, A., and Y. Gao, "Traffic scheduling in multiple WDM star systems," *Proc. ICC '92*, Chicago, IL, June 1992, Paper 349.4, pp. 1468–1472.

[17] Tanenbaum, A. S., *Computer Networks*, 2nd edition, Englewood Cliffs, NJ, Prentice Hall, 1988.

[18] Abramson, N., "The ALOHA system – Another alternative for computer communications," *AFIPS Conf. Proc.*, 1970 Fall Joint Computer Conference, Vol. 37, AFIPS Press, Montvale, NJ, 1970, pp. 281–285.

[19] Dowd, P. W., "Random access protocols for high-speed interprocessor communication based on an optical passive star topology," *IEEE/OSA J. Lightwave Technol.*, Vol. 9, No. 6, June 1991, pp. 799–808.

[20] Habbab, I. M. I., M. Kavehrad, and C. E. W. Sundeberg, "Protocols for very high-speed optical fiber local area network using a passive star topology," *IEEE/OSA J. Lightwave Technol.*, Vol. LT-5, No. 12, Dec. 1987, pp. 1782–1794.

[21] Mehravari, N., "Performance and protocol improvements for very high speed optical fiber local area networks using a passive star topology," *IEEE/OSA J. Lightwave Technol.*, Vol. 8, No. 4, April 1990, pp. 520–530.

[22] Ganz, A., and Z. Koren, "WDM passive star – Protocols and performance analysis," *Proc. INFOCOM '91*, Bal Harbor, FL, April 1991, pp. 991–1000.

[23] Glance, B. S., "Protection-against-collision optical packet network," *IEEE/OSA J. Lightwave Technol.*, Vol. 10, No. 9, Sept. 1992, pp. 1323–1328.

[24] Karol, M. J., and B. S. Glance, "Performance of the PAC optical network," *IEEE/OSA J. Lightwave Technol.*, Vol. 11, No. 8, Aug. 1993, pp. 1394–1399 – also in *Proc. GLOBECOM '91*, Phoenix, AZ, Dec. 1991, pp. 1258–1263.

[25] Karol, M. J., and B. S. Glance, "A collision-avoidance WDM optical star network," *Computer Networks and ISDN Systems*, Vol. 26, March 1994, pp. 931–943.

[26] Lee, H. W., "Protocols for multichannel optical fibre LAN using passive star topology," *Electron. Lett.*, Vol. 27, No. 17, Aug. 1991, pp. 1506–1507.

[27] Shi, H., and M. Kavehrad, "ALOHA/Slotted-CSMA protocol for a very high-speed optical fiber local area network using passive star topology," *Proc. INFOCOM '91*, Bal Harbor, FL, April 1991, pp. 1510–1515.

[28] Jia, F., and B. Mukherjee, "Bimodal throughput, nonmonotonic delay, optimal bandwidth dimensioning, and analysis of receiver collisions in a single-hop WDM local lightwave network," *Proc. OFC '92*, San José, CA, Feb. 1992, pp. 1896–1900.

[29] Sudhakar, G. N. M., N. D. Georganas, and M. Kavehrad, "Slotted Aloha and reservation Aloha protocols for very high-speed optical fiber local area networks using passive star topology," *IEEE/OSA J. Lightwave Technol.*, Vol. 9, No. 10, Oct. 1991, pp. 1411–1422.

[30] Sudhakar, G. N. M., M. Kavehrad, and N. D. Georganas, "Multi-control channel for very high-speed optical fiber local area networks and their interconnections using a passive star topology," *Proc. GLOBECOM '91*, Phoenix, AZ, Dec. 1991, pp. 624–628.

[31] Sudhakar, G. N. M., N. D. Georganas, and M. Kavehrad, "A multi channel optical star LAN and its applications as a broadband switch," *Proc. ICC '92*, Chicago, IL, June 1992, pp. 843–847.

[32] Chen, M., and T. Yum, "A conflict free protocol for optical WDMA networks," *Proc. GLOBECOM '91*, Phoenix, AZ, Dec. 1991, pp. 1276–1281.

[33] Humblet, P. A. , R. Ramaswami, and K. Sivarajan, "An efficient communication protocol for high-speed packet-switched multichannel networks," *IEEE J. Select. Areas Commun.*, Vol. 11, No. 4, May 1993, pp. 568–578.

[34] Chen, M. S., N. R. Dono, and R. Ramaswami, "A media-access protocol for packet-switched wavelength-division metropolitan area networks," *IEEE J. Select. Areas Commun.*, Vol. 8, No. 6, Aug. 1990, pp. 1048–1057.

[35] Lam, S. S., "Packet broadcast networks – A performance analysis of the R-ALOHA protocol," *IEEE Trans. Comp.*, Vol. C-29, No. 7, July 1980.

[36] Jeon, H. B., and C. K. Un, "Contention-based reservation protocol in fiber optical local area network with passive star topology," *Electron. Lett.*, Vol. 26, No. 12, June 1990, pp. 780–781.

[37] Senior, J. M., J. M. McVeigh, and D. Cusworth, "Multichannel slot reservation protocol for use on WDM optical fibre LAN," *Electron. Lett.*, Vol. 27, No. 20, Sept. 1991, pp. 1875–1877.

[38] Sudhakar, G. N. M., M. Kavehrad, and N. D. Georganas, "A simple contention-based reservation scheme for LAN using a passive star topology," *Proc. IEEE Photonics '92*, Montebello, March 1992.

[39] Jia, F., and B. Mukherjee, "The receiver collision avoidance (RCA) protocol for a single hop WDM lightwave network," *IEEE/OSA J. Lightwave Technol.*, Vol. 11, No. 5/6, May/June 1993, pp. 1053–1065.

[40] Chlamtac, I., and A. Fumagalli, "Quadro-star: A high performance optical WDM star network," *IEEE Trans. Commun.*, Vol. 42, No. 8, Aug. 1994, pp. 2582–2591.

[41] Chlamtac, I., and A. Fumagalli, "Quadro: a solution to packet switching in optical transmission networks," *Computer Networks and ISDN Systems*, Vol. 26, March 1994, pp. 945–963.

[42] Papadimitriou, G. I., and D. G. Maritsas, "Learning automata-based receiver conflict avoidance algorithms for WDM broadcast-and-select star networks," *IEEE/ACM Trans. Networking*, Vol. 4, No. 3, June 1996, pp. 407–411.

[43] Chipalkatti, R., Z. Zhang, and A. S. Acampora, "Protocols for optical star-coupler network using WDM: performance and complexity study," *IEEE J. Select. Areas Commun.*, Vol. 11, No. 4, May 1993, pp. 579–589.

[44] Bogineni, K., and P. W. Dowd, "A collisionless multiple access protocol for a wavelength division multiplexed star-coupled configuration: architecture and performance analysis," *IEEE/OSA J. Lightwave Technol.*, Vol. 10, No. 11, Nov. 1992, pp. 1688–1699.

[45] Sivalingam, K. M., and P. W. Dowd, "A multilevel WDM access protocol for an optically interconnected multiprocessor system," *IEEE/OSA J. Lightwave Technol.*, Vol. 13, No. 11, Nov. 1995, pp. 2152–2167.

[46] Sivalingam, K. M., and J. Wang, "Media access protocols for WDM networks with on-line scheduling," *IEEE/OSA J. Lightwave Technol.*, Vol. 14, No. 6, June 1996, pp. 1278–1286.

[47] Pieris, G. R., and G. H. Sasaki, "Scheduling transmissions in WDM broadcast-and-select networks," *IEEE/ACM Trans. Networking*, Vol. 2, No. 2, April 1994, pp. 105–110.

[48] Kovačević, M., and M. Gerla, "Analysis of a T/WDMA scheme with subframe tuning," *Proc. ICC '93*, Geneva, Switzerland, May 1993, pp. 1239–1244.

[49] Laarhuis, J. H., and A. M. J. Koonen, "An efficient medium access control strategy for high-speed WDM multiaccess networks," *IEEE/OSA J. Lightwave Technol.*, Vol. 11, No. 5/6, May/June 1993, pp. 1078–1087.

[50] Jeon, H. B., and C.-K. Un, "Contention-based reservation protocols in multiwavelength optical networks with a passive star topology," *IEEE Trans. Commun.*, Vol. 43, No. 11, Nov. 1995, pp. 2794–2802.

[51] Choi, H., H.-A. Choi, and M. Azizoglu, "Efficient scheduling of transmissions in optical broadcast networks," *IEEE/ACM Trans. Networking*, Vol. 4, No. 6, Dec. 1996, pp. 913–920.

[52] Jia, F., B. Mukherjee, and J. Iness, "Scheduling variable-length messages in a single-hop multichannel local lightwave network," *IEEE/ACM Trans. Networking*, Vol. 3, No. 4, Aug. 1995, pp. 477–488.

[53] Lee, J. H., and C. K. Un, "Dynamic scheduling protocol for variable-sized messages in a WDM-based local network," *IEEE/OSA J. Lightwave Technol.*, Vol. 14, No. 7, July 1996, pp. 1595–1600.

[54] Rouskas, G. N., and M. H. Ammar, "Analysis and optimization of transmission schedules for single-hop WDM networks," *IEEE/ACM Trans. Networking*, Vol. 3, No. 2, April 1995, pp. 211–221.

[55] Borella, M. S., and B. Mukherjee, "Efficient scheduling of nonuniform packet traffic in a WDM/TDM local lightwave network with arbitrary transceiver tuning latencies," *IEEE J. Select. Areas Commun.*, Vol. 14, No. 5, June 1996, pp. 923–934.

[56] Papadimitriou, G. I., and D. G. Maritsas, "Self-adaptive random-access protocols for WDM passive star networks," *IEE Proc.-Comput. Digit. Tech.*, Vol. 142, No. 4, July 1995, pp. 306–312.

[57] Kim, H. S., B. C. Shin, J. H. Lee, and C. K. Un, "Performance evaluation of reservation protocol with priority control for single-hop WDM networks," *Electron. Lett.*, Vol. 31, No. 17, Aug. 1995, pp. 1472–1473.

[58] Huang, N.-F., and C.-S. Wu, "An efficient transmission scheduling algorithm for a wavelength-reusable local lightwave network," *IEEE/OSA J. Lightwave Technol.*, Vol. 12, No. 7, July 1994, pp. 1278–1290.

[59] Chen, M., and T.-S. P. Yum, "Buffer sharing in conflict-free WDMA networks," *IEICE Trans. Commun.*, Vol. E77-B, No. 9, Sept. 1994, pp. 1144–1151.

[60] Ajmone Marsan, M., A. Bianco, E. Leonardi, M. Meo, and F. Neri, "MAC protocols and fairness control in WDM multirings with tunable transmitters and fixed receivers," *IEEE/OSA J. Lightwave Technol.*, Vol. 14, No. 6, June 1996, pp. 1230–1244.

[61] Green, P. E., Jr., "Optical networking update," *IEEE J. Select. Areas Commun.*, Vol. 14, No. 5, June 1996, pp. 764–779.

[62] Goodman, M. S., H. Kobrinski, M. P. Vecchi, R. M. Bulley, and J. L. Gimlett, "The LAMBDANET multiwavelength network: Architecture, applications, and demonstrations," *IEEE J. Select. Areas Commun.*, Vol. 8, No. 6, Aug. 1990, pp. 995–1003.

[63] Laude, J. P., "STIMAX, a grating multiplexer for monomode and multimode fibers," *Proc. ECOC '83*, Geneva, Switzerland, 1983, pp. 417–420.

[64] Lee, W. S., S. W. Bland, and A. J. Robinson, "Monolithic GaInAs/InP photodetector arrays for high density wavelength division multiplexing," *Electron. Lett.*, Vol. 24, No. 18, Sept. 1988, pp. 1143–1145.

[65] Marhic, M. E., "Combinatorial star couplers for single-mode optical fibers," *Proc. FOC/LAN '84*, Boston, MA, 1984, pp. 175–179.

[66] Senior, J. M., *Optical Fiber Communications—Principles and Practice*, 2nd edition, Cambridge, England, Prentice Hall, 1992.

[67] Kobrinski, H., R. M. Bulley, M. S. Goodman, M. P. Vecchi, C. A. Brackett, L. Curtis, and J. L. Gimlett, "Demonstration of high capacity in the LAMBDANET architecture: a multiwavelength optical network," *Electron. Lett.*, Vol. 23, No. 16, July 1987, pp. 824–826.

[68] Vecchi, M. P., R. M. Bulley, M. S. Goodman, H. Kobrinski, and C. A. Brackett, "High-bit-rate measurements in the LAMBDANET multiwavelength optical star network," *Proc. Opt. Fiber Commun. Conf.*, New Orleans, LA, Jan. 1988, Paper WO2, p. 95.

[69] Coquin, G., H. Kobrinsky, C. E. Zah, F. K. Shokoohi, C. Caneau, and S. G. Menocal, "Simultaneous amplification of 20 channels in a multiwavelength distribution system," *IEEE Photon. Technol. Lett.*, Vol. 1, 1989, pp. 176–178.

[70] Binh, L. N., and H. C. Chong, "A neural network contention controller for packet switching networks," *IEEE Trans. Neural Networks*, Vol. 6, No. 6, Nov. 1995, pp. 1402–1410.

[71] Binh, L. N., and H. C. Chong, "Improved packet switch architectures with output channel grouping," *Europ. Trans. Telecommun.*, Vol. 8, No. 1, Jan.-Feb. 1997, pp. 99–109.

[72] Jue, J. P., M. S. Borella, and B. Mukherjee, "Performance analysis of the Rainbow WDM optical network prototype," *IEEE J. Select. Areas Commun.*, Vol. 14, No. 5, June 1996, pp. 945–951.

[73] Cancellieri, G., editor, *Single-Mode Optical Fiber Measurement: Characterization and Sensing*, Norwood, MA, Artech House, 1993.

[74] Mallinson, R., "Crosstalk limits of Fabry-Perot demultiplexers," *Electron. Lett.*, Vol. 21, No. 17, Aug. 1985, pp. 759–760.

[75] Humblet, P. A., and W. M. Hamdy, "Crosstalk analysis and filter optimization of single- and double-cavity Fabry-Perot lasers," *IEEE J. Select. Areas Commun.*, Vol. 8, No. 6, Aug. 1990, pp. 1095–1107.

[76] Hill, A. M., et al., "Linear crosstalk in wavelength division multiplexed optical fiber transmission systems," *IEEE/OSA J. Lightwave Technol.*, Vol. LT-3, 1985, pp. 643–651.

[77] Ramaswami, R., and P. A. Humblet, "Amplifier induced crosstalk in multichannel optical networks," *IEEE/OSA J. Lightwave Technol.*, Vol. 8, No. 12, Dec. 1990, pp. 1882–1896.

[78] Janniello, F. J., R. Ramaswami, and D. G. Steinberg, "A prototype circuit-switched multi-wavelength optical metropolitan-area network," *Proc. ICC '92*, Chicago, IL, June 1992, Paper 330.1, pp. 818–823 – also in *IEEE/OSA J. Lightwave Technol.*, Vol. 11, No. 5/6, May-June 1993, pp. 777–782.

[79] Hall, E., J. Kravitz, R. Ramaswami, M. Halvorson, S. Tenbrink, and R. Thomsen, "The Rainbow-II gigabit optical network," *IEEE J. Select. Areas Commun.*, Vol. 14, No. 5, June 1996, pp. 814–823.

[80] "High performance parallel interface," American National Standards Institute, X3T9.3, rev. 6.9, Nov. 1989.

[81] Bertsekas, D., and R. Gallager, *Data Networks*, 2nd edition, Englewood Cliffs, NJ, Prentice Hall, 1992.

[82] Goodman, M. S., "Multiwavelength networks and new approaches to packet switching," *IEEE Commun. Mag.*, Vol. 27, No. 10, Oct. 1989, pp. 27–35.

[83] Karol, M. J., M. G. Hluchyj, and S. P. Morgan, "Input versus output queueing on a space-division packet switch," *IEEE Trans. Commun.*, Vol. COM-35, No. 12, Dec. 1987, pp. 1347–1356.

[84] Kobrinski, H., M. P. Vecchi, M. S. Goodman, E. L. Goldstein, T. E. Chapuran, J. M. Cooper, C. E. Zah, and S. G. Menocal, "Fast wavelength-switching of laser transmitters and amplifiers," *IEEE J. Select. Areas Commun.*, Vol. 8, No. 6, Aug. 1990, pp. 1190–1202.

[85] Hayes, J. F., "An adaptive technique for local distribution," *IEEE Trans. Commun.*, Vol. COM-26, No. 8, Aug. 1978, pp. 1178–1186.

[86] Capetanakis, J. I., "Tree algorithms for packet broadcast channels," *IEEE Trans. Inf. Theory*, Vol. IT-25, No. 5, Sept. 1979, pp. 505–515.

[87] Capetanakis, J. I., "Generalized TDMA: the multi-accessing tree protocol," *IEEE Trans. Commun.*, Vol. COM-27, No. 10, Oct. 1979, pp. 1476–1484.

[88] Hui, J. Y., and E. Arthurs, "A broadband packet switch for integrated transport," *IEEE J. Select. Areas Commun.*, Vol. SAC-5, No. 8, Oct. 1987, pp. 1264–1273.

[89] Narasimha, M. J., "The Batcher-Banyan self-routing network: Universality and simplification," *IEEE Trans. Commun.*, Vol. 36, No. 10, Oct. 1988, pp. 1175–1178.

[90] Batcher, K., "Sorting networks and their applications," *Proc. AFIPS Spring Joint Comp. Conf.*, Vol. 32, 1968, pp. 307–314.

[91] Eng, K. Y., "A Photonic Knockout switch for high-speed packet networks," *IEEE J. Select. Areas Commun.*, Vol. 6, No. 7, Aug. 1988, pp. 1107–1116.

[92] Eng, K. Y., M. G. Hluchyj, and Y.-S. Yeh, "A Knockout switch for variable length-packets," *IEEE J. Select. Areas Commun.*, Vol. SAC-5, No. 9, Dec. 1987, pp. 1426–1435.

[93] Yeh, Y.-S., M. G. Hluchyj, and A. Acampora, "The Knockout switch: a simple modular architecture for high-performance packet switching," *IEEE J. Select. Areas Commun.*, Vol. SAC-5, No. 8, Oct. 1987, pp. 1274–1283.

[94] Wagner, S. S., H. L. Lemberg, H. Kobrinski, L. S. Smoot, and T. J. Robe, "A passive photonic loop architecture employing wavelength-division multiplexing," *Proc. GLOBECOM '88*, Tokyo, Dec. 1988, pp. 1569–1573.

[95] Linnell, L. R., "A wide-band local access system using emerging-technology components," *IEEE J. Select. Areas Commun.*, Vol. SAC-4, July 1986, pp. 612–618.

[96] Laude, J. P., and J. Lerner, "Wavelength division multiplexing/demultiplexing (WDM) using diffraction gratings," *Proc. SPIE*, Vol. 503, San Diego, CA, 1984, pp. 22–28.

[97] Okamoto, K., K. Moriwaki, and S. Suzuki, "Fabrication of 64×64 arrayed-waveguide grating multiplexer on silicon," *Electron. Lett.*, Vol. 31, No. 3, Feb. 1995, pp. 184–186.

[98] Payne, D. B., and J. R. Stern, "Transparent single mode fiber optical networks," *IEEE/OSA J. Lightwave Technol.*, Vol. LT-4, No. 7, July 1986, pp. 864–869.

[99] Olshansky, R., and V. A. Lanzisera, "60-channel FM video subcarrier multiplexed optical communication system," *Electron. Lett.*, Vol. 23, No. 22, Oct. 1987, pp. 1196–1198.

[100] Wagner, S. S., H. Kobrinski, T. J. Robe, H. L. Lemberg, and L. S. Smoot, "Experimental demonstration of a passive optical subscriber loop architecture," *Electron. Lett.*, Vol. 24, No. 6, March 1988, pp. 344–346.

[101] Bersiner, L., and D. Rund, "Bidirectional WDM transmission with spectrum sliced LEDs," *J. Opt. Commun.*, Vol. 11, No. 2, 1990, pp. 56–59.

[102] Wagner, S. S., and T. E. Chapuran, "Broadband high-density WDM transmission using superluminescent diodes," *Electron. Lett.*, Vol. 26, No. 11, May 1990, pp. 696–697.

[103] Lee, T. T., M. S. Goodman, and E. Arthurs, "A broadband optical multicast switch," *Proc. XIII Int. Switching Symposium, ISS '90*, Stockholm, Sweden, May 1990, pp. 7–13.

[104] Atkins, J., and M. Norris, *Total Area Networking*, Chichester, England, Wiley, 1995.

[105] Chiaroni, D., P. Gavignet-Morin, J. B. Jacob, J. M. Gabriagues, J. Jacquet, D. de Bouard, G. Da Loura, and C. Chauzat, "Feasibility demonstration of a 2.5-Gbit/s 16×16 ATM photonic switching matrix," *Proc. OFC/IOOC '93*, San Josè, CA, Feb. 1993, Paper WD2, pp. 93–94.

[106] Chiaroni, D., P. Gavignet-Morin, P. A. Perrier, S. Ruggeri, S. Gauchard, D. de Bouard, J. C. Jacquinot, C. Chauzat, J. Jacquet, P. Doussière, M. Monnot, E. Grard, D. Leclerc, M. Sotom, J. M. Gabriagues, and J. Beinot, "Rack-mounted 2.5 Gbit/s ATM photonic switch demonstrator," *Proc. ECOC '93*, Montreux, Switzerland, Sept. 1993, Paper ThP12.7, pp. 77–80.

[107] Gabriagues, J. M., J. Benoit, D. Chiaroni, D. de Bouard, P. Doussière, T. Durhuus, P. Gavignet-Morin, E. Grard, J. B. Jacob, J. Jacquet, C. Joergensen, D. Leclerc, F. Masetti, P. A. Perrier, and K. E. Stubkjaer, "Design, modelling and implementation of the ATMOS project fiber delay line photonic switching matrix," *Optical and Quantum Electronics*, Vol. 26, No. 5, May 1994, pp. 497–516.

[108] Karol, M. J., and M. G. Hluchyj, "Queueing in high performance packet switching," *IEEE J. Select. Areas Commun.*, Vol. 6, No. 9, Dec. 1988, pp. 1587–1597.

[109] Kirkby, P. A., "SYMFONET: Ultra-high-capacity distributed packet switching network for telecoms and multiprocessor computer applications," *Electron. Lett.*, Vol. 26, No. 1, Jan. 1990, pp. 19–21.

[110] Kirkby, P. A., N. Baker, W. S. Lee, Y. Kanabar, and R. Worthington, "Multichannel grating demultiplexer receivers for high density wavelength multiplexed systems," *Proc. 7th Conf. Integrated Optical Commun.*, Kobe, Japan, July 1989, Paper 20 D1-3, pp. 178–179.

[111] Westmore, R. J., "SYMFONET: interconnect technology for multinode computing," *Electron. Lett.*, Vol. 27, No. 9, April 1991, pp. 697–698.

[112] Wagner, S. S., and T. E. Chapuran, "Multiwavelength ring networks for switch consolidation and interconnection," *Proc. ICC '92*, Chicago, IL, June 1992, Paper 340.5, pp. 1173–1179.

[113] Faulkner, D. W., D. M. Russ, D. Douglas, and P. J. Smith, "Novel sampling technique for digital video demultiplexing, descrambling and channel selection," *Electron. Lett.*, Vol. 24, No. 11, 1988, pp. 654–656.

[114] Smith, P. J., R. A. Lobbett, and D. W. Faulkner, "64-channel digital TV distribution system operating at 4.4 Gbit/s," *Electron. Lett.*, Vol. 24, No. 21, 1988, pp. 1336–1338.

[115] Tachikawa, Y., Y. Inoue, M. Kawachi, H. Takahashi, and K. Inoue, "Arrayed-waveguide grating add-drop multiplexer with loop-back optical paths," *Electron. Lett.*, Vol. 29, No. 24, 1993, pp. 2133–2134.

[116] Olsson, N. A., "Lightwave systems with optical amplifier," *IEEE/OSA J. Lightwave Technol.*, Vol. 7, No. 7, July 1989, pp. 1071–1082.

[117] Giles, C. R., and E. Desurvire, "Propagation of signal and noise in concatenated erbium-doped fiber optical amplifiers," *IEEE/OSA J. Lightwave Technol.*, Vol. 9, No. 2, Feb. 1991, pp. 147–154.

[118] Shimada, S., and H. Ishio (eds.), *Optical Amplifiers and Their Applications*, Chichester, England, Wiley, 1994.

[119] Ramaswami, R., and K. N. Sivarajan, "Design of logical topologies for wavelength-routed optical networks," *IEEE J. Select. Areas Commun.*, Vol. 14, No. 5, June 1996, pp. 840–851.

[120] Cancellieri, G., F. Chiaraluce, E. Gambi, and P. Pierleoni, "Devices based on the interaction of multi-frequency spatial solitons," *Proc. XI National Meeting on Electromagnetism*, Florence, Italy, Oct. 1996, pp. 585–588 (in Italian).

[121] Stern, T. E., "Linear lightwave networks: how far can they go?," *Proc. GLOBECOM '90*, San Diego, CA, Dec. 1990, pp. 1866–1872.

[122] Stern, T. E., K. Bala, S. Jiang, and J. Sharoni, "Linear lightwave networks: Performance issues," *IEEE/OSA J. Lightwave Technol.*, Vol. 11, No. 5/6, May/June 1993, pp. 937–950.

[123] Banerjee, D., and B. Mukherjee, "A practical approach for routing and wavelength assignment in large wavelength-routed optical networks," *IEEE J. Select. Areas Commun.*, Vol. 14, No. 5, June 1996, pp. 903–908.

[124] Chlamtac, I., and A. Ganz, "Lightpath communications: an approach to high bandwidth optical WAN's," *IEEE Trans. Commun.*, Vol. 40, No. 7, July 1992, pp. 1171–1182.

[125] Chlamtac, I., A. Faragó, and T. Zhang, "Lightpath (wavelength) routing in large WDM networks," *IEEE J. Select. Areas Commun.*, Vol. 14, No. 5, June 1996, pp. 909–913.

[126] Westlake, H. J., P. J. Chidgey, G. R. Hill, P. Granestrand, L. Thylen, G. Grasso, and F. Meli, "Reconfigurable wavelength routed optical networks: A field demonstration," *Proc. ECOC '91*, Paris, Sept. 1991, pp. 753–756.

[127] Kaminow, I. P., C. R. Doerr, C. Dragone, T. Koch, U. Koren, A. A. M. Saleh, A. J. Kirby, C. M. Özveren, B. Schofield, R. E. Thomas, R. A. Barry, D. M. Castagnozzi, V. W. S. Chan, B. R. Hemenway, D. Marquis, S. A. Parikh, M. L. Stevens, E. A. Swanson, S. G. Finn, and R. G. Gallager, "A wideband all-optical WDM network," *IEEE J. Select. Areas Commun.*, Vol. 14, No. 5, June 1996, pp. 780–799.

[128] Bala, K., T. E. Stern, D. Simchi-Levi, and K. Bala, "Routing in a linear lightwave network," *IEEE/ACM Trans. Networking*, Vol. 3, No. 4, Aug. 1995, pp. 459–469.

[129] Darcie, T. E., "Subcarrier multiplexing for multiple-access lightwave networks," *IEEE/OSA J. Lightwave Technol.*, Vol. LT-5, No. 8, Aug. 1987, pp. 1103–1110.

[130] Shankaranaranayan, N. K., S. Elby, and K. Lau, "WDMA/sub-carrier-FDMA lightwave networks: Limitations due to optical beat interference," *IEEE/OSA J. Lightwave Technol.*, Vol. 9, No. 7, July 1991, pp. 931–943.

[131] Deng, R. H., A. A. Lazar, and W. Wang, "A probabilistic approach to fault diagnosis in linear lightwave networks," *IEEE J. Select. Areas Commun.*, Vol. 11, No. 9, Dec. 1993, pp. 1438–1448.

[132] Bae, J. J., and T. Suda, "Survey of traffic control schemes and protocols in ATM networks," *Proc. of the IEEE*, Vol. 79, No. 2, Feb. 1991, pp. 170–189.

[133] Wong, P.-C., and K-H Chan, "Multi-fiber linear lightwave networks – Design and implementation issues," *IEICE Trans. Commun.*, Vol. E77-B, No. 8, Aug. 1994, pp. 1040–1047.

[134] Kovačevíc, M., and M. Gerla, "Rooted routing in linear lightwave networks," *Proc. INFOCOM '92*, Florence, Italy, May 1992, pp. 39–48.

[135] Lee, K.-C., and V. O. K. Li, "A wavelength rerouting algorithm in wide-area all-optical networks," *IEEE/OSA J. Lightwave Technol.*, Vol. 14, No. 6, June 1996, pp. 1218–1229.

# *Chapter 3*

# *Multihop Optical Networks*

## 3.1 PRELIMINARY REMARKS

### 3.1.1 Basic Characteristics of Multihop Networks

We have seen in Section 1.8 that optical networks can base their virtual topologies on a fully broadcast or mesh-connected media. In any case, according to the open systems interface (OSI) [1] approach, the network architecture should be almost independent of both the alternative structures.

We are going to examine a number of well-known and promising architectures, proposed and studied to support multihop virtual topologies. All of them are characterized by the fact that data packets emitted by a source are forced to cross, in general, some intermediate nodes before reaching their final destination. As will be shown below, multihop networks have been widely studied since they offer the possibility of interconnecting numbers of nodes larger than those typically handled by single-hop networks. This can be done since multihop networks scale up better their size, by exploiting the intrinsic features of mesh virtual topologies.

Moreover, they are oriented to utilize preferably a small number of fixed transmitter and receiver devices, thus greatly simplifying the hardware requirements and the costs too [2]. In fact, a given pair of adjacent nodes is expected to be always connected by means of the same wavelength channel, since the tuning of transmitters and receivers could change only in case of network rearrangement due to link or node malfunction, or when new stations are admitted in the network. For this reason, in multihop networks the tuning times are negligible in practice, because they are different from zero only when the whole network is experiencing a dramatic situation, whereas they are fundamental parameters in single-hop networks. In other words, the topology in the considered case is rather static [3].

As stated above, different multihop networks can be supported by the same physical communication system. Obviously, different architectures are not equivalent in terms of performance and cost. First, the virtual structure should be based on realistic hypotheses as regards the properties of the optical devices necessary to implement the network. In fact, many very promising solutions proposed in the literature are sometimes destined to remain on paper for long time since, at present, the state of the art of opto-electronics cannot always guarantee the performance needed by the considered systems. The evolutionary scenario of advanced integrated optics permits to assume the possibility of reaching ambitious results in the near future but, in any case, a prudent and careful dimensioning and organization of new WDM systems is highly preferable. For this reason, in the subsequent pages of this chapter we describe only multihop networks that have a certain probability of being developed, giving some references in the bibliography for the readers interested in knowing other solutions proposed in the recent years.

The problem of compatibility with the available technology concerns the hardware and software complexity of the node architecture too. To exploit the enormous amount of bandwidth made available by optical media to the most, the intermediate nodes, generally present in end-to-end communication realized in multihop networks, should not play the role of bottleneck, so dramatically reducing the maximum speed theoretically achievable in the considered systems. A certain increase in the transmission delay is due to opto-electric and electro-optic conversions that have to be realized at the nodes. In fact, at present, a sort of extended memory at optical level is not available and the buffering capacities of fiber-loop cells are not always sufficient to face all the real necessities. Other solutions have been proposed to overcome the problem of optical buffering. They are based on the principle of deflection routing that implies the intentional misrouting of packets (which in any case increases the path length), so that the optical links act like buffers. At the same time, it is important to reduce the delays caused by the processing of routing algorithms in any case. If simple and effective routing schemes are employed, short processing times are required and the presence of intermediate nodes slightly affects the overall network capacity.

### 3.1.2 Meaning and Importance of Some Performance Parameters

As stated above, multihop networks do not need fast tunable transceivers and their number not have to be high. Nevertheless, as a counterpart, packet delivery is carried out using intermediate nodes, which forward the messages until they reach their final destinations. This happens because, in general, one-hop source-destination connections are not always available. Therefore, the time invariability of multihop topologies emphasizes the necessity of defining good structures for multihop systems, as regards, for example, the average packet delay, the average node distance, the network and user throughputs, the node complexity, and so on [3]. In this sense, suitable parameters have to be used, in order to provide general criteria for evaluating the quality of different solutions. These parameters can be classified with respect to their impact on the users or on the network manager. With regard to these parameters, we think that some practical considerations are necessary. Even though they pertain in general to both single- and multihop networks, they are evidenced here since the design of multihop networks is particularly critical, because of the above-mentioned problems; thus every possible implementation must be carefully analyzed in advance to verify its feasibility, performance, and reliability.

In general, users ask for high-quality services to the telecommunication (TLC) companies, as a counterpart of the charge due for access the network and for the utilization of its services and facilities. In a competition market, where at least two TLC systems can operate, the clients choose their service providers looking at the reliability, quality, and cost of the offered services. People expect to be able to use the network and the TLC services, but they are not interested in knowing the difficulties and the technological efforts involved in this job. They are satisfied when they can obtain what they need without trouble (long access times, high cutoff probabilities, etc.) and when the relative expenses are reasonable. Consequently, the potential providers try to beat this competition designing, installing, and managing their networks according to these targets. At the same time, their commercial policy should be supported by technological choices which can offer good performance at low cost. A fundamental parameter that plays a central role in this sense is the network's efficiency (or, analogously, the network throughput) that relates to the number of potential supported users or to the telecommunication services that can be offered by the network itself. Even though, as stated above, this parameter (the network efficiency or throughput) is absolutely transparent to the users, it influences the relationship between the architecture and the cost of a system, so that it can significantly change the cost of every single connection. It is easy to understand that, if it is possible to increase the maximum level of traffic managed by the network by some percentage without significantly modifying the system structure, it is then possible to raise the proceeds of sales with no additional cost, thus partially reducing the telephone rates. This operation is not directly evident to the service purchasers, but they can benefit by this effect when present.

There are also quality parameters that are considered by both user and provider; for example, network reliability. Customers, particularly when they access the network for business activities, are greatly annoyed by troubles that can occur during the connections. This effect, as well as services costs, can induce people to subscribe to a different TLC company. The process that causes a TLC network to lose customers is progressive, sometimes slow, but continual. It depends on unsatisfactory behavior of either hardware or software. Considering once again a competition environment, where different companies fight to enlarge their market percentage, it is worth observing that the cost of gaining a new customer is much higher than the cost of retaining one [4]. This means that reliability has a great influence on the network commercial impact; therefore money investments devoted to satisfying the acquired user are more cost effective than efforts dedicated at getting new users.

How can a network ensure efficiency? Many elements can influence efficiency, but only few have a macroscopic effect on it. We can indicate three aspects: hardware complexity, network structure, and staff costs. The latter essentially depends on the number of employers in the TLC company and on the degree of automation of network management processes, but it does not deal with the scope of this book. Hardware complexity is strictly correlated with the functions assigned to the network and to the design requirements. Together with the evolution of VLSI technology and, now, with the new perspectives offered by opto-electronics, the same operation and performance can be obtained on devices characterized by lower and lower costs. In any case, this element is less important than the installation costs, that itself depends on many factors, which are subsumed in the factor of network structure optimization. This problem is analyzed in depth in this chapter: it involves topology, routing, flow control, etc. It will be demonstrated that, through suitable choices, network efficiency can be greatly increased, with cost savings and easier management procedures.

Until now we have distinguished and separated the role played by providers and customers; it is evident that this situation characterizes the systems operating on wide area networks or metropolitan area networks. But, in the case of local area networks (LANs) [1], the user and the network manager can coincide. In this particular condition, finally, all the quality parameters are visible and significant for the all the people using the communication system. In this section we have introduced some criteria to justify the attention, evident in the subsequent pages, given to the analysis of the network behavior according to suitable quality parameters. The adoption and interpretation of such parameters is not possible if their meanings in the real TLC world are not clear. Now we can go back to our original goal: the analysis of multihop WDM networks.

## 3.2 MANHATTAN STREET NETWORKS

### 3.2.1 Network Architecture

The first multihop network examined in this chapter is the so-called Manhattan street network (MSN) [5]. It is based on a regular virtual toroidal topology, composed by unidirectional links, which connects $n \times m$ nodes, as shown in Figure 3.1. The links form unidirectional rings and adjacent rings travel in opposite directions. In general, the two-connected grid could be formed by a number of rows different from the number of columns, in order to increase the flexibility and scalability of the network dimensions, even though squared MSNs are preferred for their characteristics of symmetry. Moreover, though not necessary, an even number of rows and columns is suggested [6].

The MSN is a synchronous communication system where every node transmits and receives packets, of fixed length, slotted in time. The node architecture is depicted in Figure 3.2, where, for the sake of clarity, the example of node (0,0) is considered, assuming the same network structure shown in Figure 3.1. Through two incoming links, data emitted by adjacent nodes reach the considered station, whereas packets produced by the local source enter from the external input. Both the entering slots are delayed by D1 and D2 in order to synchronize their arrivals at the switching section placed downstream. If the flowing packets are addressed to the considered node, they are extracted by the switch and sent toward the local output, or else they are pushed forward, according to routing strategies that will be discussed below. Before leaving the node to continue their propagation in the network, the same packets can be sometimes temporarily stored in buffers B1 or B2. In fact, considering that only one packet per slot can get out in each link, when both packets are addressed toward the same output, one of them is randomly selected to be stored in the buffer.

These memories are usually filled and emptied according to a FIFO procedure. As will be shown in the following sections, internal buffers are not always present, since their role can be replaced by particular routing schemes that, through misrouting of packets, avoid any need for storage.

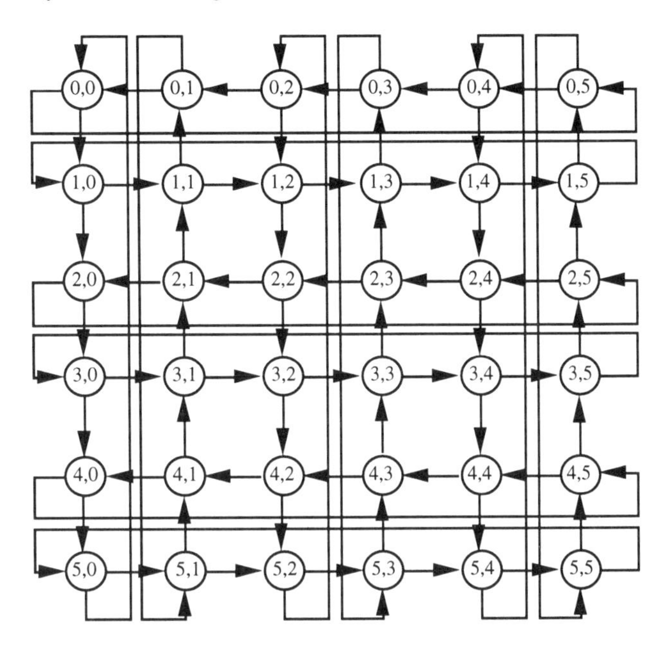


**Figure 3.1** 6×6 Manhattan street network.

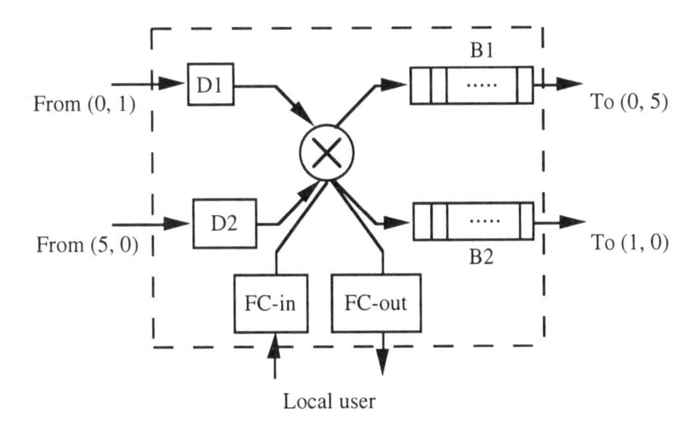


**Figure 3.2** Node architecture in a Manhattan street network.

Two devices, called FC-in and FC-out, are placed at the input and output of the node, respectively, to interface the network with the external users. They play the role of flow control devices, by regulating the speed of the data interchange process at the edge of the system. In particular, a local source can insert its messages in the node only when the node is crossed by at least one empty slot or when at least a packet is extracted from incoming links. This fact reduces progressively the bandwidth made available to each user in presence of higher and higher traffic, but it guarantees a high probability of reaching the destination to every packet accepted and transported by the network. On the other hand, to avoid that the network deliver packets faster than the local user can accept them, a buffering capacity is allocated in FC-out. Every node is identified by the so-called absolute address, represented by a pair of integer numbers that are the number of the row and the number of the column where the node is placed; they are obtained by a sequential numbering of rows and columns [6].

Scalability is one of the features characterizing multihop networks and MSNs. Therefore, the possibility of expanding the network by adding new nodes should be considered in the design and a suitable procedure of renumbering nodes, rows, and columns has to be defined a priori, maintaining the opposite directions of adjacent rings. A first solution can consist in the use of particular renumbering protocols, activated in presence of a network rearrangement. An alternative to changing the node address is to plan possible future expansions during the initial design of the network itself. In other words, to avoid the saturation of the possible addresses, rows and columns could be labeled leaving a certain number of intermediate integers unused. For example, if in the initial implementation of the MSN the rows are numbered by 00, 11, 22, etc., at least 10 new rows can be added successively.

The grid planning is less critical with respect to network expansion if fractional addressing is adopted [6]. In this case, in fact, an arbitrary number of rows and columns can be inserted anywhere in the network with no particular constraint. Fractional addressing implies that rows added in pairs are numbered as two fractions, 1/3 of the way between the two other rows. For example, let us consider an insertion between rows 0 and 1. The first new row is labeled 1/3, while the second is labeled 2/3. If a further extension occurs between rows 2/3 and 1, the first new row is labeled 7/9, and the second 8/9. Fractional addressing affords the opportunity of selecting the orientation of each row or column simply by looking at its label; all the rows or columns having an odd numerator have the same direction, whereas the even ones are oriented in the opposite direction. Moreover, fractional addressing does not constrain the number of rows and columns that can be added in the network, so that it does not change in case of network reconfiguration, whereas absolute addresses must be updated in the same conditions.

In multihop networks, frequently the path connecting two remote nodes is not unique, since several alternatives are available. The problem of the choice of the best path, and the routing of packets from source to destination, is particularly important in this sense, so that the impact of a given topology is often evaluated in terms of routing performance. Thanks to the cyclic architecture of MSN, the routing problem can be solved by just taking into account the relative position of the source with respect to the destination and considering the latter always approximately at the center of the grid. In this way, relative addressing can substitute for absolute addressing and the routing rules can be applied always in the same manner, independently of the placement of the two nodes. If, in a $n \times m$ MSN, $(r_s, c_s)$ is the absolute address of the source node, its integer relative address, with respect to the destination node having $(r_d, c_d)$ as absolute address, is given by [6]

$$r' = \frac{m}{2} - \left[\left(\frac{m}{2} - D_c\left(r_s - r_d\right)\right) \operatorname{mod} m\right] \tag{3.1a}$$

$$c' = \frac{n}{2} - \left[\left(\frac{n}{2} - D_r\left(c_s - c_d\right)\right) \operatorname{mod} n\right] \tag{3.1b}$$

with $r' \in ]-m/2, m/2]$ and $c' \in ]-n/2, n/2]$ whereas, using a fractional address, we have

$$r'' = 1 - \left[\left(1 - D_c\left(r_s - r_d\right)\right) \operatorname{mod} 2\right] \tag{3.2a}$$

$$c'' = 1 - \left[\left(1 - D_r\left(c_s - c_d\right)\right) \operatorname{mod} 2\right] \tag{3.2b}$$

where $r'' > -1$ and $c'' \leq 1$ and with

$$D_c = \begin{cases} +1 & \text{for } c_d \text{ even} \\ -1 & \text{for } c_d \text{ odd} \end{cases} \tag{3.3a}$$

$$D_r = \begin{cases} +1 & \text{for } r_d \text{ even} \\ -1 & \text{for } r_d \text{ odd} \end{cases} \tag{3.3b}$$

From the previous equations, four quadrants are defined

$$\begin{array}{ll} Q_1 & \text{for } r > 0 \text{ and } c > 0 \\ Q_2 & \text{for } r > 0 \text{ and } c \leq 0 \\ Q_3 & \text{for } r \leq 0 \text{ and } c \leq 0 \\ Q_4 & \text{for } r \leq 0 \text{ and } c > 0 \end{array} \tag{3.4}$$

which can be used as reference system for the orientation of the packet propagation inside the grid. However, note that, using fractional addressing, the destination node can sometimes be displaced from the center of the network, so that shortest path routing (which operates using information given by the above-mentioned quadrants) could fail. In this case, packets can flow along longer paths between source and destination. An example of relative addressing is shown in Figure 3.3, considering the same network of Figure 3.1. In this way, it is possible to consider the source in a central position, so that destination address can be calculated consequently.

The regular structure of MSNs is based on an even number of rows and columns. Nevertheless, in order to ensure the possibility of adding new nodes after the initial network setup, irregular configurations have to be considered. They can appear when the nodes to be added are not sufficient to complete a row or a column, or in case of node or link failure. The inclusion of a node in a previously regular topology can be realized as shown in Figure 3.4, where the links to be changed are drawn in dashed lines. The indicated procedure can be repeated for successive insertions of single nodes, until a new row or column is complete.

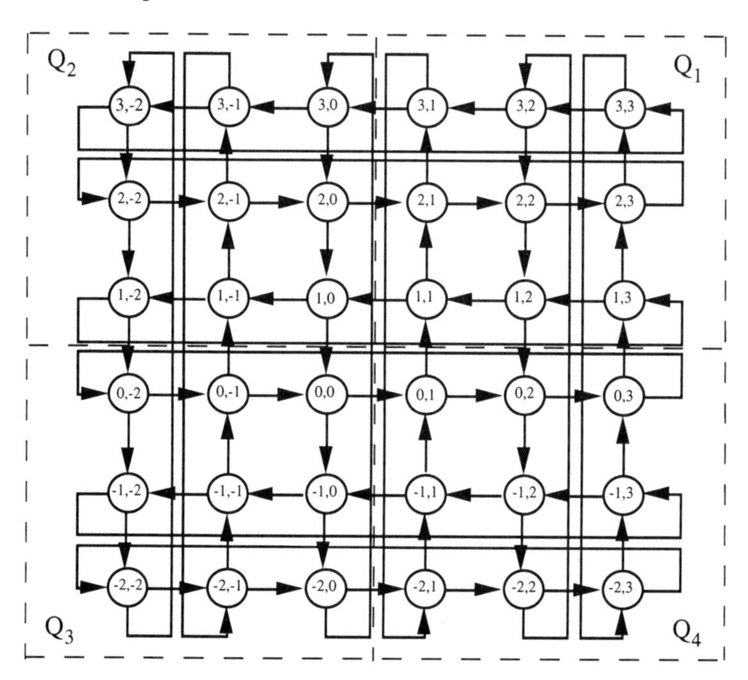


**Figure 3.3** Using relative addressing in a MSN.

Analogously, it is sometimes necessary to delete a given node that is failing or whose performance is worsening. This can cause interruption of the loops present in the network. However, thanks to the adoption of suitable relays, the failed node can be bypassed [7], so that the fault does not affect the packet propagation in the grid.

Node inclusion and failure recovery can be implemented automatically, on condition that a high-level protocol is available to inform all the stations about the modifications occurring in the topology. In this way, the risk of unsuccessful delivery of packets, due to sudden variation of previously installed connections or to the disappearance of some disconnected nodes, is avoided.

Finally, irregularities can also appear in case of link failures. They can be detected by the absence of input signal at the node placed downstream. In order to avoid packet losses, we need to activate a suitable procedure, one that is able to inform the couple of nodes previously connected by the failed link and to create an alternative path between them. This has to be done maintaining the in-degree and the out-degree of each node equal. A simple rule for accomplishing this is to avoid transmission on a row if a malfunction has occurred on a column and vice versa. An example is given in Figure 3.5, where the failed link is shown by a bold arrow, whereas the links avoided are shown by dashed lines.

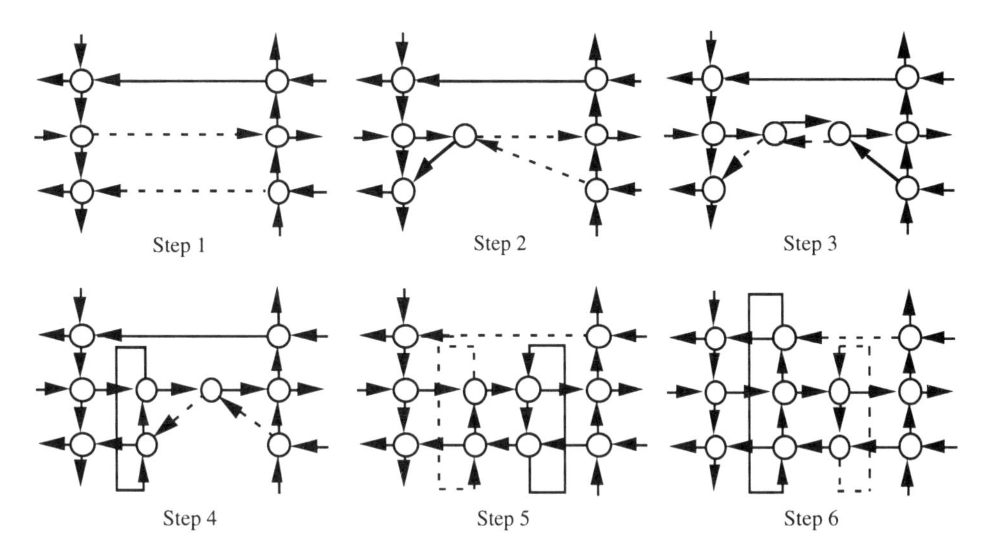


**Figure 3.4** Adding new nodes in an existing MSN topology. (*Source*: [6]. © 1987 IEEE.)

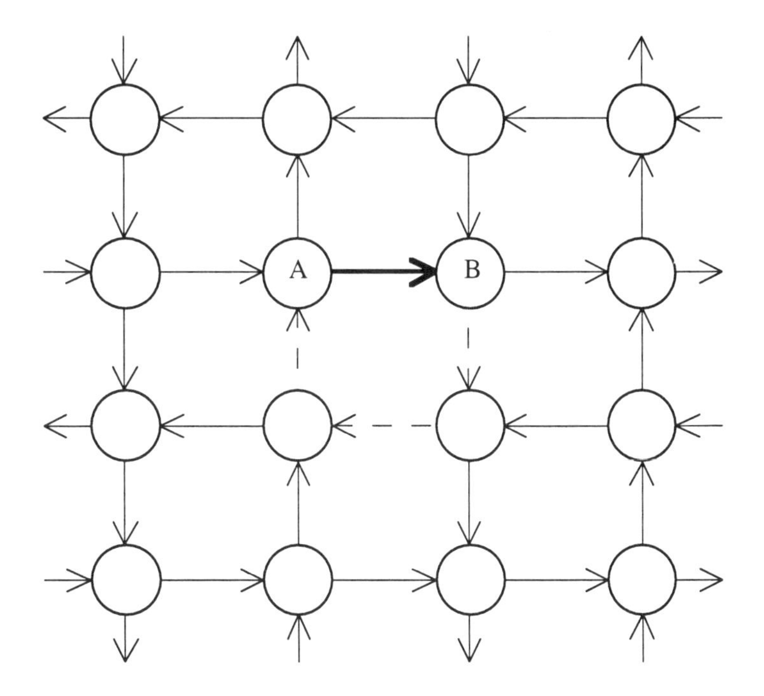


**Figure 3.5** Reaction to a link failure occurred between nodes A and B.

In practice, a single link malfunction erases four connections (one input and one output per node at the most). More complex situations, which may imply the removal of operable nodes from the network, can occur in presence of simultaneous multiple failures. Nevertheless, the joint probability relative to this event can be greatly reduced if failed links are rapidly repaired. When reactivated, the use of the link is forced by the node placed upstream. Further analysis on the reliability of MSN can be found in [8,9].

### 3.2.2 Some Topological Characteristics of a MSN

As stated above, in order to guarantee that the network is composed by two sets of horizontal and vertical pairs of adjacent counter-oriented rings, the number of rows and columns of a regular MSN must be even.

The connectivity degree $p$ of a MSN is equal to 2, as evident from previous description; therefore, in a $n \times n$ MSN the overall number of links equals $W = pn^2 = 2n^2$. In a regular topology the above-mentioned connectivity degree influences the network performance and cost as well as its capacity and overall throughput, that is, the fraction of the network capacity utilized for data delivery from sources to destinations.

Other two important parameters that define a regular network, in this case a MSN, are the so-called diameter and average distance. The former is given by the maximum value of the shortest distance existing between every pair of nodes. Because of the network symmetry, it can be considered as the greater length of the shortest paths connecting a given node with all the other nodes in the network.

For a $n \times m$ MSN, through some graph theory theorems [10], the network diameter $D$ is given by

$$D = \begin{cases} \dfrac{m}{2} + \dfrac{n}{2} + 1 & \text{for both } \dfrac{m}{2} \text{ and } \dfrac{n}{2} \text{ even integers} \\[2ex] \dfrac{m}{2} + \dfrac{n}{2} & \text{otherwise} \end{cases} \tag{3.5}$$

On the other hand, the average distance $\tau$ is the average path length from a node to any other node of the network along the shortest path. It also depends on the network dimensions [10]

$$\tau = \begin{cases} \dfrac{\frac{N}{4}(m+n+4) - n - 4}{N-1} & \text{for } \dfrac{n}{2} \text{ odd and } \dfrac{m}{2} \text{ even} \\[3ex] \dfrac{\frac{N}{4}(m+n+4) - 4}{N-1} & \text{for both } \dfrac{n}{2} \text{ and } \dfrac{m}{2} \text{ even} \\[3ex] \dfrac{\frac{N}{4}(m+n+4) - m - n - 2}{N-1} & \text{for both } \dfrac{n}{2} \text{ and } \dfrac{m}{2} \text{ odd} \end{cases} \tag{3.6}$$

If $n = m$, that is, we have a square MSN, so that $n = m = \sqrt{N}$, $D$ and $\tau$ can be written in the following forms:

$$D = \begin{cases} \sqrt{N} + 1 & \text{for } \frac{\sqrt{N}}{2} \text{ even} \\ \sqrt{N} & \text{for } \frac{\sqrt{N}}{2} \text{ odd} \end{cases} \tag{3.7}$$

$$\tau = \begin{cases} \frac{N^{3/2} + 2N - 8}{2(N-1)} & \text{for } \frac{\sqrt{N}}{2} \text{ even} \\ \frac{N^{3/2} + 2N - 4\sqrt{N} - 4}{2(N-1)} & \text{for } \frac{\sqrt{N}}{2} \text{ odd} \end{cases} \tag{3.8}$$

Another way of analyzing the connectivity graph of our MSN can be adopted. As an example, in Table 3.1 we have reported the number of nodes placed $h$ hops far from a given source in a square MSN with $n/2$ even [11].

**Table 3.1**
Node Placement in a Square MSN with $n/2$ Even

| $h$ | *Number of Nodes at Distance h* |
|---|---|
| 1 | 2 |
| $2 \le h \le \sqrt{N}/2 \quad h \ne 4$ | $4(h-1)$ |
| 4 | 11 |
| $h = \sqrt{N}/2 + 1$ | $2\sqrt{N} - 4$ |
| $\sqrt{N}/2 + 2 \le h \le \sqrt{N}$ | $4(\sqrt{N} - h + 1)$ |
| $h = \sqrt{N} + 1$ | 2 |

### 3.2.3 Distributed Routing Rules in MSNs

In this section we focus our attention on some of the proposed strategies that can be adopted at the nodes to route the packets through the MSN grid.

The routing rules represent a criterion for selecting one of the two outputs that are always present in each node. In fact, when a packet is crossing a node, to continue the propagation directed to its final destination, the node itself switches it toward an output link. This choice can be made on the basis of deterministic or random routing rules. In the first case, the chosen output is suitable to make the path short; whereas, in the second case, the output is selected randomly. It will be demonstrated that also in this particular condition the delivery of packets can be completed correctly.

However, let us begin our analysis from deterministic schemes.

### *3.2.3.1 Deterministic Routing Strategy 1*

This strategy is based on the two following rules:

- Select the preferred path if there is one preferred path from a node;
- Select either path if there are zero or two preferred paths from a node.

This strategy requires that every node knows its adjacent ones and their addresses. In Figure 3.6, the arrows show the preferred direction from the relative position in the network to the destination, considered to be located in the center of the grid. By $r_1$, $r_2$, $r_3$, $r_4$ and $c_1$, $c_2$, $c_3$, $c_4$, we have indicated the rows and the columns at the edge of the quadrants respectively. From the current position, the relative addresses of the node occupied by the packet $(r, c)$, of the next node along the column $(r_{nxt}, c)$ and of the next node along the row $(r, c_{nxt})$ are calculated to determine the preferred path. While it is worth noting that a node in $Q_4$ is in row $r_4$ if $r = 0$ and a node in $Q_2$ is in column $c_2$ if $c = 0$, we think that the directions of the arrows from various locations in the network are intuitive, therefore they can be understood without the necessity of a detailed explanation. By $Q_1$-, $Q_2$-, $Q_3$-, and $Q_4$-, we have indicated the remaining parts of the four quadrants, once we have identified columns $c_1$, $c_2$, $c_3$, $c_4$ and rows $r_1$ , $r_2$ , $r_3$ , $r_4$.

It can be demonstrated that this method allows to select the shortest path from source to destination in an integer addressed network [6].

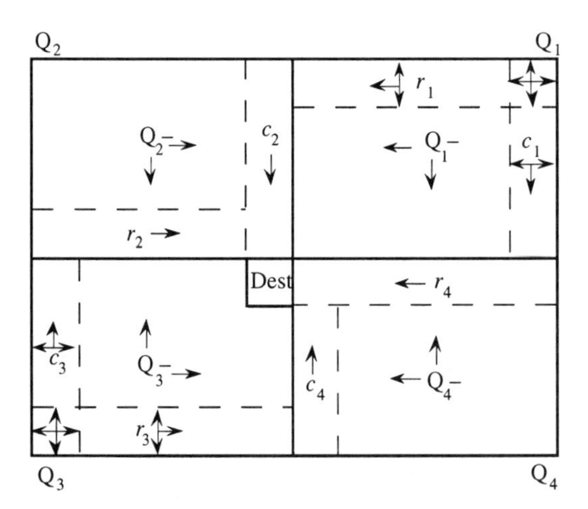


**Figure 3.6** Preferred travel directions from the relative positions in the network to destination, according to strategy 1.

*3.2.3.2 Deterministic Routing Strategy 2*

Strategy 2 is similar to strategy 1, but the preferred paths are those indicated in Figure 3.7 instead of those in Figure 3.6. It is evident that the particular cases represented by nodes placed in $r_1$, $r_3$, $c_1$ and $c_3$ disappear. This implies a certain reduction of calculations due to preferred path identification. On the other hand, this method causes some increase in the path lengths. This effect will be experienced only by nodes far from the destination, so that few packets will be affected by path length increasing, since any node closer to destination does not route its packets towards those zones.

In addition, it is a requirement of this strategy that every node knows its adjacent nodes and their addresses.

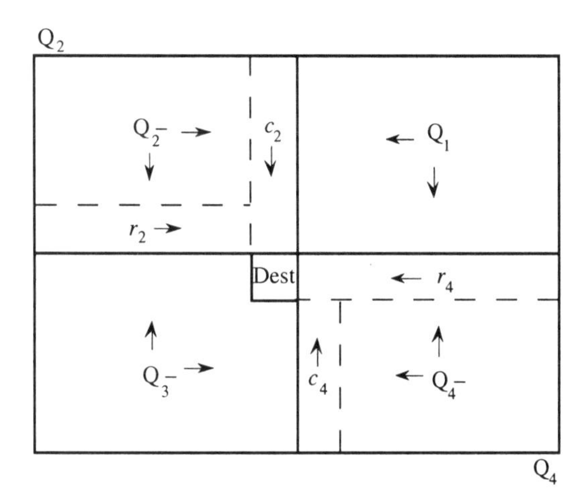


**Figure 3.7** Preferred paths in strategy 2.

*3.2.3.3 Deterministic Routing Strategy 3*

This strategy is based on the three following rules:

- Select the preferred path if there is one preferred path from a node;
- Select the alternate path if there is no preferred path and one alternate path from the node;
- Select either path if neither path is a preferred or alternate path or if both paths are preferred.

The preferred and alternate paths are shown in Figure 3.8, by solid and dashed arrows, respectively. This method also implies a reduced amount of calculations compared to strategy 2 and, furthermore, it is not dependent on the knowledge of the addresses of adjacent nodes, unlike strategies 1 and 2.

The disadvantages of strategy 3 essentially concern the possibility of generating node conflicts, since there are fewer cases, with respect to strategy 2, in which either path can be selected (see quadrants $Q_2$ and $Q_4$). Moreover, strategy 3 shows problems in handling connections in incomplete networks because, sometimes in this strategy packets cannot reach their destination [6]. In this situation the adoption of strategy 2 is preferable.

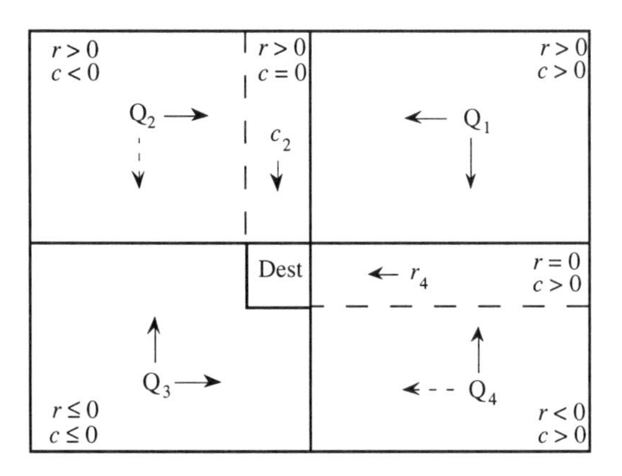


**Figure 3.8** Preferred paths in strategy 3 (preferred paths: solid arrows; alternate paths: dashed arrows).

### *3.2.3.4 Random Routing*

This procedure is completely different with respect to those previously described. In fact, in this procedure, routing choices are made randomly without any analysis of the relative distance from the current node to the destination [12].

In practice, the two possible outputs are selected randomly at every node, until the considered packet meets its destination, where it is extracted from the network. According to this procedure, every node reads the destination address field of incoming packets to check if it is their final destination. If not, the node transmits such packets toward its adjacent nodes randomly. This means that they go all over the network following a random path until they reach their destination.

It is evident that random routing heavily affects the network throughput, because of the significant increase of the average number of hops needed to reach the destination from every source. Nevertheless, thanks to the absence of a routing logic, it is very easy to implement, as well as suitable for reaching very high switching frequency. Moreover, random routing is robust and tolerant with respect to network irregularities due to link interruptions, node failures, or incomplete node insertions or extractions.

Finally we remind the reader that, to reduce the inefficiency of random routing procedure, a quasi-random technique can be adopted. In the latter case, every node knows its adjacent ones. If the destination is only one node away from the node currently occupied by the considered packet, the packet is directed there, otherwise it is switched and transmitted toward an adjacent node, randomly selected as previously indicated. This method improves the performance of this strategy without overly increasing node complexity.

### *3.2.3.5 Deflection Routing*

The above-described strategies are oriented to route the packets through the grid to their destinations. Deterministic schemes often face this problem by choosing the shortest path, whereas random ones imply longer propagations in the network. However, these methods do not solve the problem of packet delivery completely, since every node has to satisfy the necessity of routing more than one packet per slot and this can cause contentions when two packets should be addressed toward the same output link. If buffers are available at the nodes, one packet gets out whereas the other is queued. But, considering bufferless networks, it follows that every node has to be able to apply a suitable procedure to obtain a correct packet delivery without conflicts.

In this sense, a routing algorithm should provide two main functions: the basic routing (BR) function and the contention resolution (CR) function. The first is devoted to assign routing preferences, that is to say, it selects the best node outputs without considering any conflict with other colliding packets. Contention resolution, instead, plays its role when collision occurs, assigning the packets to the outputs according to a given criterion. Packets that cannot be satisfied in their preferences are deflected.

Deflection routing [13] is a good solution for regular networks, and therefore also for MSNs. In fact the latter is regular, two-connected, and it has nodes with equal in-degree and out-degree. Moreover, both paths at the outputs of many nodes have the same length to the destination (these nodes are called "don't care nodes"). Deflection routing, which is similar to "hot-potato" [14,15] and "queue plus bias" routing [16], operates as follows:

- Packets in transit are served with absolute priority with respect to those that are trying to access from the external input;
- If there is no storage available at the node and there is contention for the same outgoing channel, some packets are temporarily misrouted.

It is important to observe that if a packet is deflected its path length is only increased by four [13]. In general, increasing the path lengths increases the link utilization but reduces the maximum throughput. Nevertheless, this little penalty does not worsen the network efficiency too much. The above-mentioned good matching between MSN and deflection routing characteristics are here summarized:

- A packet that can take either link to the destination does not force another packet that arrives during the same slot to be deflected;
- When a packet is deflected onto the wrong link, at the next node it can take either link;
- As the network becomes larger, the fraction of time a packet must turn decreases since a packet is never more than three turns from its destination.

Since deflection only occurs when one packet must go straight and another must turn, the last characteristic implies that the larger the network, the smaller the deflection probability.

An accurate analysis of deflection routing in MSNs is reported in [13,17]. To evaluate the performance of this strategy, let us assume the following hypotheses:

- The sources insert packets in empty slots only, otherwise the packets are lost;
- The packet to be deflected is chosen randomly;
- The traffic is uniformly distributed toward all the possible destinations and all the nodes generate packets with the same statistic.

Let us define the absolute throughput $T^{(1)}$ as the average number of slots that a source acquires at a node and the relative throughput $T^{(2)}$ as the ratio between the absolute throughput and the maximum throughput (that depends on the inverse of the average number of hops). The last parameter is suitable to compare networks having different dimensions.

As shown in Figure 3.9, where the maximum value of the relative throughput $T^{(2)}{}_{\max}$ is plotted, the presence of buffers $b$ elements long can improve the throughput itself by about 20% to 25% for $b = 1$ and by 4% to 7% for $b = 2$. For $b > 2$ a saturation effect is rapidly reached. In practice, the impact of buffer dimension is weak and it decreases as the network dimension gets larger [18]. On the other hand, it can be observed that, in general, when the load is augmented, the number of deflections increases significantly whereas the throughput is not particularly affected by this fact. Assuming that, during a slot, a node may insert packets in empty slots with probability $p_a$ [13,18–20], the behavior of $T^{(1)}$ and $T^{(2)}$ versus $p_a$ is plotted in Figure 3.10.

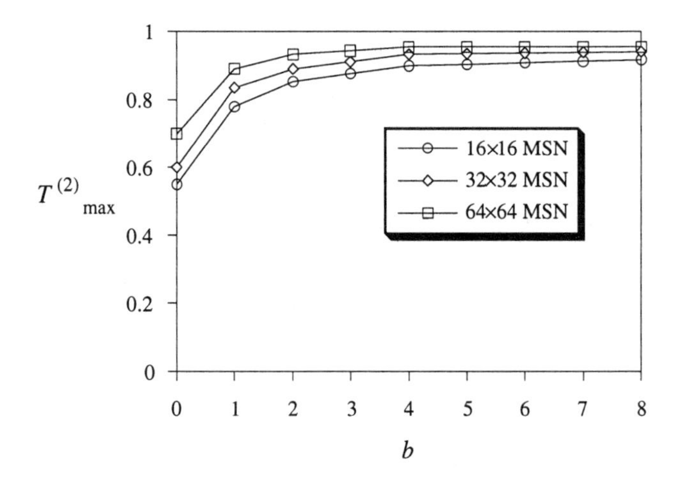


**Figure 3.9** Maximum relative throughput versus the buffer dimension.

Until now, we have assumed that the packet to be deflected will be chosen randomly. Actually, other solutions have been proposed [19]. In particular, deflection decisions can be made on the basis of the past history of the packet, on expectation of its future transit through the network, or on a combination of both. In the first case, the deflection decision takes into account the number of hops already made by the packet; therefore old packets have priority in case of contentions with other ones, so that the deflection probability decreases with time hop after hop. Alternatively, priority can be given to packets that have already experienced a high number of deflections. However, these techniques require the presence of a counter, placed in the header of every packet, where a value is stored (that is the age of the packet [19,21]), incremented at each hop or deflection. Both methods are characterized by similar performance, oriented to limit the end-to-end delay.

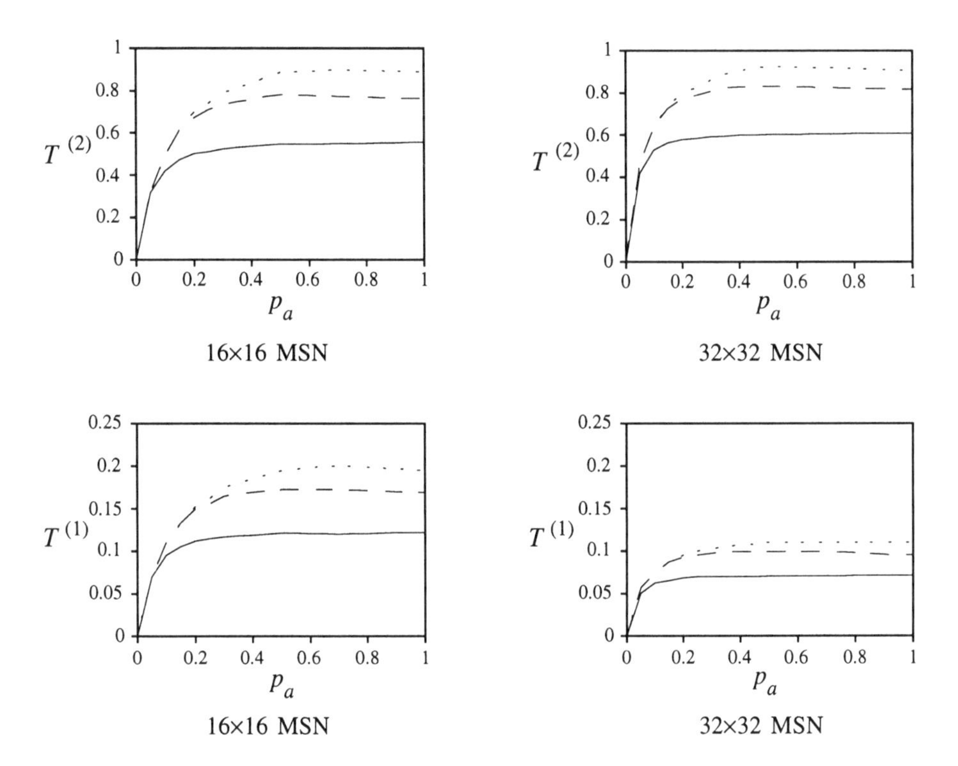


**Figure 3.10** Relative and absolute throughput in MSNs versus the arrival probability (continuous lines: $b = 0$; dashed lines: $b = 1$; dotted lines: $b = 4$). (*Source*: [13]. © 1989 IEEE.)

Other decision criteria take into account the concept of distance. The static distance strategy, in fact, gives a packet a priority equal to the distance from the source node to the destination, whereas the dynamic distance strategy equals the priority to the distance from the current node to the destination. In the first case the priority remains constant, whereas the dynamic strategy changes this value during the propagation of the packet. In both cases, the packets generated by sources very close to destinations will be given the advantage. Nevertheless, the distance algorithms perform worse than algorithms based on the previous history of the packet, in terms of packet delay, reducing the probability relating to long end-to-end paths. A detailed analysis and explanation of this effect is discussed in [19,22].

Finally, hybrid strategies are also possible [19] which take into account distance or age criteria. Their performance are comparable with those of previously examined methods.

#### *3.2.3.6 Dead Reckoning*

In this section we briefly introduce an example of a self-routing technique for primarily regular ultrahigh speed networks, which does not need landmarks for delivering packets to the destinations. It has been proposed in [23] and has been applied to packet routing in MSNs.

To reduce the delay due to routing processing at the nodes, and to simplify the process itself so that it can be implemented by ultrafast photonic devices [24], dead reckoning extracts the navigation information from only 4 routing bits. Before being inserted in the grid, a packet to be transmitted is loaded, in its header, by the destination address and by two additional bits that provide general information about the relative position of the source with respect to the destination and the two axes of the topology. The other 2 bits are used to indicate whether or not the destination row, or column, matches the node row, or column. For example, at its insertion, the packet knows that its destination is placed somewhere "south and west." At each hop, the intermediate node checks the possible correspondence between the packet destination address and its own one. This operation can be performed in only one very short step [25]. If the two address fields match, the packet is extracted, otherwise the intermediate node forwards it considering only the bits where its preferred propagation direction are stored. No optimal (and time consuming) routing process is utilized.

The considered bits are changed when the navigation should be redirected. In MSNs this happens when the packet arrives at an intermediate node located on the same row or column of the destination, or when it is forced to leave its propagation along a destination row or column, by the contention resolution procedure of some intermediate node.

The advantages offered by the dead reckoning scheme can be summarized as follows:

- Naming of rows and columns can be arbitrary;
- Nodes are not requested to know the network dimensions;
- Network irregularities due to node or link failures are tolerated;
- Routing scheme is very simple and fast.

### 3.2.4 All-Optical Implementation of MSNs

The simplicity and regularity of MSN topologies, together with the limited number of connections used to link adjacent nodes, afford the opportunity of defining efficient routing algorithms and of adopting simple architecture that can be easily implemented at electronic level. Nevertheless, the most critical problem that affects networks using optic fibers as transmission medium comes from the necessity of opto-electronic and electro-optic conversion stages, placed at the input and output interfaces of each node. In fact, at present, complex processing and buffering functionalities can be offered with good performance only by VLSI technique, since the capabilities of the state of the art optical components are not always sufficient to support advanced node functions [26].

In particular, the present technology of optical buffers is not still compatible with every extensive application [27]. This is because it seems to be well suited for adoption for parallel information processing, but not for the storage of a large mass of data [28]. However, according to the different techniques developed so far, an optical memory can be implemented by a fiber-loop cell buffer [29], or using a vertical-cavity surface-emitting laser [30]. Moreover, optical flip-flops integrated on a single chip are also available [31], operating like 1-bit dynamic memory.

The performance of optically controllable microresonators and electronically controllable vertical to surface transmission electrophotonic devices (VSTEP) are respectively reported in [32] and [33], whereas experimental applications of photonic buffering technologies are presented in [34], where fiber delay lines are used, in [35], where a composite opto-electric buffer has been utilized, and in [28], where traveling-type buffers and recirculating-type buffers are considered.

Besides optical buffering, routing techniques are also affected by the limited performance of photonic technology. It follows that, when a class of routing algorithms is too complex for its all-optical implementation, it can be applied only at the electronic level as usual.

Unfortunately, the presence of opto-electronic conversion stages at the nodes introduces a bottleneck whenever the transmission speed tends to become higher than a few gigabits per second. In this sense, the impressive transmission capacity made available by optical media is underutilized and the traffic access to the network is drastically limited. For this reason, many efforts have been made by the scientific community to realize all-optical networks, where all the processing and the routing rules can be executed at optical level, without using any conversion stage from source to destination. The only way to obtain correct traffic management in all-optical systems is represented by a drastic simplification of the above-mentioned procedures, usually adopted in electronic packet switching networks. Consequently, the residual functionality can be implemented on passive optical devices, which are very fast but have poor intelligence, and these devices can be applied in passive broadcast structures.

Many solutions have been proposed for all-optical networks implementations, essentially based on TDMA [36–38], CDMA [39,40], and wavelength/frequency division multiple access (WDMA/FDMA) [41–43]. Here, as an example [44], we consider a $n \times n$ MSN, where space diversity is combined with frequency and code division techniques. We assume the following conditions (in this section we use the term *frequency* as a synonym of wavelength):

- The nodes transmit data on the horizontal rings, whereas they receive packets on the vertical ones;
- Every node receives data at a given frequency that univocally identifies the final destination address;

- Every node transmits coded data using a codeword of a CDMA code that univocally characterizes every source.

Obviously, the number of frequencies and codewords, needed in this case, is equal to $n^2$. As a consequence of the above listed rules, a given node $(i, j)$ receives data at its proper frequency $f_{ij}$ (with $i, j = 0, 1, \ldots, n-1$), defined by the following equation:

$$f_{ij} = f_0 + [i + jn]\Delta f \tag{3.9}$$

where $f_0$ is the fundamental frequency and $\Delta f$ is the distance between adjacent frequencies. It follows that nodes belonging to the same $j$th column are tuned at adjacent frequencies, which can be grouped in the $G_j$ set

$$G_j = \{f_0 + jn\Delta f \, , f_0 + [(j+1)n - 1]\Delta f\} \tag{3.10}$$

At a generic $(x, y)$ node, the incoming optical signal is filtered to identify and remove the $G_y$ set from the horizontal ring. The extracted signal is then injected in the vertical ring, whereas at the other frequencies the data continue their propagation along the horizontal ring. In the same way, the given node receives the packets addressed to it from the vertical ring, simply filtering the input signal through a passband filter centered at the frequency $f_{xy}$. Subsequently, using a group of $n^2$ decoders, the local station can identify the transmitting node and recover the message. Let us observe that in every horizontal ring all the $n^2$ frequencies are used whereas, along the generic $j$th vertical ring, only the frequencies belonging to the set $G_j$ are present.

To implement the proposed algorithm, we have to utilize $n^2$ wide-band filters and $n^2$ narrow-band filters (to extract the $G_j$ frequency groups and the $f_{ij}$ frequencies respectively), so that the corresponding node architecture can be schematically represented as shown in Figure 3.11, where $\overline{G}_j$ and $\bar{f}_{ij}$ are, respectively, the not-extracted groups and frequencies.

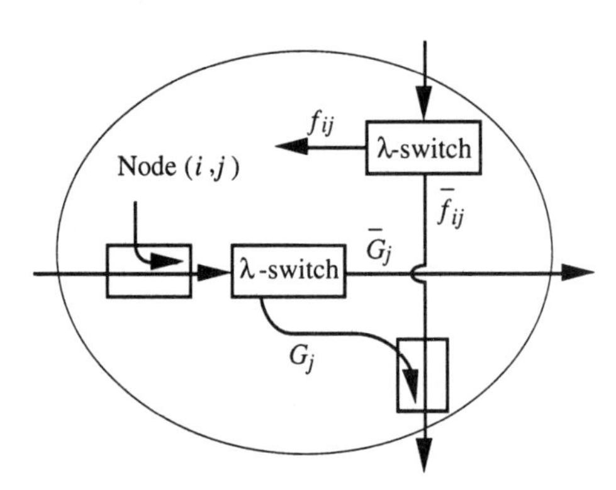


**Figure 3.11** Basic node architecture.

### 3.2.4.1 *Multiple Horizontal Rings*

Despite the relative simplicity of the above-described scheme, it must be noted that the number of needed frequencies becomes too high when the number of users rises. For this reason, the original architecture should be slightly changed (and made more complicated) to reduce the channels requirements.

Let $K$ be an integer divisor of $n$. Grouping the columns in $K$ sets, each composed by $n/K$ columns which can be assumed adjacent for simplicity sake, it results that $n^2/K$ nodes are contained in every group. Now, it is possible to organize the node transmissions so that each of them uses different horizontal rings to reach different groups of columns. In this case the total number of frequencies necessary to discriminate the possible destinations is reduced to the number of nodes belonging to every group. In other words, instead of frequency orthogonality, a suitable combination of space and frequency diversity ($K$ horizontal rings per node and $n^2/K$ frequencies) is used to separate the communication channels. The adoption of space diversity implies that, globally, $n(K + 1)$ rings should be installed in the network. Instead, the $(i, j)$ node, belonging to the $K$th group, receives data at its proper frequency, defined as follows:

$$f_{ij}^{(K)} = f_0 + \left[ i + n|j|_{\frac{n}{K}} \right] \Delta f \tag{3.11}$$

having used the notation $|x|_y$ to represent the remainder of the $x/y$ division. Also, in this case, the frequencies associated to each group are adjacent, so that every column is characterized by a particular set of frequencies, which is given by

$$G_j^{(K)} = \left\{ f_0 + |j|_{\frac{n}{K}} \cdot n\Delta f \, , \, f_0 + \left[ \left( |j|_{\frac{n}{K}} + 1 \right) n - 1 \right] \Delta f \right\} = G_{|j|_{\frac{n}{K}}} \tag{3.12}$$

The $(a, b)$ node, transmitting on the $a_k$ horizontal ring (with $k = 1, 2, \ldots, K$) traversing the node itself, uses the $f_{ij}^{(K)}$ frequency of (3.11) to reach the destination placed at row $i$ and column $q$, where

$$q = (k-1)\frac{n}{K} + |j|_{\frac{n}{K}} \tag{3.13}$$

On the contrary, the receiving section of each node operates as follows. In every $(x, y)$ node, being

$$(k-1)\frac{n}{K} \leq y < k\frac{n}{K} \tag{3.14}$$

the received signal is filtered to obtain the $G_y^{(K)}$ set of frequencies. The latter is then injected in the vertical ring. All the other nodes of the row $x$ do not handle the bandwidth partition of the signal considered, traveling on the horizontal $x_k$ ring. A node can extract the messages addressed to itself, received from the vertical ring, by filtering its proper

$f_{ij}^{(K)}$ frequency and by decoding them through an array of $n^2$ decoders. An example of schematic node architecture using this solution is represented in Figure 3.12.

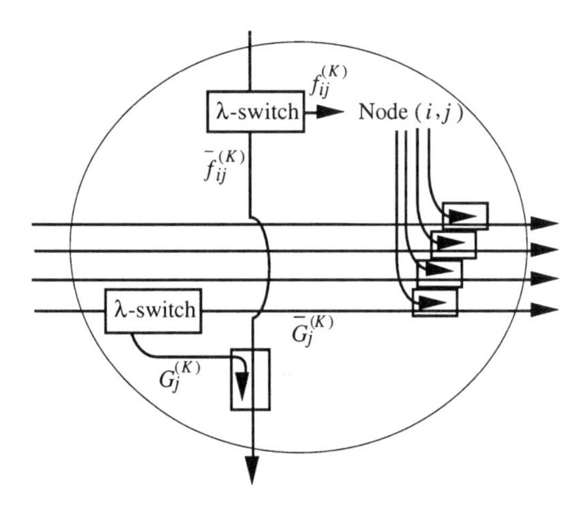


**Figure 3.12** Node architecture in presence of $K = 4$ parallel horizontal rings.

### 3.2.4.2 *Multiple Vertical Rings*

With respect to the original architecture, and as an alternative to the proposed modification presented in the previous section, it is possible to consider a structure using a reduced number of codewords, which identify the source of every packet. Using $R$ vertical rings per node and subdividing in $R$ groups the $n$ rows (assuming $R$ as an integer divisor of $n$), the network needs to use $n^2$ frequencies as in the basic configuration. Every group uses a different vertical ring to reach a given destination, so that the sources are separated in space and the codewords can be reused. In total, $n^2/R$ codewords and $n(R + 1)$ rings are needed. Once again, the nodes insert the packets to be transmitted in the horizontal rings, whereas the reception is realized on all the vertical rings.

The $(x,y)$ node, after filtering the incoming signal to select the group of frequencies $G_y$ given by (3.10), correspondent to column $y$, transmits the latter in one of the vertical rings. The $y_r$ vertical ring traversing nodes in column $y$ (with $r = 1, 2, \ldots, R$), is used by every node placed on the row $x$, being

$$(r-1)\frac{n}{R} \le x < r\frac{n}{R} \tag{3.15}$$

Every node filters the optical signals coming from the $R$ vertical links to receive its messages, by centering every passband filter at its proper frequency, given by (3.9); at the same time, the $n^2/R$ decoders placed downstream identify the sources. Therefore, in this case the network needs $Rn^2$ decoders and $Rn^2$ selective filters. The corresponding node architecture is depicted in Figure 3.13.

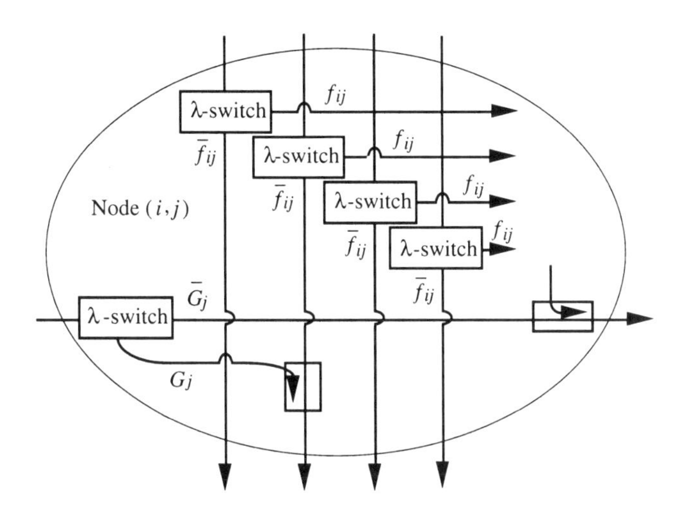


**Figure 3.13** Node architecture in the presence of $R = 4$ parallel vertical rings.

### *3.2.4.3 Frequency Diversity on a Single Vertical Ring*

The partitioning of sources in groups, whose object is to reduce the overall number of codewords, can also be obtained thanks to the adoption of frequency diversity, instead of space diversity considered in the previous section. In fact, increasing $Z$ times the number of frequencies multiplexed on a single vertical ring in the original structure, the previous partition of the $n$ rows in $Z$ subsets can be obtained. As a counterpart, the complexity of the node receiving section increases. In this case, every group of $n/Z$ rows utilizes a different set of frequencies to address the nodes placed along a given column. In this way, the sources are separated in frequency and the codewords are reused.

The $(x, y)$ node emits packets on its unique horizontal ring. Analogously to previously described architectures, the $G_{xy}$ frequency group is obtained from the horizontal ring and retransmitted in the vertical ring. $G_{xy}$ depends either on the origin row $x$ or on the destination column $y$.

Every $(x, y)$ node having

$$(z-1)\frac{n}{Z} \leq x < z\frac{n}{Z} \tag{3.16}$$

(with $z = 1, 2, \ldots, Z$) uses the same frequency set to communicate with the nodes belonging to the $y$ column. In this case the network needs $n^2$ frequencies, $n^2/Z$ codewords, $2n$ rings, but $Zn^2$ filters. The corresponding node architecture is shown in Figure 3.14.

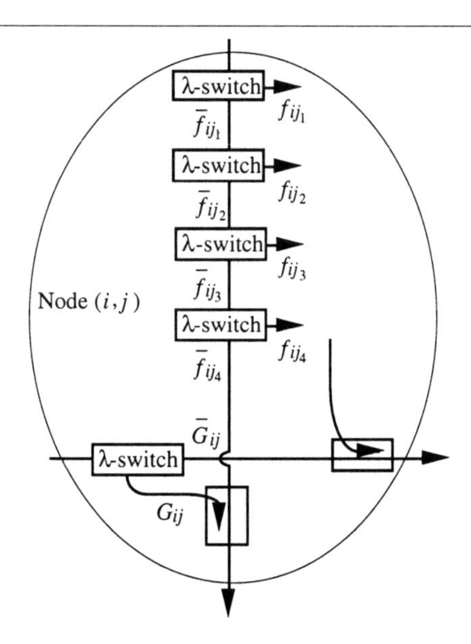


**Figure 3.14** Node architecture using frequency diversity.

### 3.2.5 Bidirectional MSNs

An evolution of MSN is the so-called bidirectional Manhattan street network (BMSN), that is, the previous MSN where all the links are bidirectional [45–47]. In it, the node positions and the grid topology remains the same but, for every previously considered connection between adjacent nodes, a reverse channel exists. In Figure 3.15, the considered topology is represented in the case of a $n \times n$ network, with $n = 4$. Every node, indicated by a circle, is placed at each cross of the grid and is directly linked with four adjacent nodes, through bidirectional connections. BMSN is therefore based on a regular mesh topology where every node is characterized by a connection degree equal to four, since there are four input and four output links per node.

Schematically, the node structure is composed of two fundamental elements: the receiving section Rx and the switching section, as shown in Figure 3.16. The access control and the extraction of packets are realized in the same way as in MSN. At the beginning of a new time slot, all the inputs are simultaneously read; the synchronous operation of nodes on each link does not require a distribution system for a centralized clock signal [48]. Therefore, the clocks can be slightly different in the different nodes and the packet alignment can be realized through suitable delay lines. After the incoming packets (having a fixed length) have entered, their addresses are examined to identify the presence of data destined to the considered node. They are then extracted and sent toward the local output. If one or some slots are empty, they can be used for sending locally generated packets and waiting for transmission in a FIFO queue. After that, all the packets that have to continue their propagation, together with the inserted ones, are transferred in the switching section, where the adopted routing scheme is applied.

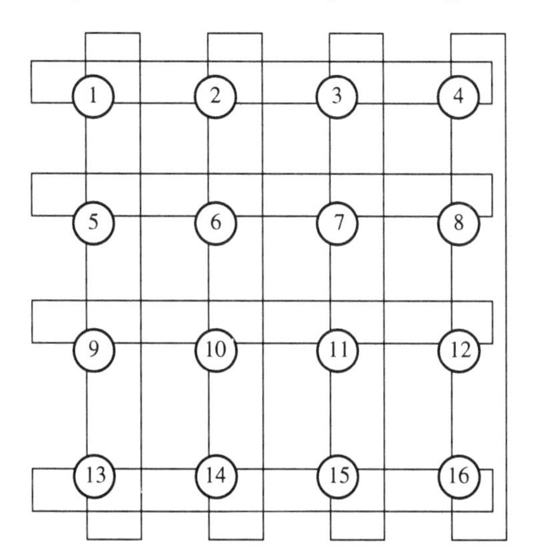


**Figure 3.15** An example of a BMSN.

In some cases, as in a MSN, the internal architecture of the switch does not contain any memory element [45]. In fact, the basic node architecture can avoid the use of buffers for any intermediate storage of packets, because this network has been specifically conceived for photonic communication. As previously discussed, the present state of the art of optic technology, in practice, sometimes makes the adoption of photonic buffering difficult in all-optical networks. For this reason, the proposed switch structure and access scheme, as well as the routing rules that will be shown below, permit us to avoid the need of memory, using the network links connecting adjacent nodes as buffers.

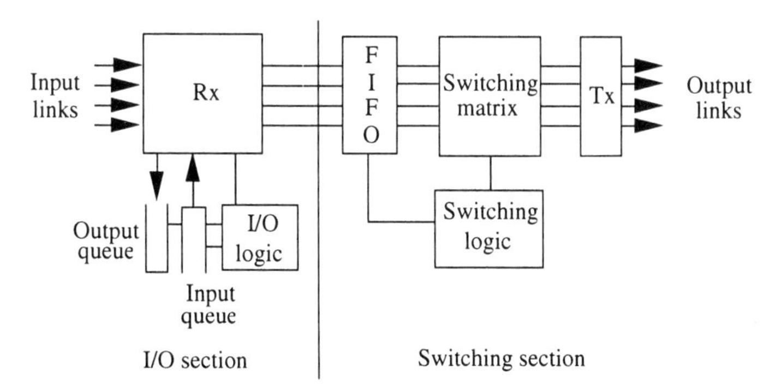


**Figure 3.16** Switching node architecture for BMSN (I/O = input/output; Tx = transmitter; Rx = receiver; FIFO = first-in first-out buffer).

The control logic, that manages the node functions, can select one of the 24 possible configurations of the 4×4 switching module (instead of the two configurations that the nodes in MSN can manage), according to the routing strategies chosen for the network operation.

The utilization of bidirectional links improves the performance of MSN topologies, increasing the connectivity of the network and its flexibility. This is evident considering that the number of alternatives, available for routing the packets at the node, is greater than the one offered by MSN stations, maintaining the complexity of the hardware architecture reasonable. Moreover, the absence of intermediate buffering greatly reduces the necessity of signal processing in the network.

Finally, with respect to the MSN, the BMSN always guarantees that the input and output degrees are equal, also in case of link failures. This means that network reconfiguration and link (or node) recovery become simpler and require minimum local logic functions.

A discussion on the all-optical implementation of BMSN, similar to that reported in Section 3.2.4, can be found in [49].

### 3.2.6 Characteristic Parameters of a BMSN

In this section some parameters useful for characterizing BMSNs are defined and utilized. The network diameter of a square $n \times n$ grid, with $n = \sqrt{N}$, is equal to [11]

$$D = \begin{cases} \sqrt{N} & \text{for } n \text{ even} \\ \sqrt{N} - 1 & \text{for } n \text{ odd} \end{cases} \tag{3.17}$$

On the other hand, the average distance turns out to be

$$\tau = \frac{1}{N-1} \sum_{h=1}^{D} h \cdot S(h) \tag{3.18}$$

where $S(h)$ is the number of nodes $h$ hops far from the considered source. In Table 3.2, some expressions of $S(h)$ are reported [11,50] where, analogously to Table 3.1, by $h$ we have indicated the minimum number of hops needed to reach the destination from the source.

Substituting the proper value of $S(h)$ in (3.18), we have

$$\tau = \begin{cases} \dfrac{\sqrt{N}}{2} & \text{for } n \text{ odd} \\ \dfrac{N^{3/2}}{2(N-1)} & \text{for } n \text{ even} \end{cases} \tag{3.19}$$

**Table 3.2**
Number of Nodes $S(h)$ Having Distance $h$ from a Given Source in a $n \times n$ BMSN, with $n=\sqrt{N}$

| | $h$ | $S(h)$ |
|---|---|---|
| $n$ even | $1 \le h < \sqrt{N}/2$ | $4h$ |
| | $h = \sqrt{N}/2$ | $4h - 2$ |
| | $\sqrt{N}/2 < h < \sqrt{N}$ | $4(\sqrt{N} - h)$ |
| | $h = \sqrt{N}$ | 1 |
| $n$ odd | $1 \le h \le (\sqrt{N} - 1)/2$ | $4h$ |
| | $(\sqrt{N} - 1)/2 < h \le \sqrt{N} - 1$ | $4(\sqrt{N} - h)$ |

A sort of normalized average distance can be defined, depending on the degree of connectivity of the nodes that equals four in BMSN, and that turns out to be

$$\Delta = p\tau = \begin{cases} 2\sqrt{N} & \text{for } n \text{ odd} \\ \dfrac{2N^{3/2}}{N-1} & \text{for } n \text{ even} \end{cases} \tag{3.20}$$

Assuming the hypothesis of uniform traffic distribution among nodes and uniform load on the links, the average traffic density $\mu$ for each connection is defined as the ratio of total traffic $L_t$ present in the network and the number of available wavelengths $W$; where, if the BMSN is realized using dedicated channels, we have $W = 4N$. Under the above-mentioned hypotheses, $L_t = \tau N(N - 1)\tilde{L}$, where $\tilde{L}$ is the average traffic exchanged between any pair of nodes. Therefore, we have

$$\mu = \frac{L_t}{W} = \frac{(N-1)\tau\tilde{L}}{p} = \begin{cases} \dfrac{\tilde{L}\left(N^{3/2} - \sqrt{N}\right)}{8} & \text{for } n \text{ odd} \\ \dfrac{\tilde{L}N^{3/2}}{8} & \text{for } n \text{ even} \end{cases} \tag{3.21}$$

On the other hand, the average traffic supported by a node, but not addressed to it, results to be $T = p\mu - (N - 1)\tilde{L}$, where $p\mu$ is the load received by the node on average and $(N - 1)\tilde{L}$ is the average traffic destined to the node itself. In general, for square BMSN,

$$T = \begin{cases} \left( \dfrac{N^{3/2}}{2} - N - \dfrac{\sqrt{N}}{2} + 1 \right) \tilde{L} & \text{for } n \text{ odd} \\ \left( \dfrac{N^{3/2}}{2} - N + 1 \right) \tilde{L} & \text{for } n \text{ even} \end{cases} \tag{3.22}$$

Finally, $r$ is the ratio of the average traffic destined to the node and the overall incoming load. This parameter is used to quantify the amount of resources spent by the nodes for routing packets. It is given by

$$r = \frac{(N-1)\tilde{L}}{p\mu} = \frac{1}{\tau} = \begin{cases} \dfrac{2}{\sqrt{N}} & \text{for } n \text{ odd} \\ \dfrac{2(N-1)}{N^{3/2}} & \text{for } n \text{ even} \end{cases} \tag{3.23}$$

### 3.2.7 Bidirectional Manhattan Topology with Uplinks

We have seen that in a BMSN the average number of hops needed to reach the destination from a given source increases with the square root of the number of nodes. This result is not completely bad, if compared with that offered by linear architectures (rings, buses, etc.), where the dependency goes with the number of nodes. Nevertheless, in this case too, the greater the network size the longer the end-to-end paths. Consequently, the throughput is affected by the enlarging of the grid dimension as well.

In order to reduce this effect, to improve the performance of BMSN, the insertion of additional links, which are called *uplinks*, has been proposed [50]. They are placed, out-of-plane, to connect not-adjacent nodes as explained, for example, in Figure 3.17. Every uplink is assigned to a pair of nodes randomly chosen.

Different solutions can be adopted:

1. One uplink originating and ending at each node;
2. The number of uplinks is smaller than the number of nodes;
3. Each node has $q$ outgoing and incoming uplinks, where $q$ is a positive integer.

The proposed technique has been tested assuming the suboptimal routing algorithm here summarized: an uplink is taken from a node if the other end of the uplink is closer to destination than any of the node neighbors. If more than one uplink per node is available, the one that ends the closest to destination is chosen. Otherwise, one of the grid links that reduces the distance is taken. As in previous analysis, the distances are considered in terms of number of hops needed to reach destination. If, in the BMSN grid, the considered distance is equal to $h$, this method ensures that the total number of hops will not be greater than $h$ [50]. This routing algorithm is not optimal but is simple and effective, since every node takes its routing decision on the basis of the destination coordinates, referring itself only to the other nodes to which it is directly connected.

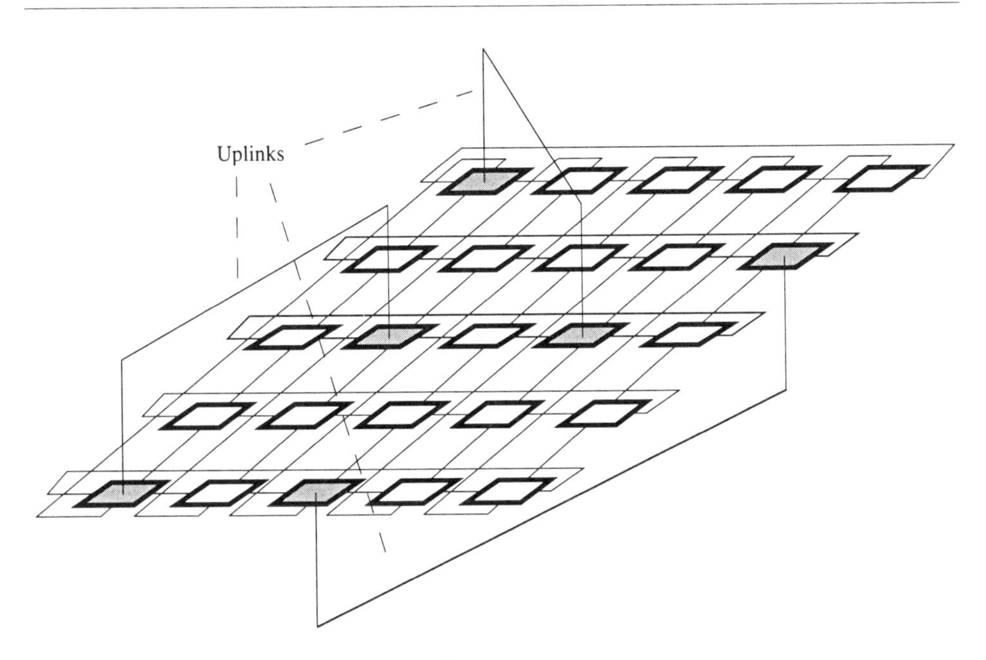


**Figure 3.17** Example of a BMSN with uplinks.

In particular, assuming the above-mentioned solution (3) applied in a $n \times n$ BMSN, with $N = n^2$, it can be shown that the expected number of hops can be approximated as

$$\tilde{h}_a \approx \sqrt[3]{\frac{N}{q}} \tag{3.24}$$

In general, the number of hops decreases as the number of uplinks per node increases, but as the cube root of $N$, instead of the square root of it, as happens in BMSNs without uplinks. Therefore, for larger values of $N$, the presence of uplinks produces significant improvement in the network throughput.

If $q = 1$, that is, one uplink per node, the degree of connectivity of every node becomes equal to 5 and the total number of links present in the network is augmented of 25% only. In this case, the throughput can be incremented by an order of magnitude for $N \approx 10^8$ [50].

### 3.2.8 Routing in BMSNs

Similarly to MSNs, many routing strategies have been proposed also for BMSNs, which exploit the availability of bidirectional links. In general they belong to the family of distributed routing algorithms, which do not require the complete knowledge of the network configuration, but only need information relative to the coordinates of destination.

Certainly, this is an interesting aspect, in terms of simplification of the algorithm implementation, above all in presence of high number of users. On the other hand, as will be shown in this section, this limited knowledge makes the optimization of the network operation impossible and this fact reduces the performance of the networks.

In general, many proposed algorithms have been designed for application in bufferless BMSNs [45–47,51,52]; they often imply the adoption of deflection techniques, particularly effective in this case, thanks to the increased connectivity degree of the BMSN nodes.

A brief discussion of some routing techniques is reported here.

#### 3.2.8.1 *Purely Random Routing*

Purely random routing (PRR) is the simplest and easiest strategy to implement compared to all the others presented in this section. In fact, the switching fabric selects one of its 24 possible configurations randomly at each slot. In this way, no information is needed to perform routing procedure and the global performance is not influenced by the load [45,53].

Although PRR is not efficient, it has the advantage of eliminating the necessity of a control unit at the node. For this reason, it makes the routing process very fast and suitable for an easy implementation in all-optical technique [54,55]. A wide analytical study of PRR behavior is reported in [56].

#### 3.2.8.2 *Hierarchical Random Routing*

Another strategy that does not require any knowledge of the network topology is the so-called hierarchical random routing (HRR) [53]. It considers the BMSNs composed by two classes of orthogonal bidirectional rings: the high-level (H) rings and the low-level (L) rings. Every L-ring is labeled by a number, that represents its address, known by all the nodes of the networks. When a node receives a packet, it compares its address with the one stored in the header of the packet itself, relative to the L-ring of the final destination. If they are equal, the packet will be preferably emitted along the L-ring, otherwise it will be preferably sent along the H-ring. The output direction along the ring is chosen randomly, but the injection of the packet in the same link used to enter the node is not allowed. In presence of conflicts among packets having the same preferred path, the configuration of the switching fabric is chosen to minimize the number of deflected packets. The packet is finally extracted from the network when it reaches its destination.

It is possible to evaluate the preferred path length $L_{HRR}$, that is, the average length of the path that packets follow if they are always routed by HRR according to their preferred outputs. It has to be noted that, because of packet collisions, the path length is in general longer than $L_{HRR}$. In a $n \times n$ BMSN it equals [45]

$$L_{HRR} = \frac{n^2}{n+1} \tag{3.25}$$

whereas the throughput limit is

$$U_{\mathrm{HRR}} = \frac{4n^2}{L_{\mathrm{HRR}}} = 4(n+1) \tag{3.26}$$

### 3.2.8.3 *Oriented Random Routing and Oriented Routing*

The HRR technique exhibits a drawback that is not negligible. Sometimes, a conflict can deflect a packet when it has just reached its L-ring. In this case the packet itself is forced to go through the entire H-ring once more, before it can reach again the desired ring. Moreover, because of the previously discussed reasons, this trip could be even longer than the ring length. To partially overcome this problem, the oriented random routing (ORR) algorithm has been proposed [45]. In this case, besides the L-ring address, the H-ring address is used to determine the position of destination. Therefore, the header of the packet contains four binary flags (H1, H2, L1, L2). When set, each of them represents the preferred output.

When the packet is injected in the network, all the flags are set on, so that every output can be chosen indifferently. The 4 bits will be varied during the propagation in the grid. When a node receives the packet, it compares the two addresses with its own addresses. If, for example, the H-addresses coincide, the L1 and L2 flags are reset, whereas the others remain unchanged. A dual situation occurs when the L-addresses are equal. In this case too, the switching configuration is chosen to minimize the number of possible deflections. After this, if the packet has been routed according to the preferred direction, for example, H1, the H2 flag is reset, so that the successive inversion of the path is not possible, whereas if the packet has been redirected, for example, along the L1 direction, the flag L2 is set, to send the packet itself toward the current node at the subsequent hop. In this way, it is possible to reach the same position again, to repeat the previous attempt, minimizing the length of the deflected paths. The preferred path length $L_{\mathrm{ORR}}$ and the throughput upper limit $U_{\mathrm{ORR}}$ obtainable by ORR are equal to the correspondent values achieved by HRR. In fact both algorithms force the same paths in absence of conflicts.

Finally, for the sake of completeness, we can say that it is also possible to consider some variations with respect to the basic ORR algorithm. For example, the preference flags can be set before inserting the packet in the grid, so that the first node can immediately know the preferred direction. In this case, we have the oriented routing (OR) algorithm. If no deflection occurs, OR routes the packets along the shortest path that connect every node pair. Therefore, preferred path length and throughput upper limit are the same of the shortest path routing (SPR), which is presented in the following section.

### 3.2.8.4 *Shortest Path Routing*

By the SPR technique, the source node selects the shortest path which connects itself with destination. This method needs the availability of a complete data base relative to the grid topology. For this reason, SPR is heavy to implement, especially in networks having a large number of nodes. The switching configuration is randomly selected between those that maximize the number of packets satisfied in their preferences. In this sense, a priority can be offered to packets having only one preference. Obviously, in order to maintain the path short, avoiding deflections, suitable buffers should be present at the nodes.

The preferred path length $L_{SPR}$ and the throughput upper limit $U_{SPR}$ are given by

$$L_{SPR} = \begin{cases} \dfrac{n^3}{2(n^2-1)} & \text{for } n \text{ even} \\ \dfrac{n}{2} & \text{for } n \text{ odd} \end{cases} \tag{3.27}$$

$$U_{SPR} \approx 8n \tag{3.28}$$

Further analysis on SPR will be developed in the following section. For the time being, to compare the performance of the routing algorithms presented here, we have reported in Table 3.3 the upper limit throughputs offered by every technique, when applied in three BMSNs of different size [45].

**Table 3.3**
Upper Limit Throughputs in Different $n \times n$ BMSNs

| | *Network Size* | | |
|---|---|---|---|
| | 5×5 | 10×10 | 20×20 |
| Purely random routing | 3.2 | 2.4 | 1.9 |
| Hierarchical random routing | 24.0 | 44.0 | 84.0 |
| Oriented random routing | 24.0 | 44.0 | 84.0 |
| Oriented routing | 40.0 | 79.2 | 159.6 |
| Shortest path routing | 40.0 | 79.2 | 159.6 |

### 3.2.8.5 *Minimum Distance Algorithm*

SPR implies some processing capability at the nodes. In practice, they have to be equipped with a routing table that indicates, for every destination $i$, the preferred output $j$ that implies the path with the minimum length $d(i, j)$, measured in number of hops [46,51].

Let us examine a possible SPR algorithm based on the use of the routing table.

Let $A$ be the reduction of distance to destination that a packet gets at each routing step ($A$ = 1, 0, −1), whose average value is indicated by E[$A$], whereas $L$ is the average length of the source-destination path. If the packets are always routed according to their preference we have $L = L_0$, being $L_0$ the shortest path length. At the equilibrium, the average global distance reduction $4N \cdot \mathrm{E}[A]$, with $N = n^2$, should be equal to the average increment due to new incoming packets, that is, $SL_0 = 4N \cdot \mathrm{E}[A]$, where $S$ is the network throughput. From this observation follows that the best strategy maximizes E[$A$]. Unfortunately, the identification of an optimum policy is not possible in this sense [46]. In fact, the reciprocal dependence between the considered parameters is very difficult to describe analytically. For this reason, a possible nonoptimal alternative method is based on the maximization, step by step, of the following cumulative quantity

$$\bar{A} = \sum_{k=1}^{4} A_k \tag{3.29}$$

where by $k$ we have numbered the four packets handled by a node. This is equivalent to minimizing the cumulative distance to destination

$$\bar{d} = \sum_{k=1}^{4} d(i_k, j_k(r)) \tag{3.30}$$

being $i_k$ and $j_k(r)$ the destination and the preferred output of the $k$th packet respectively, having labeled by $r$ one of the 24 possible switching configurations of the considered node. If several configurations satisfy the same condition, one of them is randomly selected. This policy is called minimum distance (MD) [46].

A possible increment of E[$A$] can be obtained through the following method. Assuming $n$ even, we have that at any step $A$ is either 1 or −1, depending on the fact that the packet is deflected or not. The global distance reduction in one step, that on average is equal to the throughput $S$, can also be given by

$$\sum_{i=1}^{4N} A_i = 4N - 2\delta \tag{3.31}$$

labeling by $i$ the packet coming from one of the $4N$ input links of the network and by $\delta$ the number of deflected packets. It is evident that the maximization of E[$A$] is obtained by the minimization of E[$\delta$]. In fact, an optimal routing policy should imply the minimum number of deflections.

#### 3.2.8.6 *Minimum Weights Algorithm*

Another method has been proposed in [46], which experimentally behaves very well at a local level. It minimizes the following weighted cumulative distance:

$$\bar{d}_w = \sum_{k=1}^{4} d\left(i_k, j_k(r)\right)\tilde{w}\left(i_k\right) \tag{3.32}$$

where $\tilde{w}\left(i_k\right) = n - \min d\left(i_k, j\right)$, and the minimum has to be found with respect to $j$. $\tilde{w}\left(i_k\right)$ is a weighting factor which linearly decreases as the distance between the $k$th packet and its destination is reduced. In case of conflicts this algorithm, called minimum weights (MW), gives the highest priority to packets closest to destination, whereas the others are deflected. Once again, when more than one switching configuration can produce the same results, one of them is randomly selected.

#### *3.2.8.7 The Preference Algorithms*

MD and MW belong to the family of distance algorithms. At every hop, they base the routing decisions on the packet position with respect to destination. The knowledge of the network status, that is, a complete information exchange among nodes, can improve the performance of these routing methods. On the other hand, this situation cannot be realized in practice, since it implies a too large amount of bandwidth devoted to signaling, and too heavy computing processes at the nodes. This means that, in general, only some information relative to adjacent nodes can be assumed to be available [52,57].

In a $n \times n$ BMSN two nodes can be placed in five different relative positions. Consequently, as regards the routing preferences, we can distinguish the following situations:

1. For $n$ even, the nodes can be antipodal: all of the four outputs are equivalent;
2. The nodes are placed on two different rows or columns: there are two outputs yielding minimum distance;

2.e For $n$ even, the nodes are placed on two different rows or columns and row distance or column distance equals *n/2*: there are three outputs yielding minimum distance;

3 The nodes are placed on the same row or column: there is only one output yielding minimum distance;

3.e For $n$ even, the nodes are placed on the same row or column and row distance or column distance equals *n/2*: there are two outputs yielding minimum distance.

Since there are few possible relative positions, it is possible to assign a preference value to each condition using only 3 bits, through the 0-7 preference algorithm (0-7PR) [51]. The routing is so conceived to maximize the following cumulative preference:

$$\bar{p} = \sum_{k} p\left(i_k, j_k(r)\right) \tag{3.33}$$

having denoted by $p(i_k, j_k(r))$ the above-mentioned preference, whereas the other parameters are the same as before. To explain this technique better, in Table 3.4 we have reported the correspondences between preference values and the above-listed cases.

**Table 3.4**
Preference Assignment for the 0-7 Preference Algorithm

| *Preference* | *Case* |
|---|---|
| 7 | The desired output of case 3 |
| 4 | The best output of case 2, the two equivalent outputs of case 3.e |
| 3 | The second choice of case 2, the two equivalent outputs of case 2.e |
| 2 | The third choice of case 2.e, all the outputs of case 1 |
| 0 | All the outputs leading to nonminimum paths in any case |

Routing schemes based on this method are called preference algorithms [47]. Other algorithms can be organized, using only 1 or 2 bits. They can obviously represent only a part of the cases considered in Table 3.4. In Figure 3.18, the network efficiency is plotted versus its size $n$ (with $N = n^2$), relatively to three routing algorithms considered in this section.

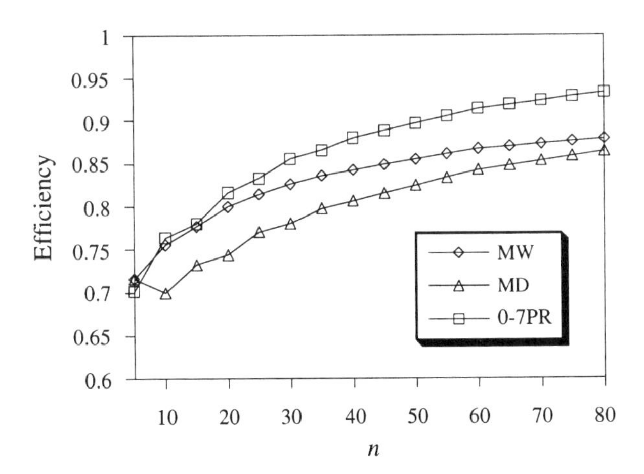


**Figure 3.18** Efficiency versus network size for some distance and preference algorithms.

The efficiency is the network throughput normalized with respect to its maximum possible value. In particular, the maximum throughput $U$ relative to $n \times n$ BMSN, that is, the network capacity, $C$ is given by [51]

$$\left.\begin{aligned} C &= \frac{4N}{L_0} \\ L_0 &\cong n/2 \end{aligned}\right\} \Rightarrow \quad C \cong 8n \tag{3.34}$$

where $L_0$ coincides in practice with the average minimum number of hops between every source-destination pair. Equation (3.34) shows that an almost linear dependence exists between the maximum throughput and the network size.

It can be observed that the efficiency is not constant, because the greater the network size the lower the probability of conflicts among packets having the same destination. The 0-7PR seems to be the best routing scheme, whereas the MW offers good results especially in small grids, since it works particularly well when the packets are close to their destinations.

Similar analysis considering BMSN loaded by nonuniform traffic can be found in [58].

#### 3.2.8.8 Routing in Time-Varying Networks

Deflection algorithms are able to bypass failed links or nodes, finding alternative paths. The optimization of routing strategy can be obtained in this case too, if suitable methods for evaluating all the new paths are adopted. This aspect is briefly examined here.

As stated above, distance algorithms require one to know the distance $d(i, j)$ to route packets through the grid. This is possible when the topology is fixed and known. On the contrary, if the structure of the network is time-dependent, every node has to evaluate and update $d(i, j)$ for every $i$ and $j$, continuously.

The backward-learning technique [14], which avoids the risks of unfairness in the access or station lock-out [59,60], offers the possibility of monitoring the network characteristics, without requiring information exchange among nodes. In fact, it estimates the distance of the $i$th node on the basis of the path lengths experienced by packets, emitted by the same node, as they come. In stationary conditions, $d(i, j)$ equals the smaller observed value but, when a grid modification occurs, an estimate updating process is activated. As an example, the estimate updating algorithm is described here [47].

Focusing our attention on a given node of the network, let $P_j$ be a packet coming from input $j$, emitted by source $V_j$. Moreover, let $D_j$ be the length of the path utilized by $P_j$, whereas $d' = d'(V_j, j)$ and $s = s(V_j, j)$ are the estimate of $D_j$ done by the considered node and a support estimate respectively.

Whenever we have that $D_j < d'$, $d'$ and $s$ are updated by $D_j$. Instead, if $D_j > d'$ and $D_j < s$, only $s$ is replaced by $D_j$. After an observation period $R$ (estimate reset period), measured by $c(V_j)$ counter, $d'$ is replaced by $s$ and the latter is set to $\infty$. In steady-state conditions, the longer $R$, the closer $s$ gets to the real distance.

It has been demonstrated [47] that, using this method, the throughput reaches the convergence rapidly, even though a good performance is not guaranteed in all the operative conditions. For example, in the presence of multiple link failures, which sometimes can disconnect some significant subsets of the network, the packets could wander all over the grid without reaching their destinations any more. In this sense, the insertion of a mechanism devoted to extract packets too old is suitable and recommended.

Other techniques can be adopted for managing routing procedure in time-varying networks [61,62]. In any case, the implementation of an integrated operation of deflection algorithms and estimation methods seems to be quite simple and effective.

***Some keypoints.*** The Manhattan street network is a regular, two-connected, multihop network that is considered to be one of the most promising architectures for future all-optical networks. In this section special attention is devoted to its topological features, also considering its performance in terms of scalability and robustness. Its regular structure makes the application of distributed routing algorithm particularly simple and, therefore, compatible with an all-optical node implementation.
Bidirectional Manhattan street networks are the bidirectional version of the considered class of nets. Their nodes are characterized by a higher connection degree that, at the expense of a moderate increase of the hardware complexity, results in general improvement in all the quality parameters of the network.

## 3.3 SHUFFLE NETWORKS

### 3.3.1 The Perfect Shuffle Topology

Shuffle network (SN) is a well-known multihop virtual topology, widely proposed for use in WDMA optical networks [63,64]. In its basic configuration, a $(p, k)$ shufflenet consists of $N = kp^k$ nodes ($k = 1, 2, \ldots; p = 1, 2, \ldots$) that are arranged in $k$ columns (numbered left to right from 1 to $k$), each of them composed by $p^k$ nodes. As an example, in Figure 3.19 we have reported the connectivity graph of (2, 2) SN. The interconnection pattern between columns is based on the $p$-perfect shuffle structure [65,66], with the left side of the first column connected with the right side of the last one. Every node of a $(p, k)$ SN is connected with $p$ nodes belonging to the subsequent column by unidirectional links, so that every column is linked with the successive one through $p^{k+1}$ channels and the last column is then connected back to the first one, as if the entire graph were wrapped around a cylinder. Therefore, every node has $p$ incoming and $p$ outgoing links and the total number of dedicated links is $W = kp^{k+1}$. In this case, we say that SN is regular with degree $p$ [67].

The main reason that justifies the introduction of this kind of architecture is the possibility of simplifying the structure of every node, so lessening the need for both transmitter and receiver tunability: each node is provided with some fixed-tuned optical transmitters and an equal number of fixed-tuned optical receivers. Here we assume that each transmitter in the network is tuned to a different wavelength [63], whereas different channel sharing techniques will be examined in the following sections. Therefore, WDM channel assignment takes the form of recirculating perfect shuffle [63] interconnections, where no $p$ transmitter writes onto the same channel and no $p$ receiver listens to the same channel, so that every node can insert in and extract messages from the network without any tuning process.

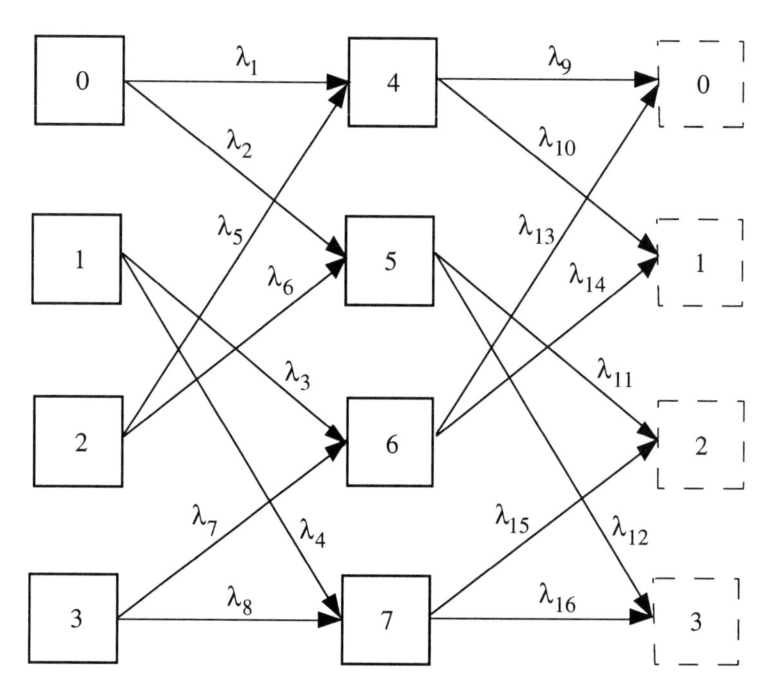


**Figure 3.19** Connectivity graph of the perfect shuffle.

The nodes are also called network interface units (NIUs), since they represent the edge of the network with respect to external users. Some solutions have been studied and proposed for an all-optical implementation of NIUs; a detailed analysis of a possible all-optical SN node architecture can be found in [68,69].

Numbering the nodes in a column from 0 to $p^k - 1$, the $i$th node is directly connected with NIUs $j, j+1, \ldots, j+p-1$, where $j = p(i \bmod p^{k-1})$ [64]. It is easy to observe that every pair of nodes can be connected through different paths. Similarly to MSNs, this feature can be utilized to bypass failed nodes or links or to avoid heavy loaded trunks. On the other hand, it has been shown that, in normal conditions, a SN may perform better than the corresponding MSN [20] or ring network [43].

The maximum distance between two nodes, that is, the network diameter, equals $2k - 1$. In fact, $k$ hops are needed to come back to the starting column and the next $k - 1$ hops are sufficient to reach all the nodes not covered during the first trip. The expected number of hops between two randomly selected nodes is given by [64]

$$\langle h \rangle = \frac{kp^k(p-1)(3k-1) - 2k\left(p^k - 1\right)}{2(p-1)\left(kp^k - 1\right)} \tag{3.35}$$

taking into account that the distribution of nodes, with respect to a given source, is the one reported in Table 3.5, where $g_h$ represents the number of nodes that are $h$ hops away from any considered node. The isotropy of SN guarantees that Table 3.5 and (3.35) hold for every starting node.

**Table 3.5**
Number of Nodes that are $h$ Hops Away from Any Given Source Node for a $(p, k)$ Shufflenet

| $h$ | $g_h$ |
|---|---|
| 1 | $p$ |
| 2 | $p^2$ |
| ... | ... |
| $k-1$ | $p^{k-1}$ |
| $k$ | $p^k - 1$ |
| $k+1$ | $p^k - p$ |
| $k+2$ | $p^k - p^2$ |
| ... | ... |
| $2k-1$ | $p^k - p^{k-1}$ |

From (3.35) it follows that the channel efficiency of a $(p, k)$ SN, in presence of a balanced traffic load on all the channels, is given by

$$\eta = \frac{1}{\langle h \rangle} = \frac{2(p-1)\left(kp^k - 1\right)}{kp^k(p-1)(3k-1) - 2k\left(p^k - 1\right)} \tag{3.36}$$

with an asymptotic value, for large $p$, equal to

$$\lim_{p \to \infty} \eta = \frac{2}{3k-1} \tag{3.37}$$

which underlines that the efficiency decreases with the inverse of the number of columns.

The same effect is shown in Figure 3.20, where the efficiency $\eta$ is plotted, varying $p$ from 1 to 10 and considering $k = 2, 3, 4$. The packet propagation is affected by the increase in the number of columns and this causes a reduction of the efficiency.

On the contrary, for constant $k$, $\eta$ slightly depends on $p$. This means that, if we want to increase the number of users maintaining a given efficiency, the nodes have to be equipped with many transmitter-receiver pairs, even though this is paid with higher costs.

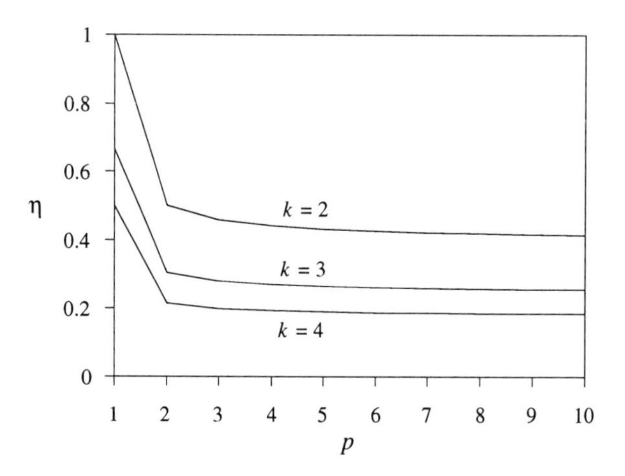


**Figure 3.20** Efficiency versus the number of transceivers per user for different number of columns.

**Table 3.6**
Dimension and Capacity of Shufflenets with Dedicated Channels, in Some Different Network Configurations

| $k$ | $p$ | $N$ | $W$ | $\langle h \rangle$ | $C'$ (Gbps) | $C_u$ (Gbps) |
|---|---|---|---|---|---|---|
| 2 | 2 | 8 | 16 | 2.00 | 8.00 | 1.00 |
| 2 | 3 | 18 | 54 | 2.18 | 24.81 | 1.38 |
| 2 | 4 | 32 | 128 | 2.26 | 56.69 | 1.77 |
| 2 | 5 | 50 | 250 | 2.31 | 108.41 | 2.17 |
| 3 | 2 | 24 | 48 | 3.26 | 14.72 | 0.61 |
| 3 | 3 | 81 | 243 | 3.56 | 68.21 | 0.84 |
| 3 | 4 | 192 | 768 | 3.69 | 208.07 | 1.08 |
| 3 | 5 | 375 | 1875 | 3.76 | 498.40 | 1.33 |
| 4 | 2 | 64 | 128 | 4.63 | 27.65 | 0.43 |
| 4 | 3 | 324 | 972 | 5.02 | 193.56 | 0.60 |
| 4 | 4 | 1024 | 4096 | 5.17 | 791.80 | 0.77 |
| 4 | 5 | 2500 | 12500 | 5.25 | 2379.82 | 0.95 |

Now, let us consider the overall network capacity of SNs using dedicated channels, that is, the network throughput. If $W$ channels operate at a fixed data rate $S$, it can be calculated as [63]

$$C = \frac{WS}{\langle h \rangle} = WS\eta \tag{3.38}$$

However, it is often useful to normalize $C$ with respect to the data rate, to show the network potentialities better. For this reason, $C' = C/S = W/\langle h \rangle = W\eta$ is also considered. Moreover, the normalized throughput per user $C_u$ can be defined as

$$C_u = C'/N = \eta p \tag{3.39}$$

in fact it depends on the number of transceivers per node and on the channel efficiency.

In Table 3.6 we have reported some results obtainable in a SN relative to the above-mentioned network parameters, assuming $p \in [2,5]$ and $k \in [2,4]$ and a 1-Gbps user transmission rate [70].

Once again we can observe that, even though the normalized capacity increases with $k$, because of the presence of a higher number of channels, the normalized throughput per user decreases as the number of columns grows.

### 3.3.2 Shufflenets with Shared Channels

In the previous section, we have considered the connectivity graph of a SN where dedicated unidirectional channels connect nodes belonging to adjacent columns. Here, instead, we are going to analyze the characteristics and performance of a similar network, but adopting a shared channel policy with single-transmitter and single-receiver per user. In this case, groups of $p$ users in each column transmit on a common channel, with a separate group of $p$ users in the next column receiving on each channel. In practice, for $i = 0, 1, \ldots, p^{k-1} - 1$, users labeled by $i, i + p^{k-1}, i + 2p^{k-1}, \ldots, i + (p-1)p^{k-1}$ in a column transmit on the same common channel that is received by users $j, j+1, j+2, \ldots, j+p-1$, where $j = p(i \bmod p^{k-1})$. It follows that there are $p^{k-1}$ channels per column, so that now we have $W = kp^{k-1}$. To explain better this new structure, a simple example of channel sharing is shown in Figure 3.21, applied to a (2, 2) SN.

It is worth observing that, with respect to the case of dedicated channels, an arrow in the graph in Figure 3.21 does not represent a separate link, but rather the possibility of connecting two different nodes in one hop. Obviously, since $p$ nodes utilize the same WDM channel, the multiple access has to be managed in a TDMA fashion. For the sake of simplicity, we can assume a fixed allocation of $1/p$ of the link capacity dedicated to every user.

For a single-transmitter and single-receiver per user SN with channel sharing among groups of $p$ users belonging to the same column, the expected number of hops between two randomly selected nodes is equal to the one obtained through (3.35), and also the channel efficiency can be calculated as indicated in (3.36). In fact, the connectivity graph does not change with respect to the dedicated channels case [70].

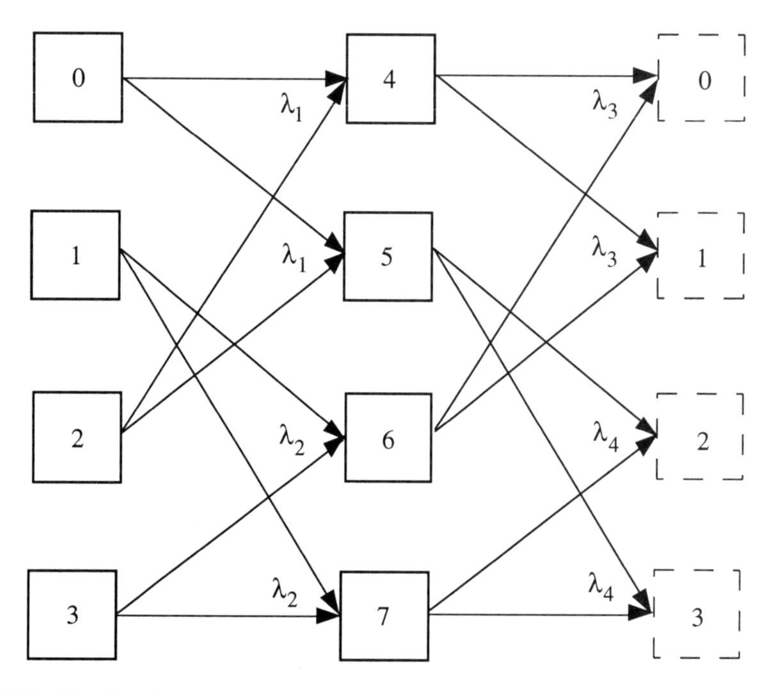


**Figure 3.21** Wavelength assignment in a (2, 2) shared channel shufflenet.

On the contrary, the network throughput and the throughput per user decrease; the former because of the reduction in the number of available links, the latter because every node can utilize only a portion of the bandwidth made available by a channel, since every link supports the multiplexed communications of $p$ users.

We have shown that, in SNs with dedicated channels, it is preferable to increase the number of users by raising $p$ instead of $k$ since the network performance, in terms of channel efficiency, weakly depends on the number of transceivers per node. Conversely, adopting channel sharing, we can observe that $W$ increases as $k$ raises. In fact, even though the channel efficiency is reduced, this is compensated by an increase in the number of links, as demonstrated in Figure 3.22. The effect of this choice on the network capacity is shown in Figure 3.23, assuming $S$ = 1 Gbps. In any case, the number of columns should be upper bounded to avoid values of the average number of hops getting too high.

In Figure 3.24, a scheme is reported that shows some possible configurations for single-transmitter and single-receiver per user shufflenets with shared-channels, varying $k$ and $p$ from 1 to 5. For $k$ = 1, SN becomes a bus, whereas for $p$ = 1 it assumes the ring connectivity. In all the other cases, SN combines the two structures, integrating the multihop feature of a ring with the shared channel feature of a broadcast bus.

Until now, we have considered nodes of shared-channel SNs provided with only one pair of transmitter and receiver devices. Nevertheless, it is also possible to use nodes equipped with multiple transceivers to increase either the network throughput or the user one. When the number of transmitters per node $T_n$ and the number of receivers per node $R_n$ are less than $p$, we maintain the channel sharing conditions. In particular, for $T_n = R_n = 1$, every channel is shared by $p$ users. On the contrary, for $T_n = R_n = p$, we have the case of dedicated channels.

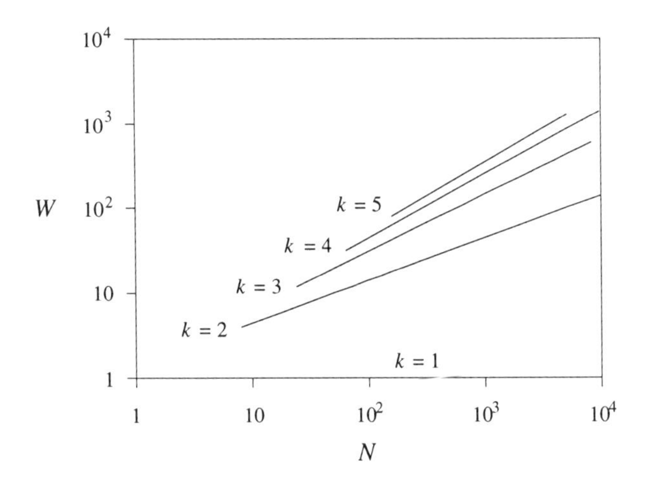


**Figure 3.22** Total number of channels in single-transmitter and single-receiver shufflenets with shared channels.

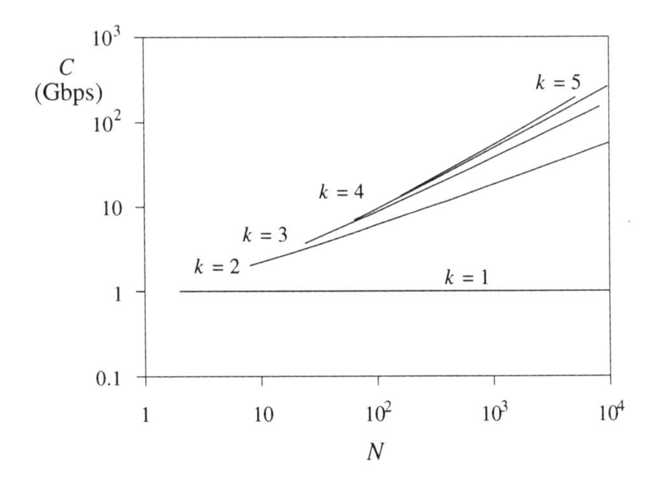


**Figure 3.23** Maximum achievable throughput in shufflenets with shared channels, assuming a user transmission rate of 1 Gbps.

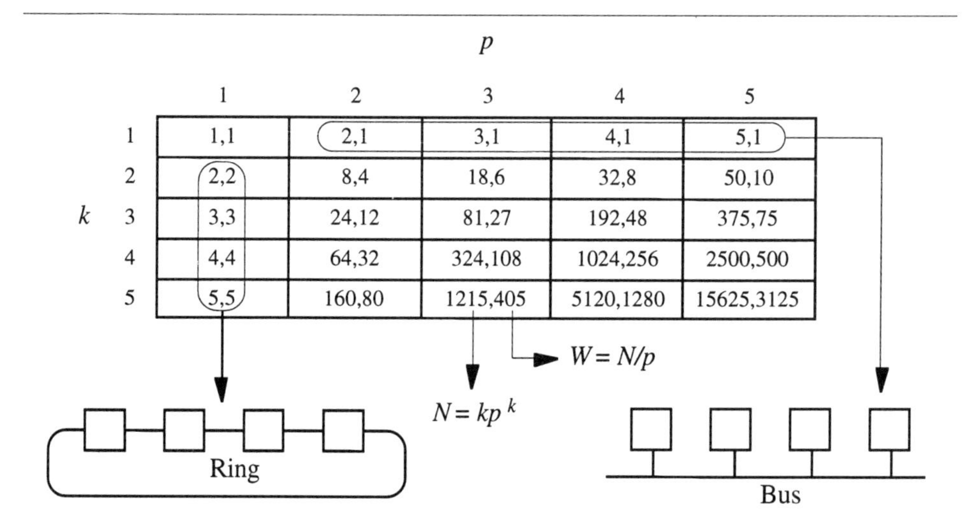


**Figure 3.24** Network configurations of single-transmitter and single-receiver per user shufflenets with shared-channels. (*Source*: [64]. © 1991 IEEE.)

It is worth observing that we can also have $T_n \neq R_n$. In fact, in Figure 3.25, we have reported two examples of a 18-user SN ($p = 3$ and $k = 2$), where for the sake of clarity, only the transmit wavelengths of nodes 0, 3, 6 and the receive wavelengths of node 9, 10, 11 are drawn. In particular, in Figure 3.25(a) $R_n$ equals $p$ whereas the nodes utilize a single transmission wavelength. The opposite configuration is represented in Figure 3.25(b). For both cases, the connectivity graph of the considered network is the same. In general, we have that $W = NT_nR_n/p$, where $T_n$ and $R_n$ can be independently set to any of the factors of $p$.

In Table 3.7, the maximum achievable network and node throughputs of a (4, 2) shufflenet, working at a 1-Gbps user transmission rate, have been listed, considering different values of $T_n$ and $R_n$ [64]. The channel efficiency depends on the connectivity graph but not on the number of transmitters and receivers per node. Moreover, the network behavior does not change if the $W/T_nR_n$ ratio remains constant.

Finally, we can observe that the results we presented above, relating to single-transmitter and single-receiver shufflenets with shared channels ($T_n = R_n = 1$), scale linearly with the product $T_nR_n$, if multiple transmitters and receivers per node are adopted [64].

The performance of dedicated-channels SNs with $T_n = R_n = p$, in terms of maximum achievable throughput per node, are compared in Figure 3.26 with the one offered by a single-channel network operating with 100% media access efficiency.

The coexistence of multiple connections on a common channel can be obtained also by time division multiplexing (TDM) of the output signals emitted by $k$ users belonging to the same row [63]. In this way, each of them, equipped with $p$ transmitters and $p$ receivers, can exploit a fraction of the available bandwidth, that equals $1/k$ in case of uniform sharing and fixed slot allocation.

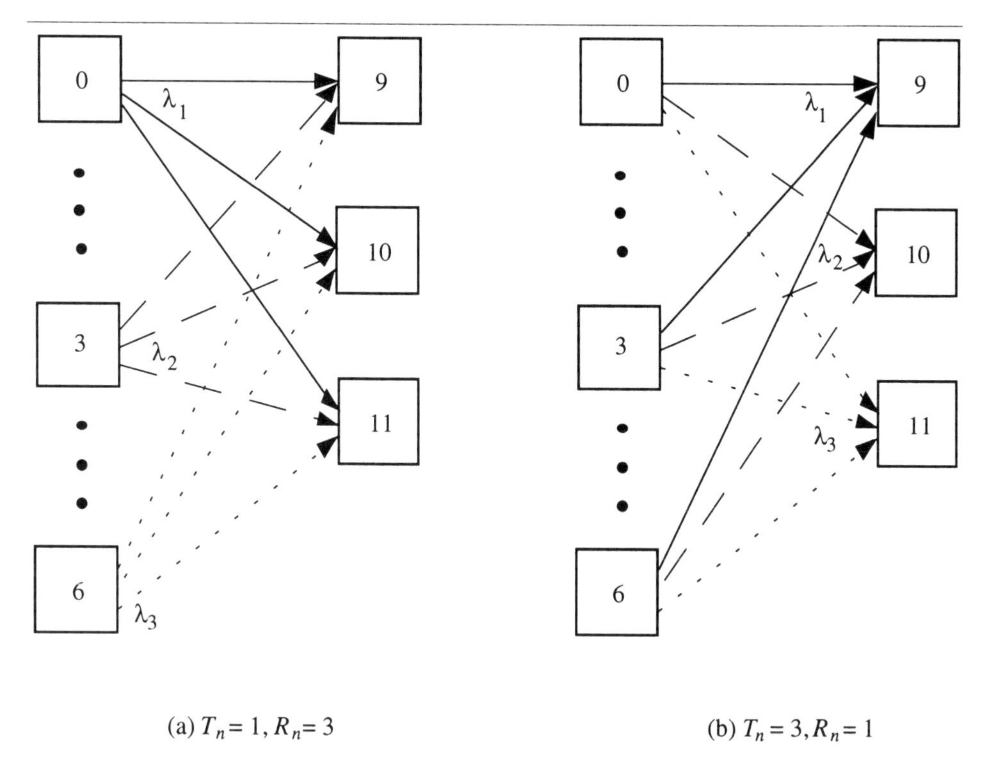


**Figure 3.25** Two examples of (a) multiple receivers and (b) multiple transmitters in a (3, 2) shared-channel shufflenet.

**Table 3.7**
Efficiency (η) and Maximum Achievable Network and Node Throughputs ($\eta WS$ and $\eta WS/N$) of a (4, 2) Shufflenet, at a 1-Gbps Transmission Rate[1]

| $T_n$ | $R_n$ | $W$ | η | $\eta WS$ (Gbps) | $\eta WS/N$ (Mbps) |
|---|---|---|---|---|---|
| 4 | 4 | 128 | 0.443 | 56.686 | 1771 |
| 4 | 2 | 64 | 0.443 | 28.343 | 886 |
| 2 | 4 | 64 | 0.443 | 28.343 | 886 |
| 2 | 2 | 32 | 0.443 | 14.171 | 443 |
| 2 | 1 | 16 | 0.443 | 7.086 | 221 |
| 1 | 2 | 16 | 0.443 | 7.086 | 221 |
| 1 | 1 | 8 | 0.443 | 3.543 | 111 |

[1] $T_n$ = transmitters per node; $R_n$ = receivers per node; $W$ = WDM channels.

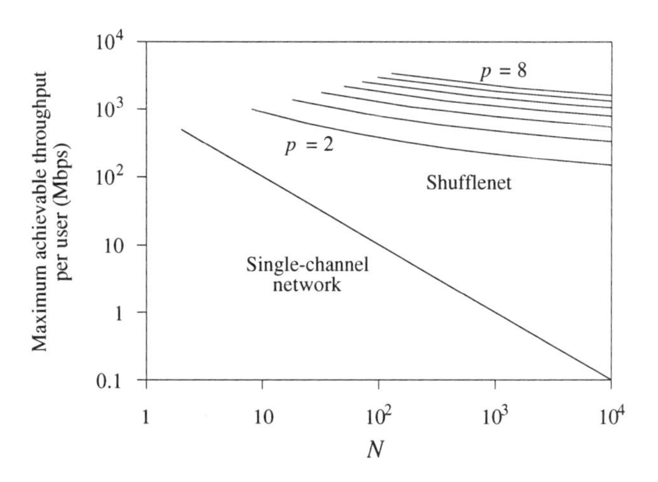


**Figure 3.26** Maximum achievable node throughput of a dedicated-channels shufflenet at a user transmission rate of 1 Gbps. (*Source*: [64]. © 1991 IEEE.)

Now, the average number of hops, assuming $p = 2$, becomes equal to [63]

$$\langle h \rangle = \frac{2 + (k-1)2^k}{2^k} \tag{3.40}$$

The latter result is lower than the one obtained through (3.35), because now $k$ users are reached in a single hop. Substituting (3.40) in (3.38) and taking into account that, in this case, the number of channels is reduced to $W = p^{k+1}$, the overall capacity for $p = 2$ turns out to be

$$C = \frac{2^{2k+1}}{2 + (k-1)2^k} S \tag{3.41}$$

In conclusion, shared channels schemes are useful when the goal is the increase of the channel efficiency, or when a given network should be implemented with a low WDM multiplexing level, that is, using a limited number of WDM channels. On the contrary, when high-performance networks have to be realized, characterized by high-speed end-to-end connections and high throughputs, the dedicated channels scheme is preferable.

### 3.3.3 Size Modifications of Shuffle Networks Based on Multistar Architecture

As shown above, the number of nodes connected by a SN is not arbitrary [3]. This means that "dummy" nodes have to be inserted to obtain a total number of nodes $N$ which equals $kp^k$, for integer values of $k$ and $p$. This fact makes the growth and expansion of SNs difficult.

A first approach to the problem is based on the observation that a $(p, k)$ SN is a subgraph of a $(p + 1, k)$ SN [43], as shown in Figure 3.27. Consequently, if we wish to raise the number of nodes by increasing the number of transmitters and receivers per user, but $k(p + 1)^k$ nodes are not needed at present, we can first deploy the nodes corresponding to the imbedded $(p, k)$ SN and switch to the target system when necessity arises [71]. However, the size of the target system has to be fixed in advance, since further network growth requires heavy hardware and software reconfigurations.

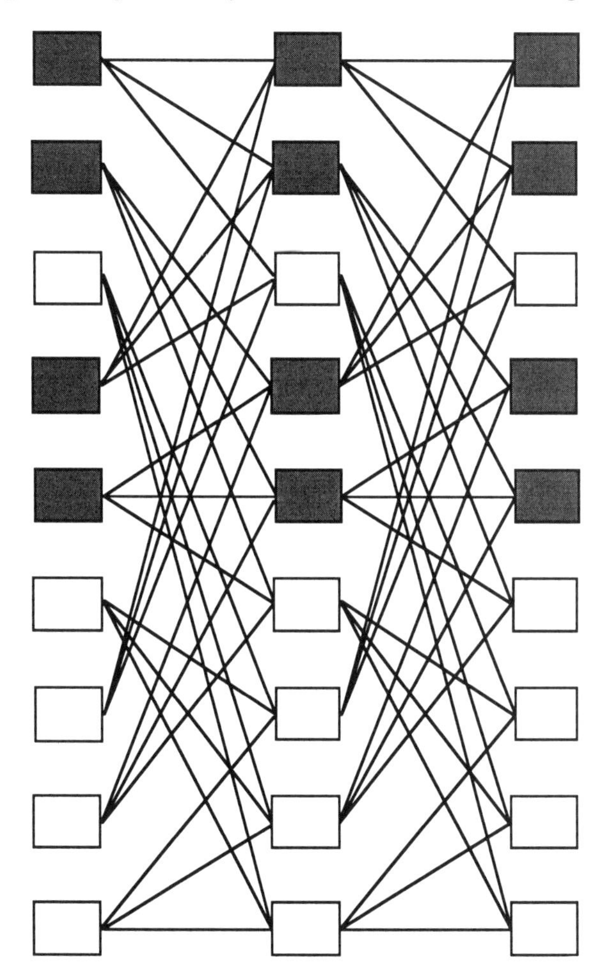

**Figure 3.27** A (2, 2) shufflenet (shaded boxes) as a subgraph of a (3, 2) shufflenet.

Alternatively, SN can be built on a backbone of multiconnected rings [72]. In this case a $(p, k)$ SN can be realized step by step, inserting groups of $k$ nodes linked by a ring one at a time. At each step, all the necessary connections are added to link the new $k$

nodes with those previously installed, maintaining the shuffle topology. In this way, we can expand a SN gradually with $p$. Every time $p$ is increased, each node has to be provided with a new pair of transmitter and receiver. This aspect can cause problems, in terms of costs and complexity, especially in case of geographically dispersed nodes.

In this section, another method of expanding the size of a SN is presented and discussed. It is based on the increment of the number of columns, so that we can focus the evolution from a $(p, k)$ to a $(p, k + 1)$ SN.

First, let us consider that a given SN can be implemented on a broadcast-and-select structure (see Chapter 2) using a single star coupler. The total number of nodes is upper bounded by the maximum number of wavelengths, that can be multiplexed and made available for dedicated channel connections [73], and by the power splitting losses. This limit introduces a constraint on the maximum size of the network. Nevertheless, in multihop networks we do not need to link a node with all the other ones directly, but only with a certain subset of them, depending on the degree of connectivity of the considered topology. Therefore, exploiting this characteristic, it is possible to organize a multihop network (a SN in this case) by connecting different smaller subnetworks. In this way, by choosing a suitable number of star couplers, asked to offer a reduced performance, a sort of integrated wavelength division multiplexing (WDM) and space division multiplexing (SDM) is adopted, which permits us to obtain either a wavelength reuse or acceptable power budget constraints. In practice, with this technique, the total number of available channels can be increased [74–76], so that the network can be enlarged according to the design necessities.

Now, let us analyze the network growth using the above-described method. Every node of the considered SN has to be connected with a first star coupler for its outgoing signals and with a second star coupler for its incoming ones. The number of available wavelengths in every star $w$ has to be at least equal to $p^2$, if $p$ is the number of inputs and outputs per node. In fact, since in a $(p, k)$ SN, a given node needs $p$ wavelengths for receiving from other $p$ nodes inside a star-subnetwork, and each of them needs $p$ distinct output channels, the minimum number of required wavelengths equals $p^2$. If the star couplers offer a larger availability of WDM channels, multiple virtual stars can be overlapped on the same physical one. Otherwise, we can assume $w = Mp^2$, where $M$ is an integer number and $1 \leq M \leq p^{k-2}$, so that the number of star couplers $n_{sc}$ is given by $n_{sc} = W/w = N/(pM)$, being $W = kp^{k+1}$ the total number of channels.

Labeling the available wavelengths from 0 to $w - 1$, every column can be partitioned in groups composed by $p$ nodes, following the rule that nodes belonging to the same group are connected with the same subset of nodes in the next column [64]. This means that, in a $(p, k)$ SN, group $i$ of any column includes nodes having the following row coordinates $\{i, i + p^{k-1}, i + 2p^{k-1}, \ldots, i + (p - 1)p^{k-1}\}$ with $0 \leq i \leq p^{k-1} - 1$ [71]. All of them are linked with consecutive nodes on the next column having the following row coordinates $\{j, j + 1, \ldots, j + (p - 1)\}$, with $j = (i \bmod p^{k-1})p$ [64].

The transmitter channel assignment, at a node or within a given star, is realized following the row major order assignment, taken from matrix theory, whereas the receiver channel assignment can be deduced directly from the connectivity graph [71]. An example is shown in Figure 3.28 considering the case of a (2, 2) SN. The first value of every pair $(x, y)$, labeling a channel, indicates the star number, whereas the second numbers the considered link within the star, so that $0 \le x \le (W/w) - 1$ and $0 \le y \le w - 1$. In Figure 3.28, we have $w = 4$.

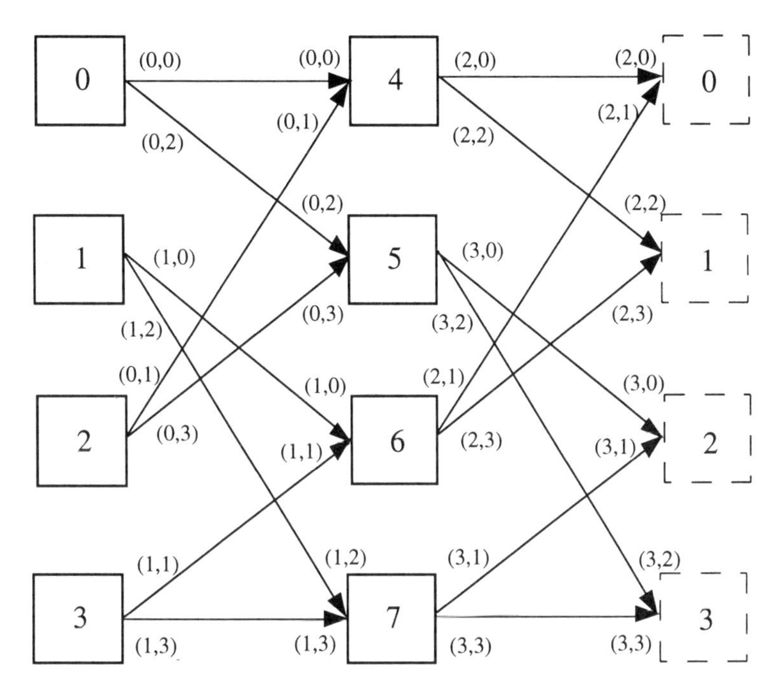


**Figure 3.28** Channels assignment for the (2, 2) shufflenet.

In a $(p, k)$ SN based on a multistar architecture with $w$ channels multiplexed on a single fiber, a generic node $n$, with $0 \le n \le N - 1$ and $N = kp^k$, is connected to the $I_n$ star for transmitting and to the $J_n$ star for receiving. These two values are obtained through a suitable channel assignment algorithm [71] which can be solved in a closed form, obtaining the following equations:

$$I_n = \left\lfloor \left( n \bmod p^{k-1} + p^{k+1} \left\lfloor n / p^k \right\rfloor \right) / \left( w / p^2 \right) \right\rfloor \tag{3.42}$$

$$J_n = \left\lfloor \left( n - p^k + N\delta\left( \left\lfloor n / p^k \right\rfloor \right) \right) / (w / p) \right\rfloor \tag{3.43}$$

where $\delta(x) = 1$ if $x = 0$ and $\delta(x) = 0$ otherwise. The expression $n \bmod p^{k-1}$ gives the group to which node $n$ belongs to within its column, whereas $\left\lfloor n / p^k \right\rfloor$, the integer result of $\left(n / p^k\right)$, is the column coordinate of node $n$ itself. Now, numbering the groups in column 0 from 0 to $p^{k-1} - 1$, in column 1 from $p^{k-1}$ to $2p^{k-1} - 1$, and so on, the number of the group $G_n$ that node $n$ belongs to is given by

$$G_n = n \bmod p^{k-1} + p^{k-1} \lfloor n / p^k \rfloor \tag{3.44}$$

Since a star can accommodate $w/p^2$ groups, (3.42) correctly gives the star coupler number which serves node $n$ for transmission.

As regards $J_n$, we can observe that the nodes using the same star coupler for receiving are adjacent along a column and numbered consecutively. Moreover, the numbering of stars used for receiving has to start from the second column since, there, nodes receive packets from groups placed on the first one. This means that the greatest value of $J_n$ corresponds to a star coupler that serves a group of nodes placed on the first column.

### 3.3.4 Modular Expansion of Shufflenets

Having considered the organization of a $(p, k)$ dedicated channels SN on a multistar architecture, that we denote here as $\Phi$, now we are able to analyze its possible modular expansion toward a $(p, k + 1)$ structure. The expansion is realized in some phases, the first of them differing from the others [71].

It is useful to define a partial $(p, k + 1, m_0)$ SN, indicated by $\Phi_e$, having a total number of nodes equal to $m_0 p^{k+1}$. Network $\Phi_e$ has $m_0$ columns, with $1 \le m_0 \le k$, so that it becomes a full $(p, k + 1)$ SN, also denoted as a $(p, k + 1, k + 1)$ SN, when $m_0 = k + 1$.

Since $\Phi_e$ contains $\Phi$, it follows that $m_0 p^{k+1} > kp^k$, and then $m_0 \ge \lfloor k / p \rfloor + 1$. Therefore, in $\Phi_e$ it is possible to insert $m_0 p^{k+1} - kp^k = p^k(m_0 p - k)$ new nodes, using $p^{k+1}(m_0 p - k)/w$ new star couplers.

The first expansion phase consists of three steps [71]:

1. Perform connectivity and channels assignment on $\Phi_e$ using the above-described methods;
2. Map each node in $\Phi$ to an equivalent node in $\Phi_e$ and update the node address accordingly;
3. Disconnect some output ports of the star couplers, add new nodes and new star couplers and connect all loose-end fibers according to the connectivity graph of $\Phi_e$.

As regards step 2, two nodes are said to be equivalent when their receiver and transmitter wavelengths are the same, even though they are connected to different star couplers. More precisely, node $\alpha$ belonging to $\Phi$ corresponds to node $\beta$ in $\Phi_e$, in accordance with the following transition formula [71]:

$$\beta = a \bmod p^{k-1} + p^k \lfloor a / p^{k-1} \rfloor + p^{k-1}(b \bmod p) + p^{k+1} \lfloor b / p \rfloor \tag{3.45}$$

where $a = \alpha \bmod p^k$ and $b = \lfloor \alpha / p^k \rfloor$.

In all the expansion phases following the first one, a new column composed by $p^{k+1}$ nodes is added to the $(p, k + 1, m_0)$ SN until it becomes coincident with a $(p, k + 1)$

SN. In practice, at the $i$th phase, a $(p, k + 1, m_0 + i - 2)$ SN is expanded to become a $(p, k + 1, m_0 + i - 1)$ SN, with $2 \leq i \leq k - m_0 + 2$, according to the following rules [71]:

1. Construct a $(p, k + 1, m_0 + i - 1)$ SN and find the connectivity and wavelength channels assignment;
2. Disconnect the output ports of the star couplers that lead to node 0, node 1, . . . , $(p^{k+1} - 1)$ in the SN of the previous phase;
3. Connect a column of $p^{k+1}$ new nodes, $p^{k+2}/w$ new star couplers and the loose-end fibers to the network, according to the connectivity graph of the $(p, k + 1, m_0 + i - 1)$ SN.

Since the network rearrangement imposed by this method guarantees that no replacement or retuning of old nodes occurs in case of new node insertion, using this approach the expansion can be cost-effective and competitive with respect to alternative techniques of network enlargement [72].

However, some problems can be experienced when routing strategies, such as static self-routing scheme [64] or dynamic routing scheme [77] have to be applied in a partial SN. In fact, because of the loss of some columns in the considered network, the nodes are no longer able to determine the output link, that has to be chosen for routing the incoming packet, only using the destination address carried by the packet header. Therefore, additional information is needed to assist routing.

Moreover, the throughput per user can increase too much, for values of $k$ too high. As discussed in the previous section, the larger $k$ the lower the channel efficiency, because of the increment in the average number of hops. For this reason, whether it is convenient to expand SN with $p$ or with $k$ depends on the specific context.

### 3.3.5 Channel Sharing in a Bidirectional Perfect Shuffle Topology

Besides the techniques discussed in Section 3.3.2, another method that implies the adoption of channel sharing in SN topology is based on the utilization of bidirectional connections between nodes. In this structure $2p$ nodes, belonging to adjacent columns, contend for a common channel. In this section, it will be demonstrated that by this solution, limiting our analysis to bidirectional SN with $p = 2$ case, the channel utilization can be improved, maintaining good performance in terms of user and network throughput.

In Figure 3.29 an example is given, for a (2, 2) SN with bidirectional channels. Nodes $i, i + 2^{k-1}$ in a column, with $i = 0, 1, \ldots, 2^{k-1} - 1$, share the same channel with nodes $j, j + 1$ in the next one, being $j = 2(i \bmod 2^{k-1})$. By $W_1$, $W_2$, $W_3$, $W_4$ we have indicated the common channels. In this way we can show that, for example, node 0 can exchange packets directly (in a single hop) with nodes 2, 4, 5 through channel $W_1$, and with nodes 1, 4, 6 through channel $W_3$. The multiple access of these nodes to the common channels can be controlled in a TDMA fashion, using a frame structure similar to that shown in Table 3.8 for the example of Figure 3.29, where time slot assignment is realized to offer the same bandwidth to each station.

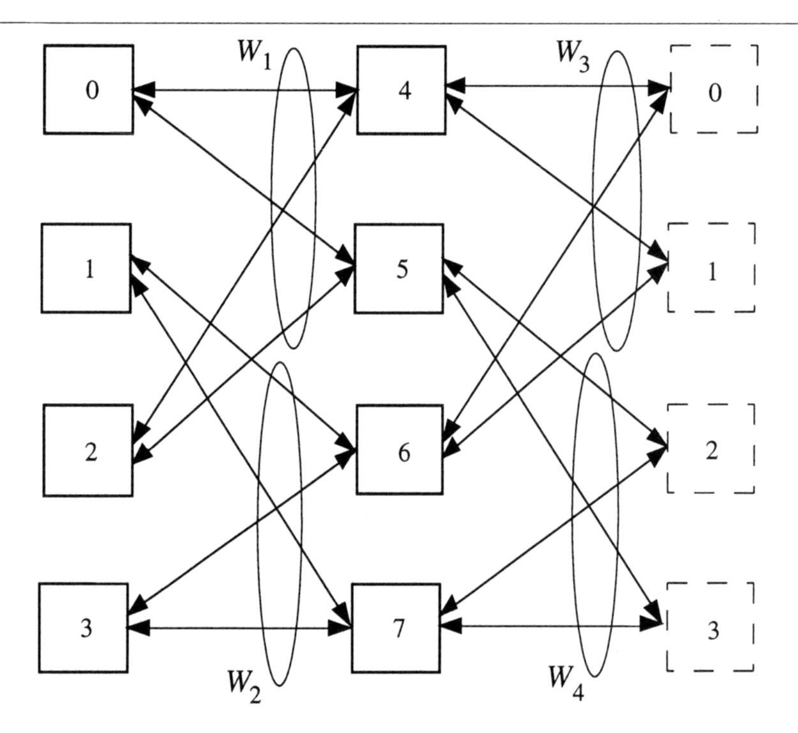


**Figure 3.29** Channel sharing in a (2, 2) shufflenet using bidirectional links.

**Table 3.8**
Slot Assignment for the Shufflenet of Figure 3.29

| | *Slot 1* | *Slot 2* | *Slot 3* | *Slot 4* |
|---|---|---|---|---|
| $W_1$ | Node 0 | Node 2 | Node 4 | Node 5 |
| $W_2$ | Node 3 | Node 1 | Node 7 | Node 6 |
| $W_3$ | Node 4 | Node 6 | Node 0 | Node 1 |
| $W_4$ | Node 7 | Node 5 | Node 3 | Node 2 |

Let us label the $(l, m)$ node by its row and column addresses, with $l = 0, 1, \ldots, 2^k - 1$ and $m = 0, 1, \ldots, k - 1$. Now, we have that:

- A f-transmission connects (forward) node $(l, m)$ to nodes $(j, m + 1)$, $(j + 1, m + 1)$;
- A b-transmission links (backward) node $(l, m)$ with nodes $(i, m - 1)$, $(i + 2^{k-1}, m - 1)$;

being $i = \lfloor l/2 \rfloor$ and $j = 2(l \bmod 2^{k-1})$. Moreover, f'-transmission and b'-transmission take place from node $(l, m)$ to node $(l + f, m)$ and node $(l + h, m)$ respectively, with $f = 2^{k-1}(-1)^{\lfloor l/(2k-1) \rfloor}$ and $h = (-1)^l$.

To explain these definitions better, let us observe that in a (2, 3) SN with bidirectional shared channels, like the one drawn in Figure 3.30, node 0 ≡ (0, 0) is connected by f-transmissions to nodes 8 and 9, by b-transmissions to nodes 16 and 20, by f'-transmissions to node 4 and by b'-transmissions to node 1.

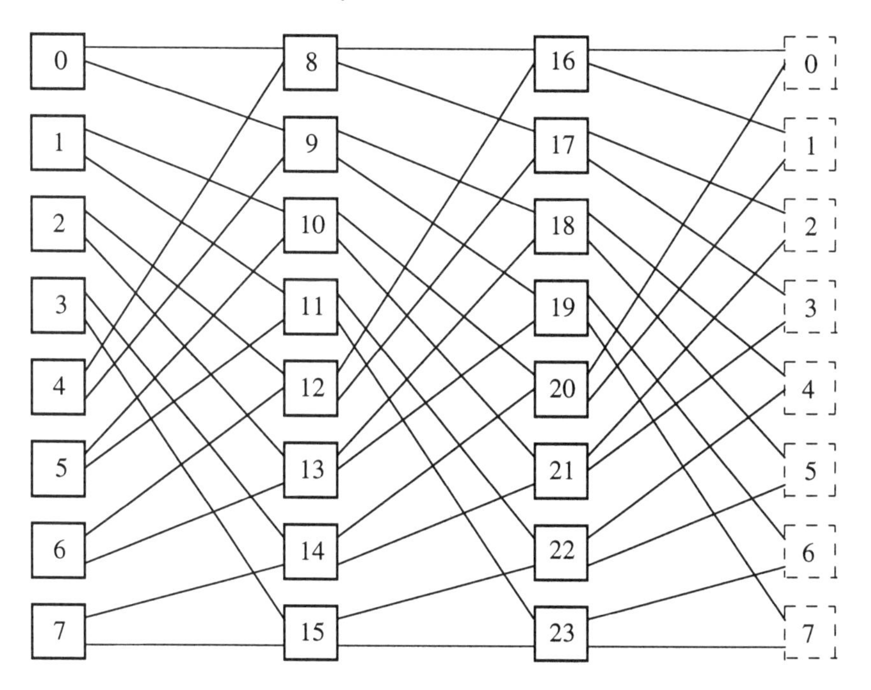


**Figure 3.30** Connectivity graph of a (2, 3) bidirectional shufflenet.

It is evident that, working on two wavelengths shared by more than $p$ nodes, as seen in Section 3.3.2, the network diameter is significantly reduced. It can be observed that the nodes placed farthest from node $0 \equiv (0, 0)$ are nodes $(2^k - 1, (k - 1)/2)$ and $(2^k - 1, (k + 1)/2)$ if $k$ is odd, and node $(2^k - 1, k/2)$ if $k$ is even. In particular, it can be demonstrated [78] that the network diameter is given by

$$D_{\text{odd}} = 2\left(\frac{k-5}{2}\right) + 2 + \frac{k+1}{2} = \frac{3k-5}{2} \qquad \text{for } k \text{ odd} \tag{3.46}$$

$$D_{\text{even}} = 2\left(\frac{k-4}{2}\right) + 2 + \frac{k}{2} = \frac{3k-4}{2} \qquad \text{for } k \text{ even} \tag{3.47}$$

In both cases, either $D_{odd}$ or $D_{even}$ are less than the equivalent value of a perfect SN with dedicated channels or with channel sharing among $p$ users belonging to the same column (see Section 3.3.2), which equals $2k - 1$. As regards the calculus of the average number of hops for the configuration here considered, we can indicate by $h(i, j)$ the minimum number of hops necessary to reach node $(i, j)$. Then, it turns out to be

$$\langle h \rangle = \sum_{i,j} \frac{h(i,j)}{N} \quad \text{for } i = 0, 1, \ldots, 2^k - 1 \text{ and } j = 0, 1 \ldots, k - 1 \tag{3.48}$$

A detailed discussion of (3.48) is reported in [78].

To evaluate the performance of the proposed architecture we will utilize the same parameters used before, such as channel efficiency $\eta$, network capacity $C$ and normalized throughput per user $C_u$; $\eta$ being the inverse of the average number of hops, whereas $C$ and $C_u$ are, respectively, defined in (3.38) and (3.39).

It can be shown [78] that in heavy-loaded traffic conditions, SNs with dedicated channels are characterized by better throughputs, with respect to bidirectional SNs with shared channels, but require a larger number of wavelengths for their implementation. On the other hand, the availability of bidirectional channels, shared among group of nodes, improves the network performance in terms of channel efficiency, even when compared with SNs with dedicated channels. In particular, to extend the performance analysis to the case of lower levels of traffic, a new definition of channel efficiency is given

$$\eta = \frac{1}{\langle h \rangle} \delta \tag{3.49}$$

where

$$\delta = \begin{cases} 1 & \text{for } \frac{1}{n} \leq \Lambda \leq 1 \\ n\Lambda & \text{for } 0 \leq \Lambda \leq \frac{1}{n} \end{cases} \tag{3.50}$$

$\Lambda$ being the average arrival rate at the output buffer of a node, normalized to the channel speed $S$, and $n$ being the number of nodes sharing a channel.

In Table 3.9, a comparison between different solutions is represented, with reference to a SN with 2,048 nodes uniformly loaded. The second row of the table refers to the scheme proposed in [64], whereas the third one concerns the scheme presented in [63], both of which are discussed in Section 3.3.2. This comparison is completed by Figures 3.31 and 3.32, where the channel efficiency and the normalized network capacity are plotted versus $\Lambda$. Figure 3.31 shows that the best channel efficiency is reached by channel sharing using bidirectional links, except for light loads, where the solution proposed in [63] seems to be preferable. On the other hand, as regards the network capacity, from Figure 3.32 we can see that the best performance is guaranteed by the utilization of dedicated channels, at least for $\Lambda \geq 0.4$, whereas the presence of bidirectional links is useful for lower traffic levels, since in this case the high channel efficiency compensates the smaller number of available channels.

**Table 3.9**
Performance of a (2, 8) Shufflenet Depending on Dedicated or Shared Channel Policies

| | $W$ | $\langle h \rangle$ | $\eta$ | $C'$ | $C_u$ |
|---|---|---|---|---|---|
| Dedicated channels | 4096 | 10.53 | $0.095 \cdot \delta$ | $389.8 \cdot \delta$ | $0.19 \cdot \delta$ |
| Channel sharing by $p = 2$ nodes in the same column | 1024 | 10.53 | $0.095 \cdot \delta$ | $97.5 \cdot \delta$ | $0.048 \cdot \delta$ |
| Channel sharing by $k$ nodes in the same row | 512 | 7.04 | $0.142 \cdot \delta$ | $73.1 \cdot \delta$ | $0.036 \cdot \delta$ |
| Bidirectional and shared channels | 1024 | 6.74 | $0.148 \cdot \delta$ | $151.6 \cdot \delta$ | $0.074 \cdot \delta$ |

*Abbreviations*: $C'$, normalized network capacity; $C_u$, normalized throughput per user; $\eta$, channel efficiency; $\langle h \rangle$, average number of hops; $W$, number of channels.

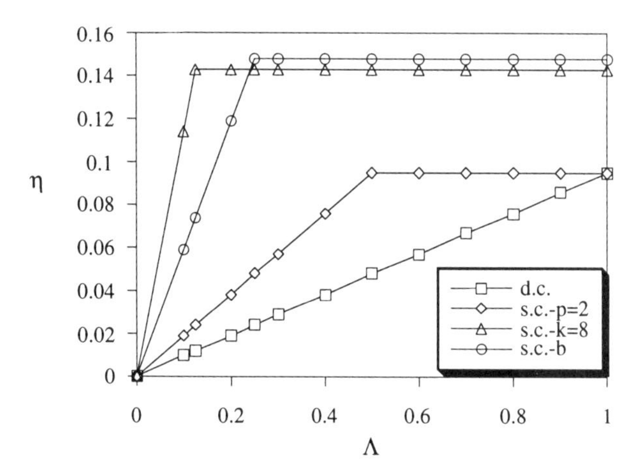


**Figure 3.31** Channel efficiency of a (2, 8) shufflenet (d.c. = dedicated channels; s.c.-b = bidirectional and shared channels; s.c.-k=8 = shared channels by $k = 8$ nodes in the same row; s.c.-p=2 = shared channels by $p = 2$ nodes in the same column).

Finally, we can extend our study to the delay analysis of the different solutions we have considered here. Assuming a TDMA multiple-access protocol and a $M/D/1$ queueing system associated with the output ports of every node, it can be demonstrated [79] that the average time delay in a node is equal to

$$t_q = \frac{\rho^2}{2\Lambda(1-\rho)} + t_t + \frac{n}{2} t_t \tag{3.51}$$

where $n$ is the number of nodes sharing a common channel, $t_t$ is the packet transmission time on the common channel, and $\rho = n\Lambda t_t$. In particular, the first term of (3.51) represents the queueing delay, whereas the third takes into account the user slot synchronization time.

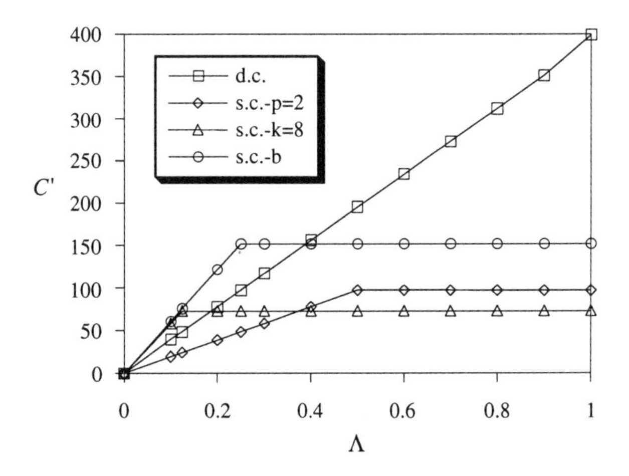


**Figure 3.32** Normalized network capacity of a (2, 8) shufflenet (d.c. = dedicated channels; s.c.-b = bidirectional and shared channels; s.c.-k=8 = shared channels by $k = 8$ nodes in the same row; s.c.-p=2 = shared channels by $p = 2$ nodes in the same column).

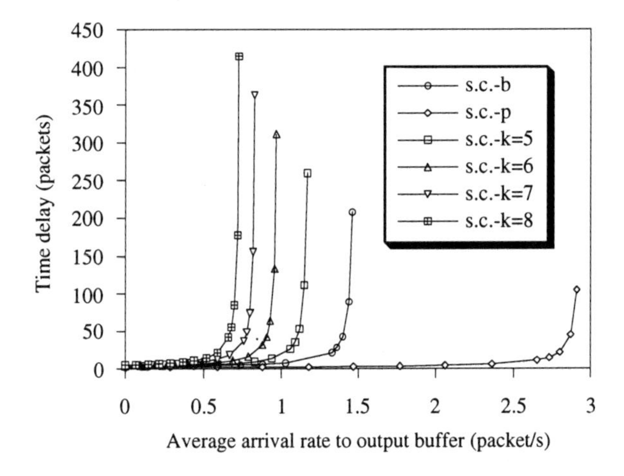


**Figure 3.33** Time delay versus the average arrival rate at the output buffer (s.c.-b = bidirectional and shared channels; s.c.-k=$x$ = shared channels by $k = x$ nodes in the same row; s.c.-p = shared channels by $p = 2$ nodes in the same column).

The time delay versus the average arrival rate at the output buffer is plotted in Figure 3.33, assuming $S$ = 2.5 Gbps, $t_t = 1.7 \cdot 10^{-7}$ sec, and 53-bytes-long packet length. The end-to-end delay depends on $t_q$ but also on the average propagation delay between nodes. It can be noted that the queueing delay obviously becomes significant with respect to the propagation delay when the load tends to get higher.

### 3.3.6 Routing in Shufflenets

We have chosen three possible routing strategies, among all the possible techniques proposed for application in SN [64,77,80,81], that seem to be useful to get the general idea of the methodologies and of the problems implied by the necessity of delivering packets in this kind of multihop networks. Their characteristics are briefly resumed in this section.

First, we are going to analyze a fixed routing algorithm and then we will also consider an adaptive scheme for datagram transmissions, both of them working in dedicated channels SN. Finally, we will present an example of fixed routing algorithm for shared channels SN.

However, even though this subject is not focused on in this section, it is worth reminding the reader that routing techniques include a contention resolution function, besides the routing one (see Section 3.2.3.5). Both of these functions contribute to the final network performance. In particular, since it has been demonstrated that, at least in SN and bidirectional SN, contention resolution based on random schemes may produce instability in traffic conditions [82], it is in general necessary to examine such strategies as a whole, to obtain a correct prediction of their effects on network behavior.

Let us adopt once again the useful notation of node labeling introduced in the previous section: in a $(p, k)$ SN every node can be identified by a pair of integers $(c, r)$, where $c \in [0, k-1]$ is the column index (having numbered the columns from left to right) and $r \in [0, p^k - 1]$ is the row index (having numbered the rows from top to bottom). Now, the address $r$ is represented in a $p$-ary notation, so that $r = r_{k-1}, r_{k-2}, \ldots, r_1, r_0$. The $r_i$ terms are the digits of the $p$-ary alphabet, consequently we have $0 \le r_i \le p-1$ and

$$r = \sum_{i=0}^{k-1} r_i p^i \tag{3.52}$$

#### 3.3.6.1 Fixed Routing

There is not a unique class of fixed routing (FR) algorithms for SN. Here we describe a possible example, which is particularly simple and easy to implement.

Denoting respectively by $(c^s, r^s)$ and $(c^d, r^d)$ the source and destination of a given sequence of packets, the column distance $\delta'$, that is, the number of columns between this node pair, is equal to

$$\delta' = \begin{cases} \left(k + c^d - c^s\right)_{\mathrm{mod}\,k} & \text{if } c^d \neq c^s \\ k & \text{if } c^d = c^s \end{cases} \tag{3.53}$$

Because of the previously examined characteristics of the SN connectivity graph, $\delta'$ is a lower bound on the number of hops between $(c^s, r^s)$ and $(c^d, r^d)$, even though it is not always possible to reach a given destination in only $\delta'$ hops.

Nevertheless a fixed routing algorithm [77] can attempt to obtain an end-to-end path $\delta'$ hops long, using the last $\delta'$ digits of $r^d$, according to the following procedure.

From $(c^s, r^s) = (c^s, r^s{}_{k-1}, r^s{}_{k-2}, \ldots, r^s{}_1, r^s{}_0)$, the packet is first routed to the first intermediate node having $((c^s + 1)_{\mathrm{mod}\,k}, r^s{}_{k-2}, \ldots, r^s{}_0, r^d{}_{\delta'-1})$ address, then toward the second intermediate node having $((c^s + 2)_{\mathrm{mod}\,k}, r^s{}_{k-3}, \ldots, r^s{}_0, r^d{}_{\delta'-1}, r^d{}_{\delta'-2})$ address. This propagation goes on until a node $\delta'$ hops far from $(c^s, r^s)$ is reached. It is destination if and only if all the following identities are simultaneously true:

$$\begin{aligned} r^s{}_{k-1-\delta'} &= r^d{}_{k-1} \\ r^s{}_{k-2-\delta'} &= r^d{}_{k-2} \\ &\ldots \\ r^s{}_0 &= r^d{}_{\delta'} \end{aligned} \tag{3.54}$$

otherwise the packet has to complete another rotation to reach $(c^d, r^d)$. Therefore, through this scheme, packet delivery can be realized in $\delta'$ or $\delta' + k$ hops at least.

In practice, when a given intermediate node $(c^x, r^x)$ receives a packet, it bases its routing choices on the following observations:

- If $(c^x, r^x) = (c^d, r^d)$ then $(c^x, r^x)$ is destination and the packet is addressed toward the local output;
- If $(c^x, r^x) \neq (c^d, r^d)$ the packet is routed toward the node $((c^x + 1)_{\mathrm{mod}\,k}, r^x{}_{k-2}, \ldots, r^x{}_0, r^d{}_{X-1})$;

where $X$ is given by

$$X = \begin{cases} \left(k + c^d + c^x\right)_{\mathrm{mod}\,k} & \text{if } c^d \neq c^x \\ k & \text{if } c^d = c^x \end{cases} \tag{3.55}$$

A similar routing scheme can be applied also to shared channels SN [64].

#### 3.3.6.2 *Adaptive Routing*

The FR algorithm presented above shows some problems, relative to the productions of traffic unbalance even when the nodes are equally loaded and when the traffic is uniformly distributed in the network. Other more involved FR schemes overcome this drawback, but they remain inefficient as regards the handling of nonuniform traffic. The effect of this fact consists in the generation of bottlenecks that greatly worsen the overall network performance.

To take into account this problem some adaptive routing procedures have been proposed. Here, we present one of them [77] that can avoid the traffic congestions, in presence either of nonuniform traffic conditions or of random packet flow peaks that can appear even if the network is uniformly loaded.

To do this, adaptive routing (AR) operates like FR when no congestion is present in the network. In this way, the packets generated by a node are usually transmitted toward the shortest path that connects the considered source-destination pair. On the contrary, when some links tend to become overloaded, AR begins to misroute a given amount of traffic toward nonoptimal paths. Obviously, these paths are longer than the optimal ones, but utilizing them can avoid the generation of bottlenecks, so that the overall network throughput increases. In other words, the saturation is reached later, and higher traffic levels can be managed in normal operative conditions, with respect to a situation where FR manages the packet delivery. Therefore, for every node pair to connect, AR exploits the presence of multiple paths made available by the SN connectivity graph.

As a counterpart, since AR can scramble the packet sequences emitted by sources, because of the different lengths of multiple end-to-end paths, it requires the availability of suitable resequencing procedures for the end users [83,84].

When a packet is misrouted, the first part of its propagation, which consists of a path $\delta'$ hops long, is "don't care," since its only goal is to deliver the packet itself to whichever node belonging to the same column of destination. Subsequently, the same packet will be routed to $(c^d, r^d)$ along a single path $k$ hops long.

AR, which uses local queue size information only [16], does not need any traffic model for its operation; furthermore, it adapts automatically to time-varying traffic conditions. Finally, it can operate either in datagram or virtual circuit mode.

Two types of packets are handled by the considered algorithm:

- *Type M packet*: it has multiple minimum-hop paths to destination and its current position is more than $k$ hops far from it;
- *Type S packet*: it has a single minimum-hop path to destination and its current position is $k$ hops far from it at the most.

Type M packets are "don't care." Consequently, nodes address them toward one of their shortest output queue, without considering the destination address contained in the packet header. Type S packets, instead, are normally transmitted according to the FR algorithm, except when the correspondent output queue is overloaded. In this case, the nodes are allowed to bump them toward the shortest output queue. Once bumped during the propagation, a proper status bit is set in the packet header, so that misroutings cannot occur any more.

It is evident that a packet can change its type during the propagation. A type S packet changes into type M packet when deflected in a nonoptimal direction, whereas a type M packet changes in type S packet when it reaches column $c^d$ after the first part of its trip.

Before analyzing the subsequent routing scheme, devoted to shared channels SN, let us compare the performance of FR and AR in some uniform traffic conditions. In Table 3.10 we have reported the results obtained by simulation [77] for FR and AR in (2, 3) and (2, 4) SNs, assuming that the external traffic is uniformly distributed toward all the possible destinations. Moreover, such a traffic is characterized by a Bernoulli arrival process at the nodes with rate $\lambda$ (measured in packets per time slot). The comparison is based on channel load and queue size (average and variance) maximum values over all the nodes and links of the considered network. It can be observed that, corresponding to these channel loads, AR accomplishes a smoothing of the traffic that tends to make the queue parameters homogeneous all over the network. This is due to the spreading of traffic created through the production of type M packets.

More significant results, as regards AR performance, can be evidenced by removing the assumption of uniform network loading or by considering time-varying traffic patterns, since under these conditions the possibility of fitting automatically the routing choices to the real conditions, offered by AR, is exploited to the highest degree [77].

**Table 3.10**
Performance of Fixed Routing and Adaptive Routing with Uniform Traffic

| *Network* | *Load* | *Parameter* | *Fixed Routing* | *Adaptive Routing* |
|---|---|---|---|---|
| | | Maximum channel load | 0.76 | 0.77 |
| (2, 3) SN | $\lambda = 0.45$ | Maximum sample mean queue size | 1.67 | 1.36 |
| | | Maximum sample queue size variance | 2.67 | 1.32 |
| | | Maximum channel load | 0.78 | 0.70 |
| (2, 4) SN | $\lambda = 0.3$ | Maximum sample mean queue size | 1.84 | 1.04 |
| | | Maximum sample queue size variance | 3.30 | 0.89 |

### 3.3.6.3 *Fixed Routing for Shared Channels Shufflenets*

In this section we briefly sum up the characteristics of a FR algorithm for the single-transmitter and single-receiver per user SN with shared channels (FR-SC) presented in Section 3.3.2. Since in this configuration $p$ users listen on the same wavelength, the algorithm must indicate the node that has to repeat the received messages.

Considering, as usual, the source-destination pair $(c^s, r^s)$-$(c^d, r^d)$ and the intermediate node $(c, r)$, this is done according to the following rules:

- If $(c, r) = (c^d, r^d)$ then $(c, r)$ is destination and the packet is not repeated;
- If $(c, r) \neq (c^d, r^d)$ then the packet is repeated if and only if $r_0 = r^d_{(k+c^d-c) \bmod k}$.

It can be demonstrated [64] that FR-SC, using only the $p$-ary digits of the received packet, delivers messages from any given source to any given destination in the minimum number of hops possible for that pair of nodes, maintaining uniform average loading on the WDM channels in presence of uniform traffic distribution. An adaptation of this scheme is proposed in Section 3.4.1 for the traffic routing in duplex shufflenets.

#### *3.3.6.4 Outline of Nonuniform Traffic Analysis Techniques*

The study of multihop networks under nonuniform traffic load, needed to test the considered communication systems in realistic conditions, requires the adoption of suitable analytical methods, that can make the interpretation of results obtained through simulation easy. For example, two possible approaches can be considered:

- The extreme-value analysis;
- The random load generation.

The former is an approximate analytical technique that assumes uniform load distribution, but also a normal distribution of traffic intensity generated by individual users. In the latter, on the contrary, the load intensity produced locally is assumed to be equal for every node, but the traffic patterns are variable. Furthermore, they are chosen randomly and their links are loaded according to the considered routing algorithm.

Using extreme-value analysis, on the basis of the overall average and the variance of the load intensity, it is possible to evaluate the overload probability of a given channel, whereas using random load generation we can estimate the capacity reduction when the degree of traffic nonuniformity is about as bad as could be expected in practice.

In [85] an extensive and detailed discussion of these two methods is reported, with several examples of their applications in SN traffic analysis. In particular, it has been demonstrated that, considering realistic load patterns, the allowable node throughput is reduced by a factor between 0.3 and 0.5 with respect to the one predicted for uniform loading. Moreover, such throughput seems to be weakly influenced by the network size.

***Some keypoints.*** Shufflenets are multicolumn networks where nodes on a column are connected to nodes in the next column by means of the so-called perfect shuffle connection pattern. Because of their low diameter and load balancing structure in uniform traffic, they are considered, together with Manhattan street networks, as a landmark for WDMA lightwave networks. Different architectures can support shufflenets' virtual topology, some architectures implying the utilization of dedicated wavelength channels, others adopting the shared channels policy. The availability of multiple end-to-end paths can be exploited by suitable and simple fixed or adaptive routing algorithms.

## 3.4 EVOLUTIONS OF THE SHUFFLE TOPOLOGY

### 3.4.1 Duplex Shufflenet

In this section, we consider some architectures that, for different reasons, can be considered as an evolution of SN structure. They have been conceived to improve overall network performance and to overcome drawbacks that affect SNs.

Even though the presence of bidirectional channels in SN has already been considered in Section 3.3.5, our analysis begins from duplex shufflenet (D-SN) [86], which can be seen in some way as a new virtual topology.

In D-SN the logical connection is similar to the original SN. It represents the evolution of the reverse channel augmented multihop SN proposed in [87]. It is based on duplex links, each composed of the unidirectional channel also present in basic SN and a reverse channel that guarantees the bidirectionality of the considered connection between a given pair of nodes. It follows that the implementation of a D-SN requires a double number of transceivers per node with respect to the basic SN architecture. Nevertheless, it will be shown that its throughput increases more than twice, if compared with the SN throughput, and other significant advantages are obtained as regards blocking probability and total delay. Moreover, by using D-SN, we can overcome the problem of asymmetric distance between nodes which affects SN [88]. In fact, if we consider two nodes belonging to a given SN, namely, A and B, and their relative distances $\delta_{AB}$ and $\delta_{BA}$, in general we have $\delta_{AB} \neq \delta_{BA}$. In Figure 3.34, a (2, 2) D-SN is represented.

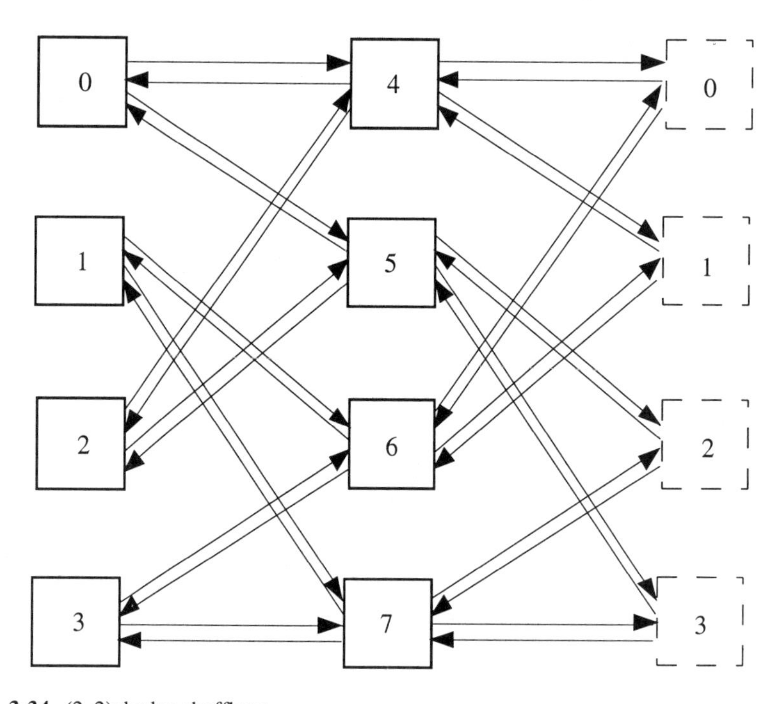


**Figure 3.34** (2, 2) duplex shufflenet.

Similarly to SN, as shown in Section 3.3.6, in a $(p, k)$ D-SN, every node can be labeled by a pair of integer numbers $(c, r)$, the first indicating the column address and the second standing for the row address. The latter counts the position of the considered node in column $c$, on the base of a $p$-ary alphabet. Therefore, the source node and destination of a given packet can be identified, respectively, by the following pairs $(c^s, r^s) = (c^s, r^s_{k-1}, r^s_{k-2}, \ldots, r^s_1, r^s_0)$ and $(c^d, r^d) = (c^d, r^d_{k-1}, r^d_{k-2}, \ldots, r^d_1, r^d_0)$.

We can consider two different routing strategies, conceived to exploit the link bidirectionality made available by D-SN: the unidirectional routing and the shortest path routing.

In unidirectional routing, when a node receives a packet it always emits it using a transmitter that is not placed on the same side where the packet comes from (except the packets addressed to the considered node). Assuming a uniform distribution of traffic toward the nodes and a uniform traffic load, the mean number of hops between any pair of source-destination nodes, for $k > 2$, is equal to [89]

$$\langle h \rangle_{\text{odd}} = \frac{p^2\left(4k + p^k\left(5k^2 - 1\right) + 2\left(k^2 - 1\right)\right) - k^2(2p-1)\left(5p^k + 2\right)}{4(p-1)^2\left(kp^k - 1\right)} + \frac{8p^{k/2}\left(p^{3/2} + p^{1/2}\right) - p^k(14p+1) - 4(k-p) - 2}{4(p-1)^2\left(kp^k - 1\right)} \quad \text{for } k \text{ odd} \tag{3.56a}$$

$$\langle h \rangle_{\text{even}} = \frac{k^2(p-1)^2\left(5p^k + 2\right) - 16p^{k+1} + 16p^{k/2+1} + 4kp^2 - 4k}{4(p-1)^2\left(kp^k - 1\right)} \quad \text{for } k \text{ even} \tag{3.56b}$$

Now, we define the so-called forward distance $D_f$ and the backward distance $D_b$, respectively, by these expressions [86]:

$$D_f = (c^d - c^s) \bmod k \tag{3.57}$$

$$D_b = (c^s - c^d) \bmod k \tag{3.58}$$

The choice of the channel and of the side where transmitting the packet is made on the basis of the following equations:

$$r^s_{k-1-D_f}\, r^s_{k-2-D_f} \ldots r^s_1\, r^s_0 = r^d_{k-1}\, r^d_{k-2} \ldots r^d_{D_f+1}\, r^d_{D_f} \tag{3.59}$$

$$r^s_{k-1}\, r^s_{k-2} \ldots r^s_{D_b+1}\, r^s_{D_b} = r^d_{k-1-D_b}\, r^d_{k-2-D_b} \ldots r^d_1\, r^d_0 \tag{3.60}$$

If only condition (3.59) is satisfied, the right-hand side channel is utilized for transmitting; on the contrary, if only condition (3.60) is satisfied, the left-hand side channel is used. If both or none of these conditions are satisfied, the right-hand side channel is used if $D_f < D_b$, whereas the left-hand side channel is used if $D_f > D_b$. Finally, if $D_f = D_b$ and both conditions are satisfied or neither is satisfied, a channel is selected randomly. Then, the packet will be sent along the path that is characterized by the minimum number of hops to reach its destination. As regards the routing choices, we can distinguish between forward and backward routing. In particular, when the $(c, r)$ node receives a message addressed to $(c^d, r^d)$ from the left-hand side, a forward transmission is realized according to the following rules [64]:

- If $(c, r) = (c^d, r^d)$ then $(c, r)$ is destination and the packet is not repeated;
- If $(c, r) \neq (c^d, r^d)$ then the packet is repeated if and only if $r_0 = r^d_{(k+c^d-c) \bmod k}$.

On the contrary, backward routing is adopted when the packet comes from the right-hand side. In this case we have two other cases [86]:

- If $(c, r) = (c^d, r^d)$ then $(c, r)$ is destination and no user repeats the packet;
- If $(c, r) \neq (c^d, r^d)$ only the node with $r_{k-1} = r^d_{k-1-(c-c^d) \bmod k}$ will repeat the packet.

Shortest path routing, instead of considering the side where the packets come from, as performed by unidirectional routing, admits that a node broadcasts them toward all of its channels, without considering their origins. In this way, it can provide the optimal shortest path for a D-SN [86]. It requires the availability of two fields in the packet header: the directional field and the transmission field. After identifying the shortest path, the source node writes in the directional field the direction of transmission of the packet itself. Then, in the transmission field, it includes information utilized by intermediate nodes to repeat the packet during its propagation across the network.

In Figure 3.35, the channel efficiencies of SN and of D-SN are compared.

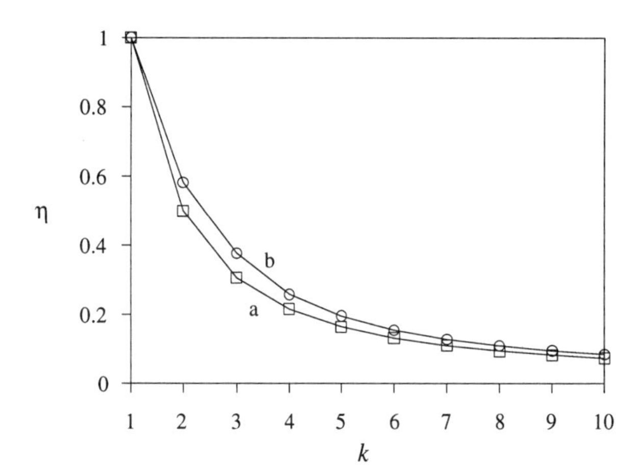


**Figure 3.35** Channel efficiency for $p = 2$: (a) shufflenet and (b) duplex shufflenet.

They have been obtained adopting the original routing [64,86] in SN and the shortest path routing in D-SN. Here we can observe a slight improvement due to the utilization of bidirectional links, but more significant advantages are achieved as regards network throughput.

In fact, in Figure 3.36, the node throughput achievable using shortest path routing and unidirectional path routing in D-SN and original routing in SN, is plotted versus the arrival probability. By shortest path routing the throughput can be approximately doubled and the effect of saturation due to the increase in traffic load happens later. Analogously, other significant benefits are implied by the bidirectionality as regards end-to-end delay and packet loss probability [86].

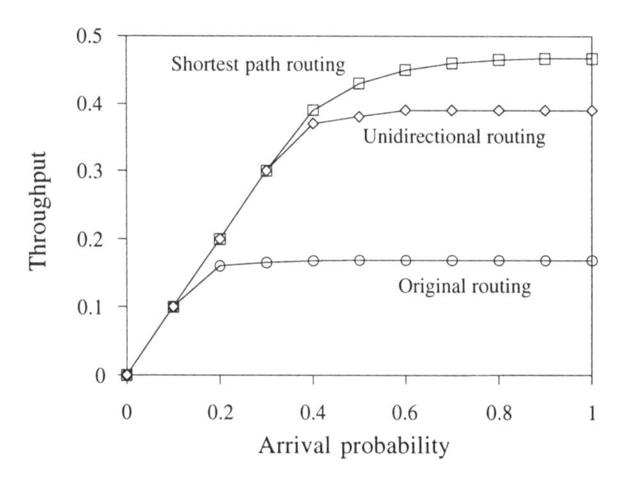


**Figure 3.36** Node throughput versus the arrival probability in a (2, 5) SN (original routing) and in a (2, 5) D-SN (shortest path and unidirectional routing).

### 3.4.2 GEMNET

GEneralized shuffle-exchange Multihop NETwork (GEMNET) [90] is a regular multihop architecture, that represents a generalization of shuffle exchange network, and includes shufflenet and de Bruijn (DB) graph (see Section 3.5) as members of its family.

It has been conceived taking into account the following targets:

- Small nodal degree, to reduce node costs;
- Small diameter, to reduce end-to-end delays;
- Simple routing: to simplify the processing at the nodes and to increase their speed;
- Growth capability; to remove the problems encountered in more rigid topologies.

GEMNET can be realized on a passive star topology at physical level and organized on a regular WDM multihop architecture at virtual level for wide area [91] applications.

A $(k, m, p)$ GEMNET is composed by $k \times m$ nodes, having a degree of connectivity equal to $p$, that is, the number of transceivers per node, being $k, m, p$ integer numbers with $k \geq 1$ and $m \geq 1$. The nodes are arranged in $k$ columns, with $m$ nodes per column, connected by directed links [92] and wrapped around a cylinder as in shuffle-exchange connectivity pattern. It is immediately evident that this generalization removes the constraint of $N = kp^k$ or $N = p^d$ nodes, experienced in SNs and in DB graphs (see Section 3.5), respectively. Consequently, it follows that an arbitrary number of nodes $N$ can be connected through a regular topology by GEMNET.

A given node $a$ ($a = 0, 1, \ldots, N-1$), indicated by the $(c, r)$ pair, is placed on the $r$th row and on the $c$th column, where $c = (a \bmod k)$ with $c = 0, 1, \ldots, k-1$, and $r = \lfloor a / k \rfloor$ with $r = 0, 1, \ldots, m-1$, being $\lfloor x \rfloor$ the greatest integer smaller than $x$.

It must be noted that there are as many GEMNETs as there are divisors of $N$. Therefore, for a given value of $N$, at least two different configurations exist, for $m = 1$ and for $k = 1$. In particular, for $m = 1$ GEMNET reduces to a ring with consecutive nodes connected by $p$ parallel links; for $m = p^k$ it becomes coincident with a $(p, k)$ SN; finally for $m = p^d$ and $k = 1$ it reduces to a DB graph with diameter $d$ ($d = 2, 3, 4, \ldots$).

An example of GEMNET virtual topology is reported in Figure 3.37, in the (2, 5, 2) case. Starting from a node, all the nodes of a column have been visited for the first time after $\lceil \log_p m \rceil$ hops ($\lceil x \rceil$ being the smaller integer greater than $x$), whereas some uncovered nodes are present in the previous column. The remaining nodes will be covered in the subsequent $k-1$ hops. Therefore, the diameter of a $(k, m, p)$ GEMNET is equal to

$$D = \lceil \log_p m \rceil + k - 1 \tag{3.61}$$

The column distance, defined as the minimum number of hops needed to reach a node (not necessarily coincident with destination) belonging to the same column of destination $(c_d, r_d)$, starting from the source node $(c_s, r_s)$, is given by the formula $\delta = [(k + c_d) - c_s] \bmod k$ [93]. The hop distance between source-destination pair, instead, is equal to the smallest integer $h$ that assumes the form $(\delta + jk)$ with $j = 0, 1, 2, \ldots$, and satisfies the following expression:

$$R = \left[m + r_d - \left(r_s p^h\right) \bmod m\right] \bmod m < p^h \tag{3.62}$$

$R$ is called route code, since it indicates the shortest path from source to destination, if it is expressed as a sequence of $h$ base-$p$ digits. In particular, if $R = [a_1, a_2, \ldots, a_h]_{\text{base } p}$, the packet has to be sent toward the $a_j$ output link of the current node at its $j$th hop. Finally, if $p^h$ is greater than the number of nodes belonging to a column, multiple shortest paths can exist between the given pair of nodes to be connected.

The availability of multiple shortest paths offers the possibility of routing packets towards less loaded channels, trying to balance the link flows on the whole network. In this sense, we can distinguish three different approaches [93]:

- *Unbalanced routing scheme*: the traffic tends to be unbalanced since once multiple links are available, the one indicated by base-$R$ code is chosen;

- *Partially balanced routing scheme*: this method spreads the load across multiple shortest paths, limiting the choice to no more than $p$ alternatives;
- *Random routing scheme*: between all the available shortest paths, one is randomly chosen.

As regards link loading, we must take into account that random routing performs better than unbalanced and partially balanced routing scheme [93], being able to balance nonuniform traffic also.

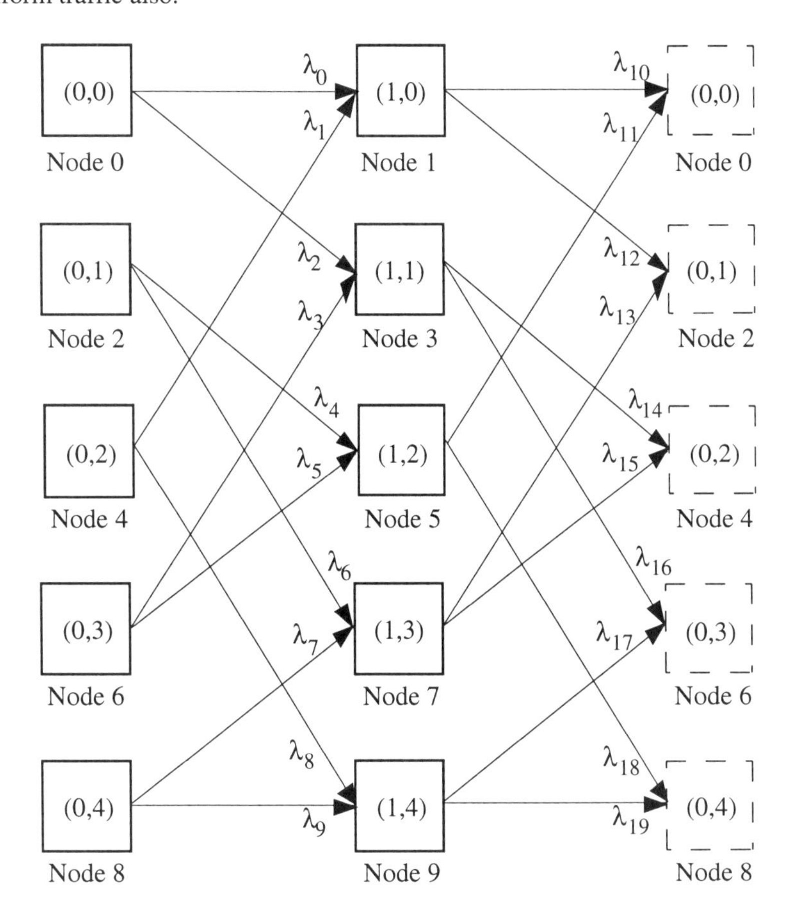


**Figure 3.37** Virtual topology and channels assignment of a (2, 5, 2) GEMNET.

A closed-form expression of the average distance between any pair of nodes in a GEMNET has not been found so far. Nevertheless, a method to evaluate the network performance has been used in [90], based on the calculus of the minimum and maximum average number of hops. It can be shown that they are given by the following formulas, respectively:

$$\bar{h}_{\max} = \frac{mk\left(D + \frac{1-k}{2}\right) - k\left(\frac{p^{D-k+1}-1}{p-1}\right)}{mk-1} \tag{3.63}$$

$$\bar{h}_{\min} = \frac{1}{N-1}\left[p\left(\frac{(L+1)p^{L}(p-1)-\left(p^{L+1}-1\right)}{(p-1)^2}\right) + \frac{mk(2L+k+1)}{2} - \delta(0, G-1)\right.$$

$$\cdot\frac{\left(p^k\right)^G - 1}{p^k - 1}\left(p^{Q+1}\frac{(k-Q)p^{k-Q-1}(p-1)-\left(p^{k-Q}-1\right)}{(p-1)^2} + (L+1)\frac{p^k-1}{p-1}\right.$$

$$\left. + \frac{p^Q\delta(k-Q, k-1)}{(p-1)^2}\left\{(p-1)\left(k-(k-Q)p^{-Q}\right) - p\left(1-p^{-Q}\right)\right\}\right)$$

$$- p^{Gk}\delta(Gk+k, L+k)\left((Gk+k)\frac{p^{L-Gk+1}-1}{p-1}\right.$$

$$\left.\left. + p\frac{(L-Gk+1)p^{L-Gk}(p-1)-\left(p^{L-Gk+1}-1\right)}{(p-1)^2}\right)\right] \tag{3.64}$$

where $G = \lfloor (L+1)/k \rfloor$, $Q = (L+1) \bmod k$ and the parameter $L$ indicates the last column in which exactly $p^i$ new nodes are reached. It can assume the following values:

$$L = \begin{cases} D-k & \text{if } \dfrac{\left(p^k\right)^{F+1}-1}{p^k-1}\, p^{(D-k) \bmod k} \le m \\ D-k-1 & \text{otherwise} \end{cases} \tag{3.65}$$

where $F = \lfloor (D-k)/k \rfloor$. The expression of $\bar{h}_{\max}$ is obtained considering the longer paths that connect a given source with all the other nodes of the network. This means that the spanning tree covering the GEMNET from that source is organized in order to minimize the new nodes reached at every hop, by overlapping (when possible) previously crossed nodes at each step.

On the contrary, to calculate $\bar{h}_{\min}$, we have to consider the tree that maximizes the number of new nodes reached with every hop. In general, step by step, we must visit only nodes that have never been reached before. If the values of $p$, $k$, $N$ are inserted in (3.64) and (3.65) properly, the former equation matches the equivalent formula for SNs and the lower bound for DB graphs [90].

The minimum average hop distance can be given also by the following more compact form

$$\bar{h}_{\min} = \frac{1}{N-1}\left[\sum_{i=0}^{L} ip^{i} + \sum_{i=L+1}^{L+k} i\left(m - p^{i \bmod k}\sum_{j=0}^{\lfloor i/k \rfloor - 1} p^{jk}\right)\right] \tag{3.66}$$

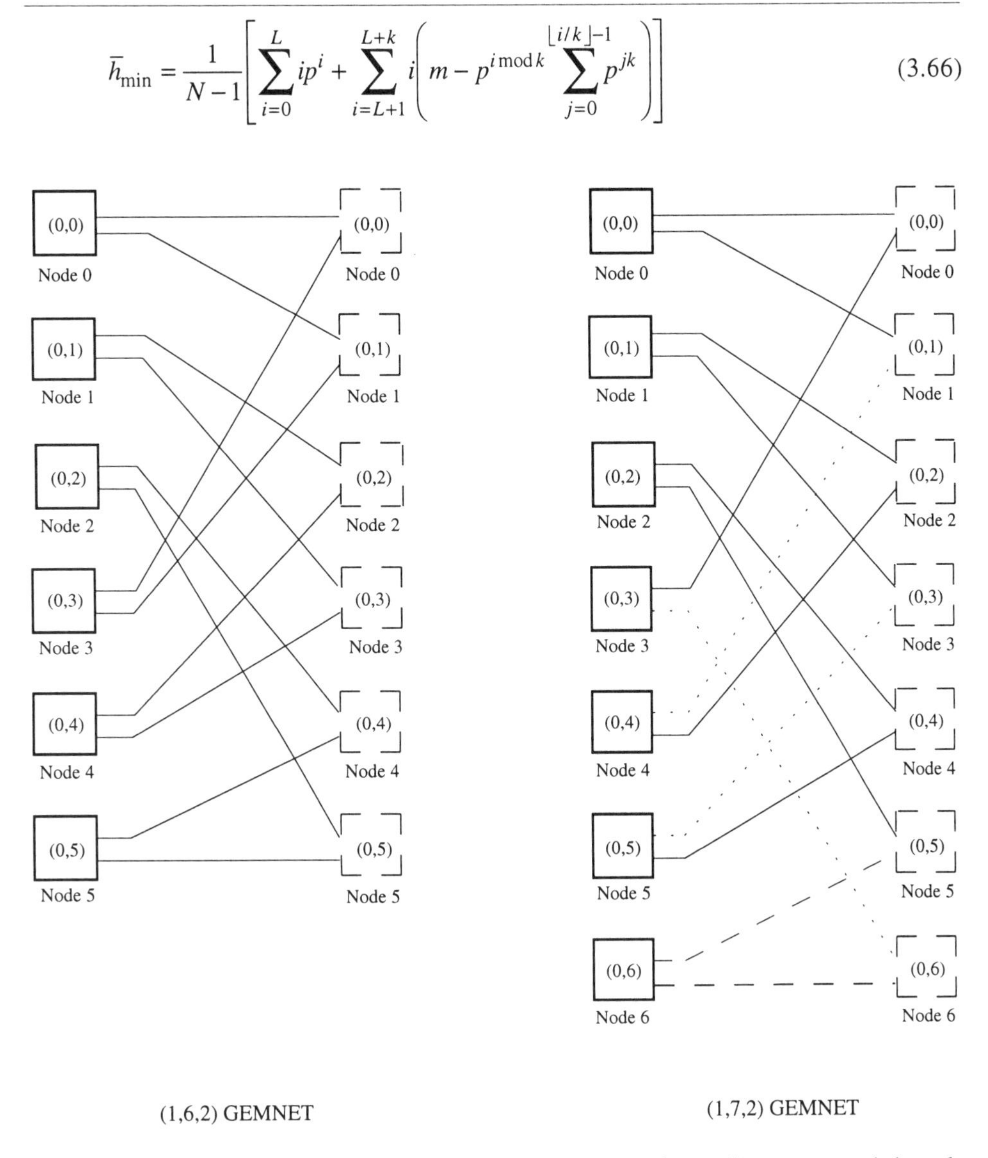


**Figure 3.38** Insertion of a new node in a (1, 6, 2) GEMNET (continuous lines = nontuned channels; dotted lines = retuned channels; dashed lines = pretuned channels).

We have the maximum uncertainty on the true value of the average number of hops in correspondence with the largest difference $\Delta h$ between $\bar{h}_{\max}$ and $\bar{h}_{\min}$. It can be verified that this condition occurs when $k = 1$, $p = 2$, and for high values of $N$; in this particular case it is interesting to observe that $\Delta h < 1$ for both cases of $L$.

In general, especially for $p \geq 3$, with $k = 1$ we can obtain the lowest average hop distance, even though in particular conditions ($p = 2$) the structure can be organized on few columns maintaining good performance [93]. Furthermore, when $k = 1$, some advantages are evident that are relative to network reconfiguration flexibility also.

In fact, beside low average distance among nodes and balanced link loading capabilities, another important feature is the network scalability. In particular, GEMNET structure is compatible with a one node at a time growth, when it is provided by just one column. This opportunity matches with the fact that one-column architecture is characterized, in general, by minimum average distance. Therefore, the best conditions as regards scalability and efficiency can be obtained through one-column architecture.

An example of network scaling is given in Figure 3.38, where a new node is inserted in a (1, 6, 2) GEMNET to obtain a (1, 7, 2) GEMNET (the wavelength reassignment is also considered in the same figure).

The insertion of new nodes in the bottom row is suggested, to simplify the rearrangement of the network [90]. Moreover, multiple node insertion can also be performed by adding new rows of nodes progressively, until all the new nodes have been connected. The number of transceiver retunings $r_k$, needed in case of $k$ nodes insertion in a $N$-node network, is given by [93]

$$r_k = \sum_{i=2}^{p} \left\lceil \frac{N}{p} \right\rceil (i-1) \geq \frac{N(p-1)}{2} \tag{3.67}$$

In conclusion, we can say that GEMNET is a very promising architecture, able to guarantee good properties both in terms of load balancing and easy scalability. It can represent an interesting structure for practical implementation of future WDM networks. Unfortunately, GEMNET is not always isotropic and this fact can penalize the average number of hops, as will be shown in Section 3.4.3.1.

To face the problem of limited availability of WDM channels, as previously done for SN, channel sharing may be adopted also in GEMNET. Besides that, it can be shown that channel sharing may be particularly effective when the network considered has to support multicast traffic. An analytical method for the study of this technique has been proposed in [94].

### 3.4.3 Enlarged Shufflenet Architecture

In this section we present another modification of SN topology, which allows one to connect a larger number of nodes with a negligible penalty in network efficiency, without increasing the complexity at node level. It is shown that the new architecture, called enlarged shufflenet (ESN) [95], since it is based on an optimized connectivity graph, also compares favorably with the GEMNET architecture presented in Section 3.4.2. At a parity of connected nodes, in fact, it allows to reduce the average number of hops.

Instead of $p^k$ nodes, each column of the enlarged shuffle structure consists of $p^k + 1$ nodes, so that the total number of connected stations becomes

$$N_{\text{ESN}} = k\left(p^k + 1\right) \tag{3.68}$$

In order to have the same network diameter of the original SN (i.e., the same maximum number of hops) ESN must exhibit a different connectivity graph, especially as regards the closure connection between the last and the first columns. Some examples are shown in Figures 3.39 to 3.41, for different values of $p$ and $k$. To preserve the intelligibility of the graphs, the wavelengths assigned to the various virtual connections have been explicitly shown only in Figure 3.39. The same assumption of one distinct wavelength per each link, however, must be considered applied to Figures 3.40 and 3.41, as well as to any other, even more general, enlarged topology.

The idea of enlarging the structure arose from the observation that, in practical cases, the traffic matrix is characterized by diagonal elements equal to zero; it is in fact rather unusual that a node uses the network to receive messages it transmitted itself. In this condition, the modified architecture allows to reach $p^k$ nodes after $k$ hops (instead of $p^k - 1$, as in the original SN) thus better utilizing, in some respects, all the connections.

Although a rigorous mathematical formulation of the criteria here presented could be given [96], in searching for the optimized enlarged structures above, its conclusions can be summarized in a few simple rules, immediately usable for design purposes. In practice, two kinds of virtual connections can be identified in the enlarged architecture:

1. Those involving the columns from 1 to $k$;
2. Those between the $k$th and the first column (closure connections).

In case (1), the wavelengths are assigned following the same principle used for the perfect shuffle: the first node of each column transmits to the first $p$ nodes of the next column, the second node of each column transmits to the second group of $p$ nodes of the next column, and so on. When all nodes of the next column have been linked once, they are connected once more proceeding from top to bottom. In case (2) (which is the novelty of this approach) this connection order is inverted. Thus, the first node of the $k$th column transmits to the last $p$ nodes of the first column, the second node of the $k$th column transmits to the last group but one of $p$ nodes of the first column, and so on. When all nodes of the first column have been linked once, they are connected once more proceeding from bottom to top. In other words we can say that the algorithm used for connecting forward the nodes belonging to the $k$th column is antisymmetric with respect to that used for connecting forward the nodes of all the other columns. This difference has important implications, for example, as regards the average number of hops between two randomly selected users.

If the network is *isotropic* (which means that the routing trees look the same regardless of source node) the average number of hops can be computed, as usual, by using the following general expression:

$$\langle h \rangle = \frac{1}{N-1} \sum_{h=1}^{2k-1} h g_h \tag{3.69}$$

where, as assumed in previous sections, $g_h$ represents the number of nodes that are $h$ hops away from any considered node.

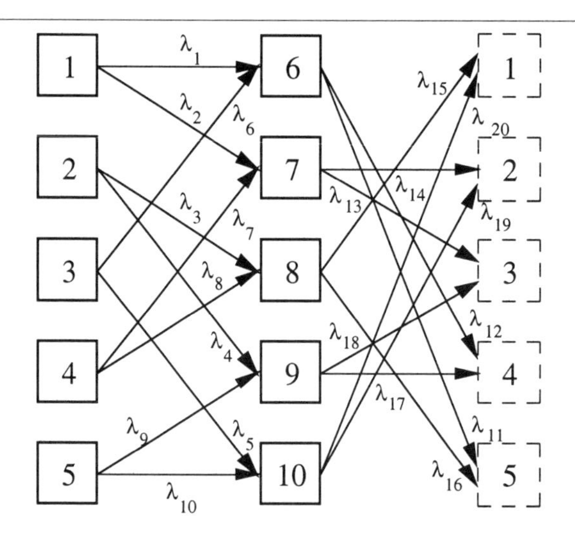


**Figure 3.39** Enlarged shufflenet for the case $p = 2$ and $k = 2$.

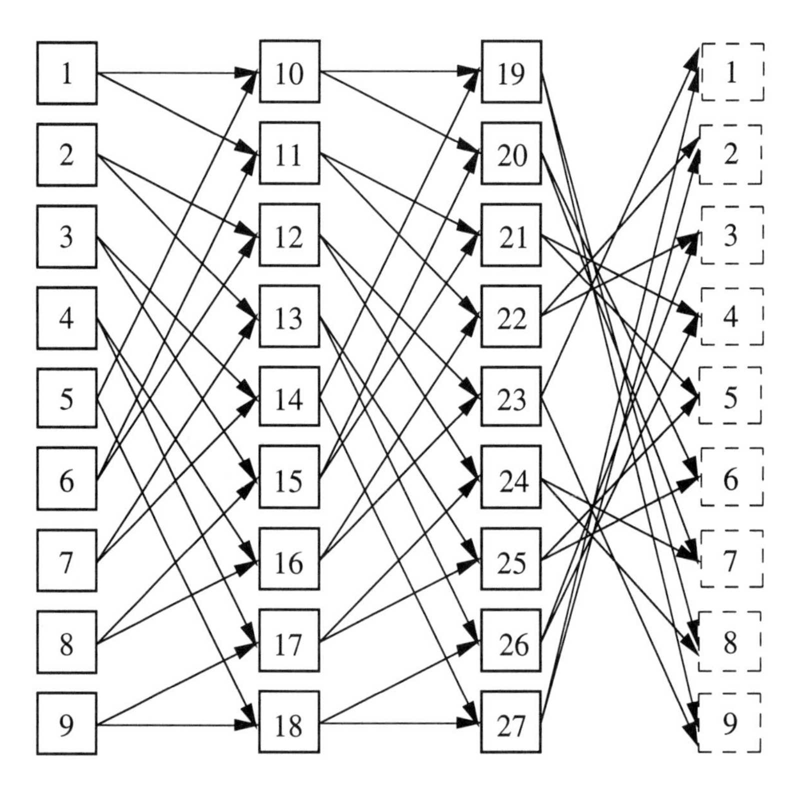


**Figure 3.40** Enlarged shufflenet for the case $p = 2$ and $k = 3$.

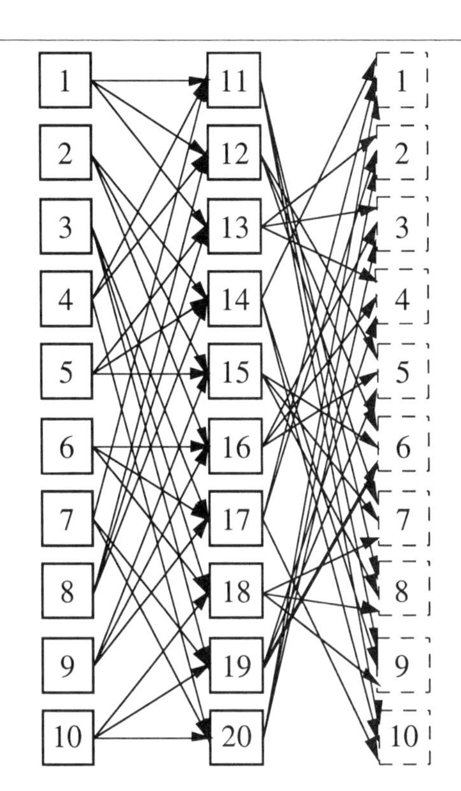


**Figure 3.41** Enlarged shufflenet for the case $p = 3$ and $k = 2$.

The isotropic property, which is verified for both SN and ESN, ensures that the value of $g_h$ is the same for all source nodes. In particular, the couples $h$-$g_h$ for ESN, directly derivable from the connectivity graph, are reported in Table 3.11 (see for comparison Table 3.5). Inserting them into (3.69) we obtain, with simple algebra,

$$\langle h_{\mathrm{ESN}} \rangle = \frac{k}{k\left(p^k + 1\right) - 1}\left[\frac{3}{2}\left(p^k + 1\right)(k-1) + p\left(\frac{p^k - 2p^{k-1} + 1}{p-1}\right)\right] \tag{3.70}$$

This expression can be compared with the one valid for a shufflenet and reported in (3.35), which results also in the form [97]

$$\langle h_{\mathrm{SN}} \rangle = \frac{k}{kp^k - 1}\left[p^k\left(\frac{3k-1}{2}\right) - \frac{p^k - 1}{p-1}\right] \tag{3.71}$$

A more significant analysis can be developed on the basis of the network efficiency $\eta$ that, assuming a uniform traffic pattern and a routing algorithm that balances the traffic load on all the WDM channels, is simply given by the reciprocal of the average number of hops. Thus, we have $\eta_{\mathrm{SN}} = 1/\langle h_{\mathrm{SN}} \rangle$ and $\eta_{\mathrm{ESN}} = 1/\langle h_{\mathrm{ESN}} \rangle$, respectively, with in general, $\eta_{\mathrm{ESN}} < \eta_{\mathrm{SN}}$.

**Table 3.11**
Number of Nodes that Are $h$ Hops away from any Given Source Node for a $(p, k)$ Enlarged Shufflenet

| $h$ | $g_h$ |
|---|---|
| 1 | $p$ |
| 2 | $p^2$ |
| . | . |
| . | . |
| $k-1$ | $p^{k-1}$ |
| $k$ | $p^k$ |
| $k+1$ | $p^k - p + 1$ |
| $k+2$ | $p^k - p^2 + 1$ |
| . | . |
| . | . |
| $2k-1$ | $p^k - p^{k-1} + 1$ |

A direct comparison between the original and the modified structures is not possible, since the networks always have a different number of nodes. Nevertheless, we can give an idea of the performance obtainable by plotting the opposite of the percent variation in the network efficiency $\Delta\eta / \eta_{SN}\% = (\eta_{ESN} / \eta_{SN} - 1) \cdot 100$ versus the percent variation in the number of nodes $\Delta N / N_{SN}\% = (N_{ESN} / N_{SN} - 1) \cdot 100$. This is shown in Figure 3.42 for several $(p, k)$ pairs.

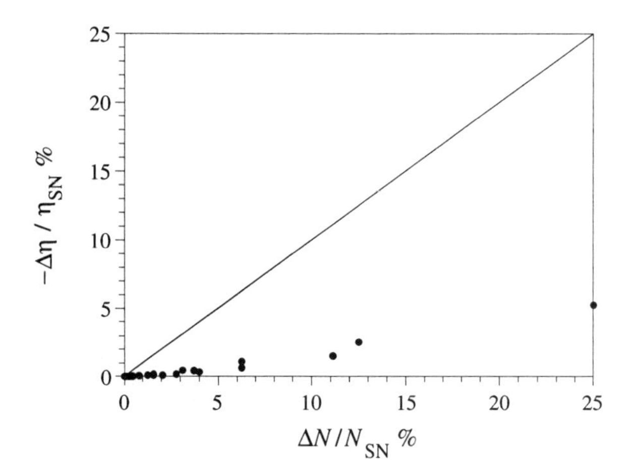


**Figure 3.42** Percent variation of the efficiency as a function of the percent variation of the number of nodes.

Since the dots are all far below the line $-\Delta\eta / \eta_{SN}\% = \Delta N / N_{SN}\%$ (drawn for the sake of clarity), we can say that the positive effect of increasing the number of nodes largely compensates the negative effect of decreasing the network efficiency. More specifically, we can see that there are several situations for which ESN ensures a relevant (in percent) increase in the number of connected nodes, while preserving almost the same value of efficiency of the original SN. For very high values of $k$ and $p$ (interesting in theory but difficult to realize in practice) all the dots collapse in the origin, since for networks of great size ESN and SN become practically indistinguishable.

#### *3.4.3.1 Comparison with GEMNET*

The idea of extending the SN structure has also been pursued recently, as discussed above, in the so-called GEMNETs. The main object of GEMNET is to increase scalability, that is, the capability for the network to grow without the severe constraints imposed by SN architecture. In GEMNET, contrary to SN, there is no restriction in the number of nodes, and the topology represents a very wide family of arbitrary-sized networks including, as special cases, SN as well as other structures (like the DB graph). Thus, we have a chance to compare the performance of GEMNET with that of ESN; in fact, this can be done by setting the parameters of GEMNET in such a way as to have $m = p^k + 1$ nodes per each column.

To construct the connectivity graph, GEMNET uses the same rules of the perfect shuffle. In this sense, it is not necessarily optimized, and the corresponding ESN behaves better, for example, as regards the average number of hops. A simple explanation of this conclusion can be found in Table 3.12, where we have reported the couples $h$-$g_h$ for the graphs of Figure 3.37 and Figure 3.39, respectively. The GEMNET with 10 nodes arranged in $k = 2$ columns is not isotropic, and 8 nodes have one station at a distance of 4 hops. This obviously penalizes the average number of hops, which is expected to be higher than the one resulting from ESN, in the same conditions. The latter is isotropic, with a maximum distance of 3 for all the nodes.

**Table 3.12**

Couples $h$-$g_h$ for $p = k = 2$: (a) (2, 5, 2) GEMNET; (b) (2, 2) Enlarged Shufflenet.

(a)

Node $i \neq 3, 8$

| $h$ | $g_h$ |
|---|---|
| 1 | 2 |
| 2 | 3 |
| 3 | 3 |
| 4 | 1 |

Node $i = 3, 8$

| $h$ | $g_h$ |
|---|---|
| 1 | 2 |
| 2 | 4 |
| 3 | 3 |

(b)

| $h$ | $g_h$ |
|---|---|
| 1 | 2 |
| 2 | 4 |
| 3 | 3 |

Because of its anisotropy, the value of $\langle h_{GN} \rangle$ for GEMNET must be computed as

$$\langle h_{GN} \rangle = \frac{1}{N(N-1)} \sum_{j=1}^{N} \sum_{h=1}^{h_{max}(j)} h g_h(j) \tag{3.72}$$

where $g_h(j)$ represents the value of $g_h$ for the $j$th source node and $h_{max}(j)$ is the corresponding maximum number of hops. On the basis of Table 3.12(a), we have $\langle h_{GN} \rangle = 2.289$ while applying (3.70), on the basis of Table 3.12(b), we find $\langle h_{ESN} \rangle = 2.111$, which is about 8% smaller.

In conclusion, ESN gives an example of a different arrangement of the connectivity graph (with respect to the basic perfect shuffle) that is potentially able to improve performance. In particular, it allows the average hop distance to be maintained sufficiently low while increasing the number of nodes. The choice of the best topology for constructing the next generation of lightwave networks, however, will also take into account other properties, such as the capacity of balancing the load on the various links [98]. For this reason, and to improve scalability, the addition of further degrees of freedom in the network design (like in GEMNET) should be considered, even if this may imply some variation with respect to the connection strategy above described.

### 3.4.4 Modification of the Shufflenet Connectivity Graph

The main characteristics of SNs have been presented in the previous sections, as regards, in particular, the total number of wavelengths and the average number of hops in the case of uniform traffic distribution. When such quantities are combined to derive the aggregate capacity or the network efficiency, however, the optimistic assumption that the traffic is balanced on all the WDM channels is often made. In reality, it is easy to prove that when a shortest path algorithm is used for routing, this hypothesis is not true, and indeed that this leads to worsening in the performance with respect to the expected values. The extent of such worsening can be remarkable, especially when the number of connected nodes $N$ is small. Let us consider the case of $N = 8$. The connectivity graph of the perfect shuffle has been shown in Figure 3.19, with $k = 2$ and $p = 2$, where $\lambda_i$, being $i \in [1,16]$, represents the generic wavelength.

Now, let us make the uniform traffic assumption: a cell that originates at any given source station is equally likely to be destined to any of the other $N - 1$ stations. Formally, this corresponds to having a uniform traffic matrix with zero diagonal elements. Furthermore, to reinforce this idea, we suppose that the network operates in overload conditions (i.e., at any instant, each node has at least one cell to transmit to any other node). We call the *transmission cycle* a time period long enough to permit, on average, each station to transmit one cell to any other node of the network and to receive one cell from all of them as well.

The aggregate capacity $C_T$ can be computed as the ratio between the number of cells delivered in a transmission cycle, multiplied by the bit rate per channel $S$, and the number of time slots required to complete the transmissions. Both these quantities can be easily derived for the network shown in Figure 3.19. Let us consider, to this purpose, Figure 3.43. It represents a possible slot assignment, in a transmission cycle, under the hypothesis of using the shortest path as the routing algorithm; $T_h$ represents the $h$th time slot and $x/y$ denotes a cell emitted by node $x$ (source) and addressed to node $y$ (final destination). Because of the multihop mechanism, the cells using more than one wavelength appear with their own multiplicity in the slot assignation table of Figure 3.43. The shaded cells correspond to the links univocally determined by the shortest path algorithm, while the others (not shadowed) correspond to "don't care" paths (two different paths, connecting a given couple of nodes, are don't care when they consist of the same number of hops), which are randomly selected. The slot assignation table considered is not unique, in the sense that the order of the assignation and, most of all, the assignation of the don't care paths can be obviously rearranged without altering the features of the table itself. In any case, in fact, the conclusion is the same: half the wavelengths support, in each transmission cycle, two cells more than the other half. Thus the load is not balanced, and it is interesting to note that the unbalancing is due to the cells univocally attributed by the shortest path algorithm and not to the don't care paths.

| | $\lambda_1$ | $\lambda_2$ | $\lambda_3$ | $\lambda_4$ | $\lambda_5$ | $\lambda_6$ | $\lambda_7$ | $\lambda_8$ | $\lambda_9$ | $\lambda_{10}$ | $\lambda_{11}$ | $\lambda_{12}$ | $\lambda_{13}$ | $\lambda_{14}$ | $\lambda_{15}$ | $\lambda_{16}$ |
|---|---|---|---|---|---|---|---|---|---|---|---|---|---|---|---|---|
| $T_1$ | 1/6 | 1/8 | 3/2 | 3/4 | 5/6 | 5/8 | 7/2 | 7/4 | 2/5 | 2/7 | 4/1 | 4/3 | 6/5 | 6/7 | 8/1 | 8/3 |
| $T_2$ | 1/3 | 2/5 | 3/1 | 2/7 | 4/1 | 5/7 | 4/3 | 7/5 | 2/4 | 1/6 | 4/2 | 1/8 | 3/2 | 6/8 | 3/4 | 8/6 |
| $T_3$ | 3/2 | 6/5 | 1/6 | 6/7 | 8/1 | 3/4 | 8/3 | 1/8 | 4/1 | 5/6 | 2/5 | 5/8 | 7/2 | 4/3 | 7/4 | 2/7 |
| $T_4$ | 7/2 | 1/5 | 5/6 | 3/5 | 5/1 | 7/4 | 7/1 | 5/8 | 8/1 | 1/3 | 6/5 | 4/6 | 6/2 | 8/3 | 8/4 | 6/7 |
| $T_5$ | 6/2 | 1/7 | 3/6 | 3/7 | 5/3 | 5/4 | 7/3 | 7/8 | 5/1 | 2/6 | 1/5 | 4/8 | 6/4 | 6/3 | 3/5 | 8/7 |
| $T_6$ | 1/2 | 2/4 | 2/6 | 6/8 | 4/2 | 8/4 | 4/6 | 4/8 | 2/1 | 2/8 | 4/5 | 1/7 | 3/1 | 7/3 | 8/2 | 3/7 |
| $T_7$ | | 6/4 | | 2/8 | 8/2 | | 8/6 | | | 5/3 | | 5/7 | 7/1 | | 7/5 | |
| $T_8$ | | 1/4 | | 3/8 | 5/2 | | 7/6 | | | 2/3 | | 4/7 | 6/1 | | 8/5 | |

**Figure 3.43** Example of slot assignment in a transmission cycle for a (2, 2) shufflenet; the shaded cells correspond to the links univocally determined by the shortest path algorithm.

From a different point of view, a qualitative explanation of the unbalanced nature of the load can also be given looking directly at Figure 3.19 and considering, for example, the two wavelengths, $\lambda_1$ and $\lambda_2$, used by node 1 for transmission. Confining ourselves to examining the paths with one or two hops (those with three hops are don't care and the relative traffic is always balanced), we observe that $\lambda_1$ is used:

1. To connect node 0 to node 4;
2. To forward to node 4 the cell emitted by node 0 and addressed to node 1;
3. To repeat the cells coming from node 6 but addressed to node 4.

On the contrary, $\lambda_2$ is used:

4. To connect node 0 to node 5;
5. To forward to node 5 the cells emitted by node 0 and addressed to nodes 2 and 3;
6. To repeat the cells coming from nodes 4 and 6 but addressed to node 5.

Comparing (2) with (5) and (3) with (6), we note that, in each transmission cycle, $\lambda_2$ must support two cells more than $\lambda_1$; the same remark holds for any pair of channels $(\lambda_{2j-1}, \lambda_{2j})$ with $j \in [1,8]$.

Since $N(N-1) = 56$ connections must be established in a transmission cycle and, according to Figure 3.43, the number of time slots required is equal to 8, the aggregate capacity for the case $k = p = 2$ is $C_T = 7S$.

The conclusions drawn for the network of Figure 3.19 can be extended to a perfect shuffle of any size. In general, the wavelengths with higher load must support, in each cycle, $k$ cells more than those with lower load, and these "extra cells" are always due to the paths with $k$ hops. Taking this into account, the evaluation of the duration $T$ (in time slots) of a transmission cycle ($T = 8$ in Figure 3.43) can be made as follows.

First, the number of cells (with their multiplicity in the sense above specified) circulating inside the network in a transmission cycle is computed. Such a calculus can be easily developed starting from the knowledge of the number of nodes $g_h$ which are $h$ hops away from any considered node. This information, reported in Table 3.5, is extractable directly from the connectivity graph and holds for both the perfect shuffle and the modified shuffle presented in this section.

The total number of cells is evaluated multiplying the number of nodes $N = kp^k$ by $\sum_h hg_h$; to this number, however, $k$ extra cells must be artificially added, once again multiplied by the number of nodes, to take into account the load problem.

At this point, the result can be divided by the number of wavelengths $W = kp^{k+1}$, obtaining in this way the value of $T$ we are searching for. With simple algebra, on the basis of Table 3.5, it is possible to verify that [98]

$$T = \frac{N\left(\sum_{h=1}^{2k-1} hg_h + k\right)}{W} = k\left[\frac{p^{k-1}(3k-1)}{2} - \frac{p^{k-1}-1}{p-1}\right] \tag{3.73}$$

Finally, inserting (3.73) in the definition of the aggregate capacity, we have

$$C_T = \frac{N(N-1)}{T}S = \frac{2p^k(p-1)\left(kp^k-1\right)}{p^{k-1}(3k-1)(p-1) - 2\left(p^{k-1}-1\right)}S \tag{3.74}$$

Expression (3.74) is different from the one reported in most of the previous literature [97,99] since it has been derived, with the above procedure, considering that the channels $\lambda_i$ ($i = 1, 2, \ldots, kp^{k+1}$) are not equally loaded.

The problem of improving the perfect shuffle connectivity graph, even in the most general case of nonuniform point-to-point traffic matrices was faced, for example, in [100], just for a SN with $N = 8$. The heuristic procedure developed in that case, however, substantially based on a local search, was not successful in its attempt to find a better connectivity diagram for the specific case of uniform traffic matrix with zero diagonal elements. For this reason, it is of interest to solve this problem separately, completing, in some respects, the analysis of [100] and related papers [81]. The proposed structure [98] is shown in Figure 3.44. The philosophy adopted is rather simple: each station transmits toward the stations of the subsequent column that are not directly received from the previous column. For example, station 2 transmits toward stations 3 and 5, whereas it receives from stations 1 and 7.

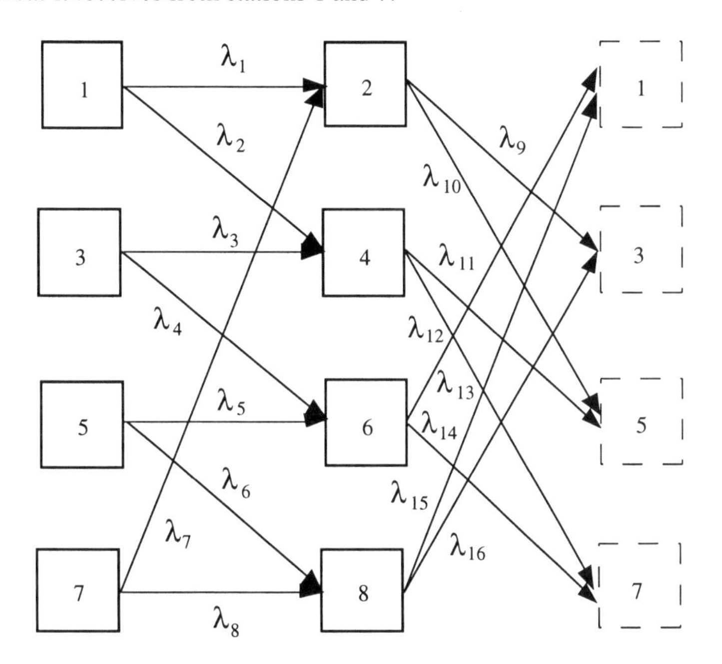


**Figure 3.44** Connectivity graph of the modified shuffle.

Maintaining unchanged the routing algorithm (shortest path with random selection of don't care paths), a possible slot assignment in a transmission cycle is the one reported in Figure 3.45. The meaning of the cells *x/y*, shadowed or not, is the same discussed in Figure 3.43. Contrary to the perfect shuffle, here we have some don't care paths with two hops also (they concern, precisely, the connections 1/5, 5/1, 2/6, 6/2, 3/7, 7/3, 4/8, 8/4). Anyway, distributing uniformly the corresponding links, Figure 3.45 evidences that all the channels are equally loaded.

Since Table 3.5 holds also for the modified structure, the duration of the transmission cycle is here given by

$$T^{(M)} = \frac{N\sum_{h=1}^{2k-1} hg_h}{W} = \frac{k}{p}\left[\frac{p^k(3k-1)}{2} - \frac{p^k - 1}{p-1}\right] \tag{3.75}$$

clearly more favorable than (3.73). Correspondingly, the aggregate capacity results in

$$C_T^{(M)} = \frac{N(N-1)}{T^{(M)}}S = \frac{2p^{k+1}(p-1)\left(kp^k-1\right)}{p^k(3k-1)(p-1)-2\left(p^k-1\right)}S = \frac{W}{\langle h \rangle}S \tag{3.76}$$

in turn larger than (3.74). In (3.75) and (3.76), superscript ($M$) states that we are referring to the modified graph; the same notation will be also adopted in the sequel. Expression (3.76) is the most familiar one in previous literature [97,99], really corresponding to the assumption of balanced traffic load, now verified thanks to the modified connectivity graph. By replacing $k = p = 2$, we obtain $C_T^{(M)} = 8S$, with an increase of more than 14% with respect to the unbalanced case. For instance, assuming $S = 1$ Gbps the aggregate capacity passes from 7 to 8 Gbps. The same percentage increase can be found in the network efficiency $\eta$; in fact

$$\eta^{(M)} = \frac{C_T^{(M)}}{WS} \tag{3.77}$$

is equal to 0.5, whereas $\eta = C_T/(WS) = 0.4375$ for the connectivity graph of Figure 3.19.

| | $\lambda_1$ | $\lambda_2$ | $\lambda_3$ | $\lambda_4$ | $\lambda_5$ | $\lambda_6$ | $\lambda_7$ | $\lambda_8$ | $\lambda_9$ | $\lambda_{10}$ | $\lambda_{11}$ | $\lambda_{12}$ | $\lambda_{13}$ | $\lambda_{14}$ | $\lambda_{15}$ | $\lambda_{16}$ |
|---|---|---|---|---|---|---|---|---|---|---|---|---|---|---|---|---|
| $T_1$ | 1/3 | 1/6 | 3/2 | 3/1 | 5/1 | 5/2 | 7/4 | 7/1 | 2/1 | 2/7 | 4/3 | 4/1 | 6/3 | 6/2 | 8/2 | 8/5 |
| $T_2$ | 1/5 | 1/7 | 3/5 | 2/1 | 2/7 | 4/3 | 6/2 | 4/1 | 1/3 | 2/8 | 1/6 | 3/2 | 3/1 | 6/5 | 8/4 | 8/6 |
| $T_3$ | 1/8 | 8/4 | 3/7 | 3/8 | 5/4 | 2/8 | 3/2 | 7/3 | 2/4 | 1/5 | 3/5 | 1/7 | 2/1 | 2/7 | 7/1 | 4/3 |
| $T_4$ | 1/2 | 1/4 | 2/4 | 8/6 | 1/6 | 5/3 | 7/5 | 7/6 | 2/6 | 1/8 | 4/6 | 3/7 | 5/4 | 3/8 | 5/2 | 7/3 |
| $T_5$ | 6/3 | 5/4 | 8/5 | 2/6 | 4/6 | 1/8 | 6/5 | 3/8 | 7/4 | 2/5 | 4/8 | 4/2 | 5/1 | 6/8 | 8/7 | 5/3 |
| $T_6$ | 8/2 | 8/7 | 7/4 | 3/6 | 5/7 | 4/8 | 4/2 | 6/8 | 6/3 | 7/5 | 8/5 | 4/7 | 6/4 | 6/7 | 8/1 | 7/6 |
| $T_7$ | 5/2 | 6/4 | 3/4 | 7/6 | 5/6 | 5/8 | 7/2 | 7/8 | 2/3 | 6/5 | 4/5 | 8/7 | 6/1 | 5/7 | 4/1 | 8/3 |

**Figure 3.45** Example of slot assignment in a transmission cycle for the modified shuffle; the shaded cells correspond to the links univocally determined by the shortest path algorithm.

The modification of the perfect shuffle topology we have illustrated in the case $N = 8$ could be extended, in principle, to greater values of $k$ and $p$. The advantage one can find from this change, however, becomes smaller and smaller for increasing $N$.

More precisely, noting by $F$ the extent of such improvement, it is possible to show that [98]

$$F = \frac{C_T^{(M)}}{C_T} = \frac{\eta^{(M)}}{\eta} = 1 + \frac{2}{p^k(3k-1) - 2\sum_{j=0}^{k-1} p^j} \qquad (3.78)$$

This expression has been plotted in Figure 3.46 for different values of the overall number of stations. For $N = 18$ the improvement reduces to about 5%, whereas for $N > 100$ it becomes smaller than 1% [98]. This behavior is qualitatively expected since, for increasing values of $N$, the effect of the zero diagonal elements in the uniform traffic matrix (which is responsible for the load unbalance) becomes less and less important.

The choice of $N = 8$ is justified by the observation that only networks with a small number of nodes look realistic on the basis of the present optical technology; supposing to use a fully broadcast physical topology (which is the simplest one to realize) the number of stations is limited by the maximum number of multiplexed wavelengths. Anyway, the basic idea that suggested the modification can also be extended to networks of greater size.

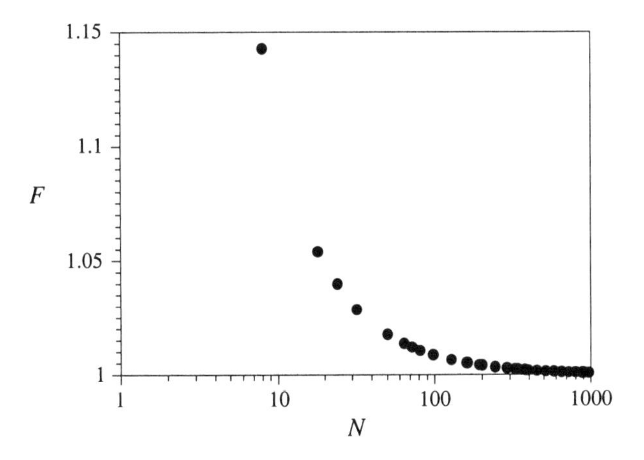


**Figure 3.46** Improvement resulting from balanced load as a function of the network size.

### 3.4.5 Banyan Net

As evidenced in the previous sections of this chapter, many attempts have been made in order to overcome the disadvantages of SN, essentially due to nonsymmetric node distance (the distance from node $i$ to node $j$ does not equal the distance from node $j$ to node $i$) and difficult network expansion.

As regards the solution of the first problem, among the other possible solutions presented so far, the bilayered shufflenet (BS) has been proposed [101]. It uses the basic SN node placement, but connects adjacent columns through a symmetric set of unidirectional links, as shown in Figure 3.47. Now, maintaining the same node number of SN, the degree of connectivity is increased at $2p$. This solution exhibits a good behavior but at the expense of higher implementation costs. Moreover, this architecture does not solve the above-mentioned problems.

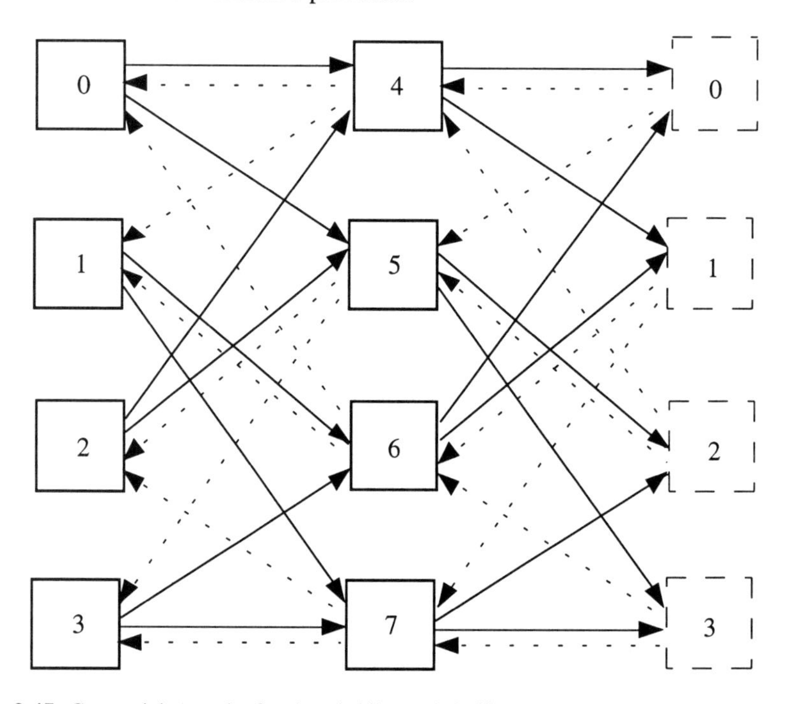


**Figure 3.47** Connectivity graph of an 8-node bilayered shufflenet.

On the contrary, in this section, we will focus on another layout that seems to offer a better performance. The network based on this particular structure, called SW-Banyan net (SW-BN), is topologically equivalent to SN [102].

There are many structures similar or equivalent to SW-Banyan networks, such as omega [103,104], delta [66], baseline [102], Banyan networks [105], presented below and others. They are collectively known as multistage interconnection networks (MINs) [103], since they have been considered for the interconnection of multiple processor and memory elements in computer or switching multistage architectures. Here, instead, we analyze the utilization of similar solution as WDM network virtual topologies.

Regular SW-BN is a subclass of the BN family, composed by $N = p^m \times k$ nodes, each of them identified by a $(x, y)$ pair, where $x \in \{0, 1, \ldots, k-1\}$ and $y \in \{0, 1, \ldots, p^m - 1\}$; in particular $y = (y_0, y_1, \ldots, y_{m-1})$ and also $y = y_0 p^{m-1} + \cdots + y_{m-1} p^0$.

A generic $(x, y)$ node is connected with the following adjacent nodes:

$$([x + 1] \bmod k, y_0, \ldots, y_{r-1}, 0, y_{r+1}, \ldots, y_{m-1})$$
$$([x + 1] \bmod k, y_0, \ldots, y_{r-1}, 1, y_{r+1}, \ldots, y_{m-1})$$
$$\ldots$$
$$([x + 1] \bmod k, y_0, \ldots, y_{r-1}, p, y_{r+1}, \ldots, y_{m-1})$$

$$([x-1] \bmod k, y_0, \ldots, y_{r-2}, 0, y_r, \ldots, y_{m-1})$$
$$([x-1] \bmod k, y_0, \ldots, y_{r-2}, 1, y_r, \ldots, y_{m-1})$$
$$\ldots$$
$$([x-1] \bmod k, y_0, \ldots, y_{r-2}, p, y_r, \ldots, y_{m-1})$$

being $r = x \bmod m$ and $y_i \in \{0, 1, \ldots, p-1\}$.

SW-Banyan topology is wrapped around a cylinder, so that the first and the last column are connected, as in SN architecture. On the contrary, unlike in SN, the number of columns $k$ is not limited to $m$, but is a multiple of it (this aspect can increase the flexibility of the network, making its expansion easier) and all the links are bidirectional. It results in a cylindrical and multiple cascaded version of the SW-Banyan proposed in [105], that from now on will be called Banyan net (BN) [106].

The network performance of BN, in terms of throughput and efficiency, can be significantly improved with respect to SN, as will be shown later. Other properties of BN, when applied in multiprocessors or in multistage switching fabrics, are reported in [107,108] and also in [109], where the hybrid family of Banyan-hypercube networks is presented and discussed. The implementation of a $p^m \times k$ BN is shown in Figure 3.48, for $p = 2$, $m = 2$, and $k = 4$.

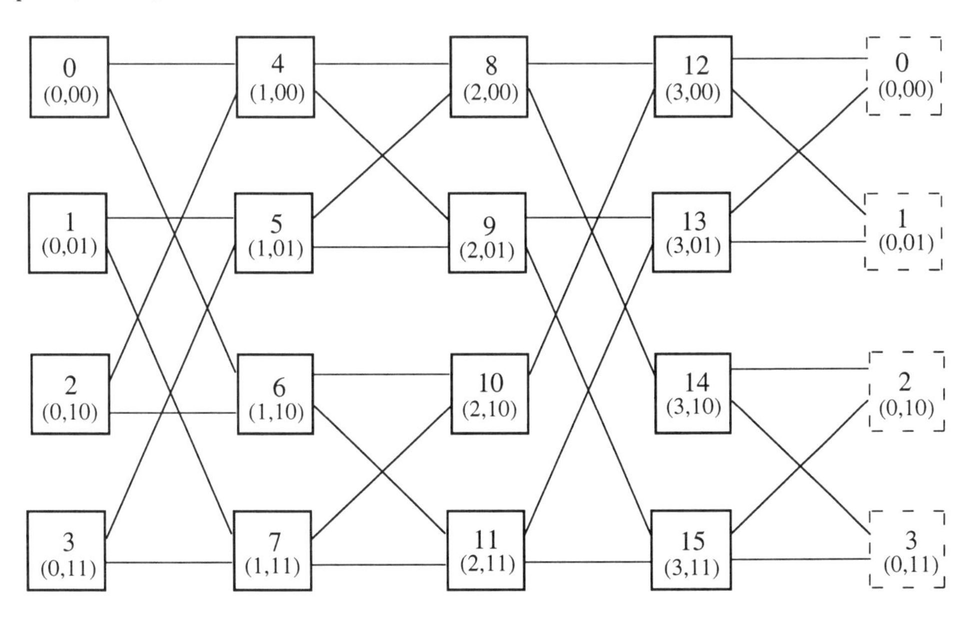


**Figure 3.48** Connectivity graph of a (2, 2, 4) Banyan net.

Because of the bidirectionality of the links, the degree of connectivity of each node equals $2p$, as in bilayered shufflenet, every transceiver pair working at its proper wavelength $\lambda_{ij}$ (with $i, j = 0, 1, \ldots, kp^m - 1$) to connect node $i$ to node $j$. Considering a generic transmitting node $(x, y)$ and assuming from now on $p = 2$, we can distinguish four types of transmission:

1. Forward straight: $(x, y)$ connected with $([x+1] \bmod k, y)$
2. Forward exchange: $(x, y)$ connected with $([x+1] \bmod k, y_0, \ldots, \bar{y}_r, \ldots, y_{m-1})$
3. Reverse straight: $(x, y)$ connected with $([x-1] \bmod k, y)$
4. Reverse exchange: $(x, y)$ connected with $([x-1] \bmod k, y_0, \ldots, \bar{y}_{r-1}, \ldots, y_{m-1})$

whereas more involved situations are managed by routing schemes when $p > 2$.

An interesting characteristic, that makes the adoption of relatively simple routing algorithm possible, is that all the nodes belonging to the same row are connected by a ring. In this sense a $p^m \times k$ BN can be seen as composed by $2^m$ interconnected rings of $k$ nodes each. The diameter of a $2^m \times k$ BN is equal to [106]

$$D_{\text{BN}} = \begin{cases} m + \lfloor m/2 \rfloor & \text{if } k = m \\ \max(2m, \lfloor k/2 \rfloor) & \text{if } k > m \end{cases} \tag{3.79}$$

Expression (3.79) results from the following considerations. If $k = m$, the distance between nodes on the same column can remain lower than or equal to $m$ [106], whereas nodes separated by $i$ columns, with $1 \le i \le \lfloor m/2 \rfloor$, are $m + i$ hops far at the most. In the latter case, in fact, $i$ hops are needed to reach the destination column and other $m$ hops are requested when the considered node is not placed on the same ring of the source. Therefore, taking into account the link bidirectionality, the packet propagation cannot be longer than $m + \lfloor m/2 \rfloor$.

If $k > m$, nodes closer than $m$ columns from the source are reached in no more than $2m$ hops, since no more than $m$ hops are needed to reach the desired ring and other $m$ hops are used to reach destination along the ring itself. For nodes separated by $i$ columns from the source one, with $m < i \le \lfloor k/2 \rfloor$, the packet propagation is $i$ hops long; the first $m$ used to reach the ring, the last $i - m$ to move along it. Thus, for $k > m$, the diameter cannot be greater than $\max(2m, \lfloor k/2 \rfloor)$.

To compare BN with SN we introduce the concept of generalized shufflenet (G-SN) [106], that is composed by $N = p^m \times k$ nodes, with $k$ multiple of $m$, arranged in k columns of $p^m$ nodes. According to this architecture, in G-SN nodes remain connected through a shuffle permutation and the network diameter becomes equal to $D_{\text{G-SN}} = k + m - 1$ [106].

In Figure 3.49, the ratio $D_{\text{G-SN}}/D_{\text{BN}}$ is plotted versus the $k/m$ ratio, for $m = 3$ and $m = 8$. It is interesting to observe that it is always $D_{\text{BN}} < D_{\text{G-SN}}$, with a maximum difference for $k/m = 4$ when $m = 8$, and for $k/m = 5$ when $m = 3$.

To extend this analysis and comparison, let us consider also the channel efficiency $\eta = 1/\langle h \rangle$ and the network throughput $C = \eta W$, where $W$ is, once again, the total number of channels in the network.

For a bidirectional $p^m \times k$ BN, the considered parameters are given by [106]

$$W_{\text{BN}} = 2kp^{m+1} = 2pN \tag{3.80}$$

$$C_{BN} = \eta_{BN} W_{BN} = 2\eta_{BN} k p^{m+1} \tag{3.81}$$

whereas the channel efficiency $\eta_{BN}$ is obtained by numerical simulation, considering the application of a given routing algorithm. In fact, an analytical expression of the average number of hops in BN is not available in closed form.

In Figure 3.50, $\eta_{BN}$ is reported versus $k/m$ and for different values of $m$. The increase of $k$ with respect to $m$ produces longer end-to-end paths in average, thus reducing the channel efficiency. Once known $\eta_{BN}$, from (3.81) the network throughput can be calculated. Its trend is represented in Figure 3.51, assuming a 1-Gbps user transmission rate, for $m \in [3,8]$.

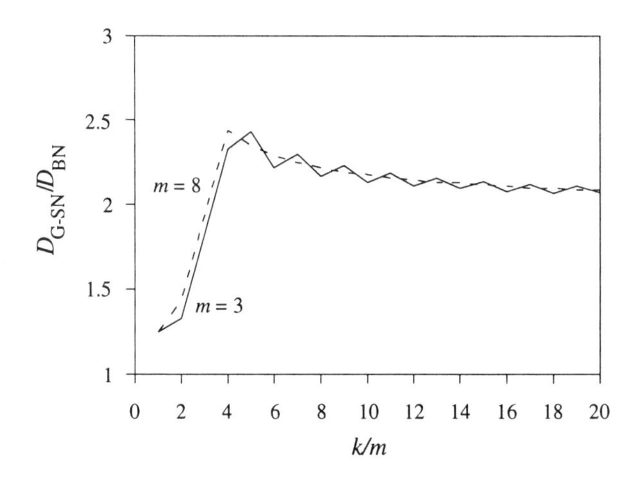


**Figure 3.49** Diameter comparison between Banyan net and generalized shufflenet.

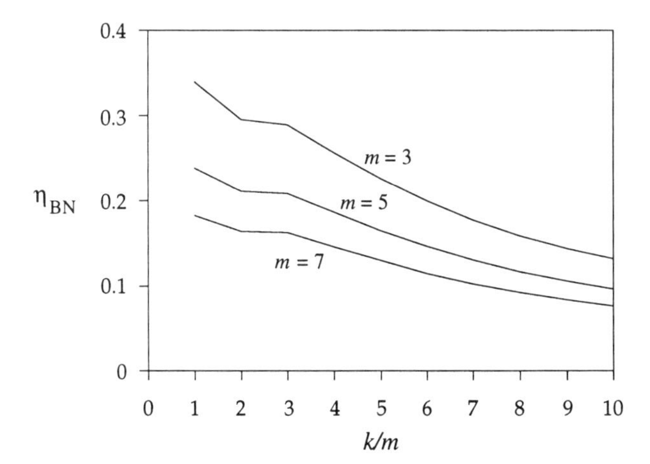


**Figure 3.50** Channel efficiency in a Banyan net.

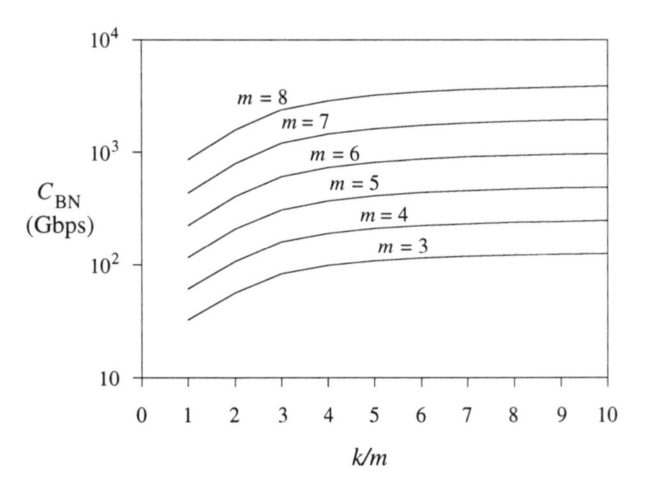


**Figure 3.51** Network throughput of a Banyan net for a 1-Gbps user transmission rate.

In Table 3.5, the number of nodes $g_h$ that are $h$ hops away from any considered source was reported for the SN topology. The extension to the $p^m \times k$ G-SN structure implies the modification of the $h$-$g_h$ scheme [110]

$$g_h = \begin{cases} p^h & \text{if } h = 1, \ldots, m-1 \\ p^m & \text{if } h = m, \ldots, k-1 \\ p^m - p^{h-k} & \text{if } h = k, \ldots, k+m-1 \end{cases} \tag{3.82}$$

It can be demonstrated that now, for G-SN, we have [110]

$$\langle h \rangle_{\text{G-SN}} = \frac{kp^m(p-1)(k+2m-1) - 2k\left(p^m - 1\right)}{2(p-1)\left(p^m k - 1\right)} \tag{3.83}$$

and then $\eta_{\text{G-SN}} = 1/\langle h \rangle_{\text{G-SN}}$. The network parameters of a G-SN are listed below [106]:

$$W_{\text{G-SN}} = kp^{m+1} = pN \tag{3.84}$$

$$C_{\text{G-SN}} = \eta_{\text{G-SN}} kp^{m+1} = \eta_{\text{G-SN}} W_{\text{G-SN}} \tag{3.85}$$

For the sake of completeness, the network parameters of a $(p, m)$ BS (where $k = m$) are also reported [101,106]

$$\eta_{\mathrm{BS}}^{(\mathrm{o})} = \frac{(1-p)^2(N-1)}{(1-p)^2 mN + (1-p)N - (1-p)p^m - (1-p)^2 p^{m-1} - 2\left(1-p^{(m-1)/2}\right)} \tag{3.86}$$

for $m$ odd, and

$$\eta_{\mathrm{BS}}^{(\mathrm{e})} = \frac{(1-p)^2(N-1)}{-\frac{2}{p} - 2 + p^{\frac{m}{2}-2}\left(m + (2-m)p + 2p^2\right) + N\left(m + \frac{3}{2} - 2mp - 2p + mp^2 + \frac{p^2}{2}\right)} \tag{3.87}$$

for $m$ even. Moreover,

$$W_{\mathrm{BS}} = W_{\mathrm{BN}} \tag{3.88}$$

$$C_{\mathrm{BS}} = 2\eta_{\mathrm{BS}} m p^{m+1} \tag{3.89}$$

From (3.81), (3.85), and (3.89), it results that

$$\frac{C_{\mathrm{BN}}}{C_{\mathrm{G-SN}}} = \frac{2\eta_{\mathrm{BN}}}{\eta_{\mathrm{G\text{-}SN}}} \tag{3.90}$$

$$\frac{C_{\mathrm{BN}}}{C_{\mathrm{BS}}} = \frac{\eta_{\mathrm{BN}}}{\eta_{\mathrm{BS}}} \quad \text{for } k = m \tag{3.91}$$

so that by comparing the network efficiencies, the throughputs are also compared.

In Figure 3.52, the $\eta_{\mathrm{BN}}/\eta_{\mathrm{G\text{-}SN}}$ ratio is plotted assuming $p = 2$. The advantage that can be achieved by adopting the BN architecture is evident; $\eta_{\mathrm{BN}}$ can be increased with respect to $\eta_{\mathrm{G\text{-}SN}}$ until the upper bound of $\eta_{\mathrm{BN}} \approx 2.15\eta_{\mathrm{G\text{-}SN}}$, that represents the saturation level reached for $k/m \geq 5$. In any case, it is worth observing that the better the performance, the higher the costs. In fact, in the considered conditions, BN needs twice as many transceivers as in the SN architecture; thus, it is useful to compare the network performance considering the same number of transceivers as done in Figure 3.53.

The comparison seems not favorable to BN. Nevertheless, it has to be considered that the best channel efficiency shown by SN with $p = 4$ and $k = m$, or by BS with $p = 2$ and $k = m$, go with the above-mentioned problems which affect both the networks: limited number of configurations or non-symmetric node distance.

In this sense, and considering the easier implementation of decentralized self-routing procedures [106], BN can be competitive, maintaining the topological equivalence with SN.

In conclusion, we want to underline that BN is not a different topology from SN, but a new layout that makes the adoption of bidirectional routing algorithms easy. This means that, since SN and BN are isomorphic, the latter can be seen as a different representation of the bidirectional version of SN.

***Some keypoints.*** Several architectures (duplex shufflenet, GEMNET, enlarged shufflenet, bilayered shufflenet, Banyan net, etc.) are considered to overcome the problems exhibited by shufflenets, essentially concerning asymmetric node distance, limited network scalability and flexibility, and unbalanced link loading. They can be seen as a generalization of shufflenets, resulting from the extension or modification of some original network characteristics. For example, some of them use bidirectional links, others adopt different closure connections or node placement layout, and so on. Their features are described and their correspondent behaviors and performance are compared with those of basic shufflenets.

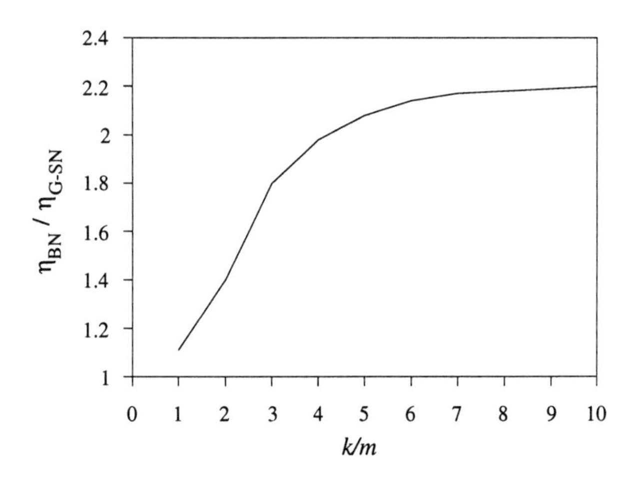


**Figure 3.52** Comparison of Banyan net and shufflenet efficiencies. (*Source*: [106]. © 1994 IEEE.)

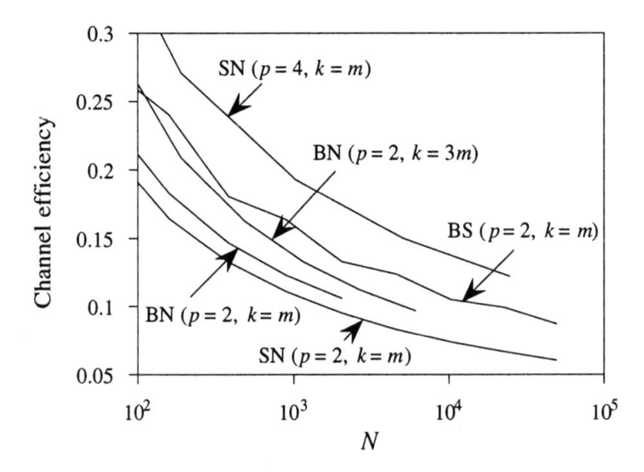


**Figure 3.53** Channel efficiency versus number of users in shufflenet (SN), Banyan net (BN), and bilayered shufflenet (BS).

## 3.5 DE BRUIJN GRAPH TOPOLOGY

### 3.5.1 The de Bruijn Graph

de Bruijn (DB) graphs [111] have been proposed as a class of logical and physical topologies for multihop networks, so that we can consider them as DB networks. As will be shown in this section, they are characterized by good properties as regards the possibility of connecting a large number of nodes, using simple addressing and routing techniques.

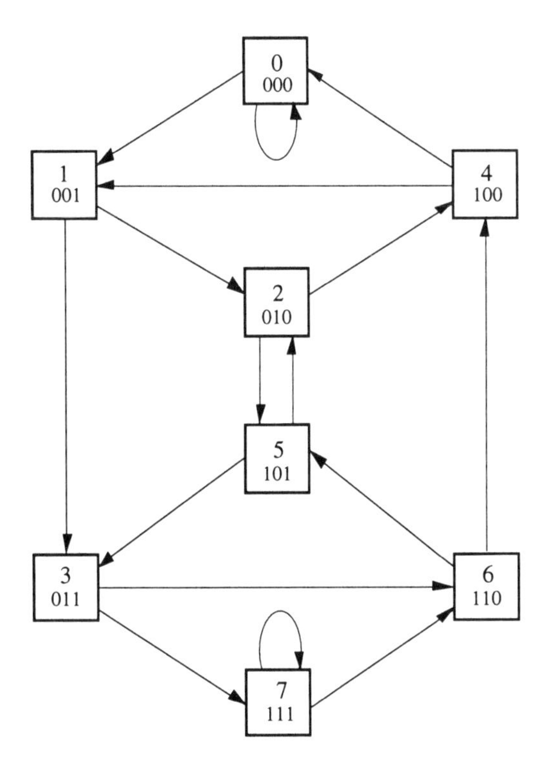


**Figure 3.54** Virtual topology of a (2, 3) de Bruijn network.

A $(\Delta, D)$ DB is a graph that connects a set of nodes $\{0, 1, 2, \ldots, \Delta - 1\}^D$ through a group of channels, in such a way that node $(a_1, a_2, \ldots, a_D)$ is linked with node $(b_1, b_2, \ldots, b_D)$ if and only if $b_i = a_{i+1}$ for $1 \le i \le D - 1$. Globally, $N = \Delta^D$ nodes are present in the network, all of them characterized by the same connection degree. In particular, considering that all the links are unidirectional, the in-degree equals the out-degree and both are equal to $\Delta$. Nevertheless it is worth observing that, according to the previous definition, $\Delta$ nodes have self-loops that are considered in the graph but are meaningless in a communication network; in particular, they are located on the nodes $(a, a, \ldots, a)$ with $a \in \{0, 1, \ldots, \Delta - 1\}$. Neglecting the self-loops, we conclude that the connection degree is equal to $\Delta$ for $N - \Delta$ nodes, whereas it equals $\Delta - 1$ for the remaining $\Delta$ ones.

An example of a (2, 3) DB network is reported in Figure 3.54; where it is easy to observe that the diameter of a DB network coincides with $D$.

There is a one-to-one correspondence between the nodes of a $(\Delta, D)$ DB and the possible states of a $\Delta$-ary shift register with $D$ stages. In other words, the DB graph results equivalent to the state transition diagram of the considered shift register. In fact, node (state) $x \equiv (x_1, x_2, \ldots, x_D)$ is connected to node (state) $y \equiv (y_1, y_2, \ldots, y_D)$ if the address of $y$ is obtained by shifting the address of $x$ of one digit and adding a new digit [112], that is to say, if $(y_1, y_2, \ldots, y_D) \equiv (x_2, x_3, \ldots, x_D, j)$.

Consequently, a link connecting node $x$ with node $y$ can be indicated by the string $(x_1, y_1, y_2, \ldots, y_D) \equiv (x_1, x_2, \ldots, x_D, y_D)$ that is a sequence of $D + 1$ digits, the first $D$ representing the address of $x$, the last $D$ equal to the address of $y$. Analogously, every path $k$ hops long that links a given pair of nodes is identified by a string composed by $D + k$ digits.

Since in a graph with maximum out-degree equal to $\Delta$ we can find no more than $\Delta$ node-disjoint paths between any pair of nodes, in a $(\Delta, D)$ DB network there are $\Delta - 1$ ones [113]. It follows that a $(\Delta, D)$ DB network maintains an end-to-end connection also in presence of $\Delta - 2$ node faults. In the latter worst case, the network diameter is increased by one at the most [113]. Other studies concerning the fault-tolerance capability of DB networks, and its improvement through scalable architectures, can be found in [114].

### 3.5.2 Routing in de Bruijn Networks

In this section we will analyze two different possibilities of packet routing in DB networks: shortest path routing and longest path routing.

#### 3.5.2.1 *Shortest Path Routing*

Let us examine the problem of node connection using the shortest path routing in a $(\Delta, D)$ DB network. Let node $A \equiv (a_1, a_2, \ldots, a_D)$ be the source of a given sequence of packets addressed to node $B \equiv (b_1, b_2, \ldots, b_D)$. By $S(i, A, B)$, with $0 \le i \le D$, we indicate a logical function on strings $A$ and $B$ that is true if and only if $(b_1, b_2, \ldots, b_{D-i}) = (a_{i+1}, a_{i+2}, \ldots, a_D)$ and is false otherwise. At the same time, $M(i, A, B)$, with $0 \le i \le D$, coincides with the string $(a_1, a_2, \ldots, a_D, b_{D-i+1}, \ldots, b_D)$ composed by $D + i$ elements.

The routing algorithm considered operates according to the procedure below [115]:

number of hops
$i = 0$ while $S(i, A, B) = \text{FALSE}$
$i = i + 1$
end while
$P = M(i, A, B)$

$P$ being the string that labels the shortest path between nodes $A$ and $B$.

As an example, let us consider once again the (2, 3) DB network of Figure 3.54, where we assume $A \equiv (0,0,1)$ and $B \equiv (1,0,1)$. In this case $S(0, A, B) = S(1, A, B) =$ FALSE, whereas $S(2, A, B) =$ TRUE. Therefore, since $M(2, A, B) = (0,0,1,0,1)$, the shortest path between $A$ and $B$ is $(0,0,1) \rightarrow (0,1,0) \rightarrow (1,0,1)$.

It can be demonstrated [116] that the average number of hops experienced in DB networks, when shortest path routing is adopted, is included between the following bounds:

$$D\frac{N}{N-1} - \frac{\Delta}{(\Delta-1)^2} + \frac{D}{\left(\Delta^D - 1\right)(\Delta - 1)} \leq \langle h \rangle \leq D\frac{N}{N-1} - \frac{1}{\Delta - 1} \tag{3.92}$$

whereas the exact value of $\langle h \rangle$ can be calculated using a recursive procedure [116].

As regards (3.92), it is worth observing that the difference between the lower and the upper bounds of $\langle h \rangle$ is $O(1/\Delta^2)$ for high values of $\Delta$, so that the average number of hops tends to the minimum bound as the number of transceivers per node increases.

When shortest path routing is used, the average link loading can be expressed as

$$\langle L \rangle = \frac{N(N-1)}{\Delta N - \Delta}\langle h \rangle = \frac{\langle h \rangle}{\Delta}\Delta^D \tag{3.93}$$

In fact, the average distance between source and destination equals $\langle h \rangle$ and there are $N(N-1)$ possible source-destination pairs, assumed to be equally likely. Therefore, (3.93) is justified considering that $\Delta N - \Delta$ links are present in the network, neglecting the self-loops.

The upper bound of $\langle L \rangle$ can be calculated by exploiting the analogies between DB graphs and shift register transition diagrams. The maximum link loading $L_{max}$ is obtained considering the maximum number of paths, traced by shortest path routing, that include a given edge and are not longer than $D$.

If $k$ is the path length, the path itself can be represented by a string composed by $D + k$ elements, whereas the considered edge, as shown before, is always determined by a string $D + 1$ digits long. Thus, the number of paths $P_k$ of length $k$ that pass along this edge is equal to the number of strings of length $D + k$ containing the relative string of length $D + 1$. For $1 \leq k \leq D$, this number is

$$P_k = k\Delta^{k-1} \tag{3.94}$$

since the shorter string can be placed in $k$ different positions and there are $\Delta$ possible values for each of the remaining $k - 1$ elements of the longer string. Now considering all the possible path lengths, we finally have [116]

$$L_{max} = \sum_{k=1}^{D} P_k \leq \frac{D\Delta^{D+1} - (D+1)\Delta^D + 1}{(\Delta - 1)^2} \tag{3.95}$$

For large values of $\Delta$, $L_{max} \approx D\Delta^{D-1}$.

### 3.5.2.2 *Longest Path Routing*

The goal of longest path routing is to obtain a fair distribution of traffic in the network, avoiding the presence of load peaks in some links, created by the application of shortest path routing. In fact, the latter always tends to minimize the average path length, but this cause high traffic levels on few links and the underutilization of the remaining available edges. Longest path length, instead, minimizes the maximum link loading, accepting the increase in the average number of hops.

Let us consider once again the connection of nodes $A \equiv (a_1, a_2, \ldots, a_D)$ and $B \equiv (b_1, b_2, \ldots, b_D)$, now using longest path routing in a $(\Delta, D)$ DB network. According to this strategy, the unique path $(a_1, a_2, \ldots, a_D, b_1, b_2, \ldots, b_D)$ of length $D$ is chosen and it can be represented by a string of length $2D$. This path can have circuits. If they are removed from the path, the length becomes shorter than $D$ and the routing scheme does not select the longest connection between $A$ and $B$ anymore.

**Table 3.13**
Performance of Shortest Path Routing and Longest Path Routing in Different $(\Delta, D)$ de Bruijn Graphs

| $(\Delta, D)$ | $\langle h \rangle$ | $N$ | $N_{max}$ | $N/N_{max}$ | $L_{max}$(SP) | $L_{max}$(LP) |
|---|---|---|---|---|---|---|
| 2,2 | 1.5000 | 4 | 5 | 0.8000 | 3 | 4 |
| 2,3 | 2.1071 | 8 | 9 | 0.8889 | 11 | 12 |
| 2,4 | 2.8333 | 16 | 19 | 0.8421 | 29 | 32 |
| 2,5 | 3.6492 | 32 | 39 | 0.8205 | 81 | 80 |
| 3,2 | 1.6667 | 9 | 10 | 0.9000 | 7 | 6 |
| 3,3 | 2.4786 | 27 | 29 | 0.9310 | 31 | 21 |
| 3,4 | 3.3861 | 81 | 88 | 0.9205 | 138 | 108 |
| 3,5 | 4.3440 | 243 | 266 | 0.9135 | 535 | 405 |
| 4,2 | 1.7500 | 16 | 17 | 0.9412 | 9 | 8 |
| 4,3 | 2.6399 | 64 | 67 | 0.9552 | 57 | 48 |
| 4,4 | 3.5985 | 256 | 269 | 0.9517 | 313 | 256 |
| 4,5 | 4.5844 | 1024 | 1079 | 0.9490 | 1589 | 1280 |
| 5,2 | 1.8000 | 25 | 26 | 0.9615 | 11 | 10 |
| 5,3 | 2.7277 | 125 | 129 | 0.9690 | 86 | 75 |
| 5,4 | 3.7059 | 625 | 647 | 0.9660 | 586 | 500 |
| 5,5 | 4.7000 | 3125 | 3234 | 0.9663 | 3711 | 3125 |
| 6,2 | 1.8333 | 36 | 37 | 0.9730 | 13 | 12 |
| 6,3 | 2.7823 | 216 | 221 | 0.9774 | 121 | 108 |
| 6,4 | 3.7694 | 1296 | 1327 | 0.9766 | 985 | 864 |
| 6,5 | 4.7665 | 7776 | 7966 | 0.9761 | 7465 | 6480 |

$\langle h \rangle$ = average number of hops.

Analogously to the previous section, we are now able to analyze the consequences of the adoption of longest path routing on the maximum link loading. Following the same procedure used before, we can say that the problem concerns the calculus of the maximum number of paths, traced by longest path routing without removing circuits, that included a given edge, that is the number of strings of length $2D$ containing a given string of length $D + 1$. Here we have

$$L_{max} \leq D\Delta^{D-1} \tag{3.96}$$

In Table 3.13 we have reported some results relative to $(\Delta, D)$ DB with $2 \leq \Delta \leq 6$ and $2 \leq D \leq 5$. By $N_{max}$ we have indicated the maximum number of nodes that can be supported in any directed graph for the same average number of hops, whereas $L_{max}$(SP) and $L_{max}$(LP) are the maximum link loading using shortest and longest path routing, respectively [116].

We can observe that for acceptable values of $\Delta$, $(\Delta, D)$ DB tends to the optimal topology. On the other hand, as expected, longest path routing always reduces the maximum link loading, except for $\Delta = 2$ and $D \leq 4$.

### 3.5.3 de Bruijn Versus Shufflenet

As seen in Section 3.3.1 in a $(p, k)$ SN the out-degree equals $p$ for every node belonging to the network. On the other hand, the network diameter is equal to $2k - 1$. To maintain the convention on symbols adopted for $(\Delta, D)$ DB, we can say that $p$ corresponds to $\Delta$ and $k = (D + 1)/2$. Therefore, for a $(\Delta, k)$ SN we have

$$N = k\Delta^k = \frac{D+1}{2}\Delta^{(D+1)/2} \tag{3.97}$$

and (3.35) can be rewritten as

$$\langle h \rangle = \frac{k\Delta^k(\Delta-1)(3k-1) - 2k\left(\Delta^k - 1\right)}{2(\Delta-1)\left(k\Delta^k - 1\right)} \tag{3.98}$$

There is a sort of analogy between SN and DB, since a $(\Delta, D)$ DB can be seen as a two-column graph, where the same $\Delta^k$ nodes are contained in both the columns and they are connected through a perfect $\Delta$ shuffle.

For this reason, their performance can be compared using the same parameters considered in Table 3.13, except for the average link loading, that is now given by

$$\langle L \rangle = \frac{N(N-1)}{W}\langle h \rangle = \frac{N(N-1)}{k\Delta^{k+1}}\langle h \rangle \tag{3.99}$$

In Table 3.14 we have reported the characteristic parameters of different SNs, for an out-degree $2 \leq \Delta \leq 6$ and a diameter $D \in \{3,5,7,9\}$. It can be seen that SN performs well for small diameters and number of nodes, but DB networks can have a larger number of stations with the same average number of hops.

**Table 3.14**
Performance of $N$-Nodes Shufflenets for Different Values of the Out-Degree ($\Delta$) and of the Network Diameter ($D$)

| $(\Delta, D)$ | $\langle h \rangle$ | $N$ | $N_{max}$ | $N/N_{max}$ | $\langle L \rangle$ |
|---|---|---|---|---|---|
| 2,3 | 2.0000 | 8 | 9 | 0.8889 | 7.0 |
| 2,5 | 3.2609 | 24 | 30 | 0.8000 | 37.5 |
| 2,7 | 4.6349 | 64 | 84 | 0.7619 | 146.0 |
| 2,9 | 6.0692 | 160 | 256 | 0.6250 | 482.5 |
| 3,3 | 2.1765 | 18 | 19 | 0.9474 | 12.3 |
| 3,5 | 3.5625 | 81 | 122 | 0.6639 | 95.0 |
| 3,7 | 5.0217 | 324 | 549 | 0.5902 | 540.6 |
| 3,9 | 6.5074 | 1215 | 3289 | 0.3694 | 2633.3 |
| 4,3 | 2.2581 | 32 | 33 | 0.9697 | 17.5 |
| 4,5 | 3.6911 | 192 | 343 | 0.5598 | 176.2 |
| 4,7 | 5.1730 | 1024 | 2192 | 0.4672 | 1323.0 |
| 4,9 | 6.6683 | 5120 | 21864 | 0.2342 | 8533.7 |
| 5,3 | 2.3061 | 50 | 51 | 0.9804 | 22.6 |
| 5,5 | 3.7620 | 375 | 784 | 0.4783 | 281.4 |
| 5,7 | 5.2525 | 2500 | 6522 | 0.3833 | 2625.2 |
| 5,9 | 6.7505 | 15625 | 97688 | 0.1599 | 21094.0 |
| 6,3 | 2.3380 | 72 | 73 | 0.9863 | 27.6 |
| 6,5 | 3.8068 | 648 | 1559 | 0.4157 | 410.5 |
| 6,7 | 5.3012 | 5184 | 16014 | 0.3237 | 4579.3 |
| 6,9 | 6.8002 | 38880 | 335971 | 0.1157 | 44064.1 |

*Abbreviations*: $\langle L \rangle$, average link loading; $N_{max}$, maximum number of nodes in any directed graph characterized by the same average number of hops $\langle h \rangle$.

It can be demonstrated [116] that, considering the user throughput too, DB can show a better performance than SN. Similarly, taking into account the average queueing delay, DB with shortest path routing behaves better than SN for lower offered loads, whereas it always works better than SN when longest path routing is adopted.

DB networks have been proposed as physical topology too, where nodes, connected by point-to-point links, adopt wavelength-routing (see Sections 2.1 and 2.4). Given the diameter $D$ and the maximum out-degree $\Delta$, we can search for the best physical topology, that is to say, the architecture that maximizes the number of connected nodes and respects the constraints imposed by power loss limits.

In this sense the Moore bound [111], that indicates the maximum number of nodes supported by any graph, can be used to evaluate the efficiency of the DB solution. It is equal to

$$N_{Moore} = \frac{\Delta^{D+1} - 1}{\Delta - 1} \tag{3.100}$$

We can normalize the total number of stations $N_{SN}$ and $N_{DB}$, relating to SN and DB networks, respectively, with respect to the Moore bound, so that $R_{SN} = N_{SN}/N_{Moore}$ and $R_{DB} = N_{DB}/N_{Moore}$.

As shown in Figure 3.55 [116], DB is not optimal but can approach the upper bound for suitable values of $\Delta$ and $D$. Moreover, it performs better than SN, when the latter is considered as a physical topology too.

As regards the fault-tolerance properties of DB and SN, a comparative study is reported in [117].

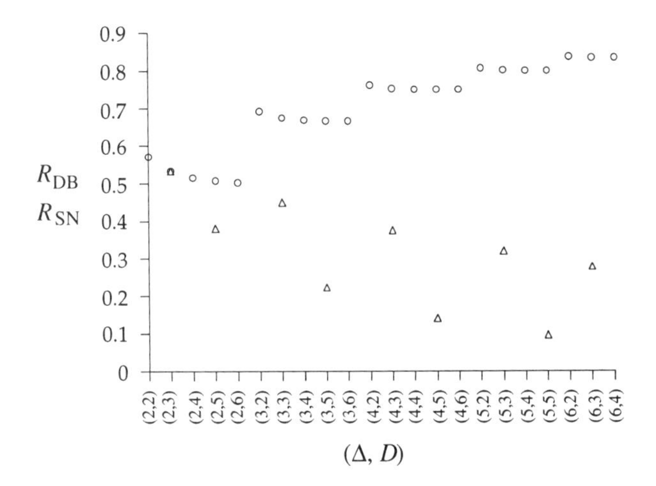


**Figure 3.55** Normalized number of stations supported by de Bruijn networks (circles) and shufflenets (triangles) for different out-degrees and diameters.

### 3.5.4 The Modified de Bruijn Topology

Let us consider a multifiber implementation of a $(\Delta, D)$ DB optical network. We indicate by $C$ the number of fibers which connect the outputs (inputs) of a given node with other stations of the network, so that $C$ represents the adopted space diversity degree. When a pair of nodes is linked by more than one fiber, groups of $F$ fibers can be arranged in bundles. If $B$ is the number of bundles (or cables), we have that $C = FB$, whereas the total number of fibers in the network equals $NC$.

Alternatively, the space diversity can also be obtained by using multiple parallel topologies, which overlap the considered network; they are completely disjointed even though they connect the same group of users. Let $P$ be the number of parallel topologies. In this case, $C = FBP$, assuming that every single topology utilizes multifiber cables too.

In a traditional DB network we have that $B = \Delta$, $F = 1$, $P = 1$ and then $C = \Delta$. Nevertheless, it is also possible to consider an extended DB graph topology where there are multiple connections between a given node and its $\Delta$ adjacent stations, in a way that $B = \Delta$ and $F = D$, to obtain $C = \Delta D$. Finally, if we consider a topology multiplicity too, with $P > 1$, we have the so-called modified de Bruijn (MDB) network, characterized by a space diversity degree $C = \Delta DP$ [26].

Wavelength-routing has been proposed in [26] for possible application in MDB. By this technique, in a different fashion from shortest path routing examined in Section 3.5.2.1, an end-to-end path is always of length $D$. In fact, its relative string is obtained by a progressive shifting of the digits that compose the destination address, into the least significant positions of the string indicating the source address. For example, if (0011) is the source address and (1001) is the destination address, in a (2, 4) MDB, the considered path traverses the following subsequent nodes: (0111), (1110), (1100), before reaching (1001). It can be verified that this strategy does not avoid self-loops, since two equal addresses can be subsequently originated by the above-described procedure.

Suitable indexing solutions are used to label the different bundles and fibers. In particular, every cable is labeled by an integer number $b \in [0, \Delta[$, whereas a fiber in the cable is indexed by an integer number $f \in [0, D[$. In [26] it is shown that $f$ is arbitrary inside a bundle, whereas $b$ is univocally determined by the addresses of the nodes placed at the ends of the bundle itself.

An end-to-end channel, and its relative wavelength, is associated with a $\Delta$-ary string $X = X(x_0, x_1, \ldots, x_{D-1})$ generated by the sequence of $b$ symbols relative to the bundles that compose the channel itself, from source to destination. At every node, $f_{in}$ and $b_{in}$, which are the fiber and bundle indexes characterizing the incoming signal, are sufficient to determine $f_{out}$ and $b_{out}$ (the fiber and bundle indexes of the output signal, respectively), that is to say, to realize the routing choices.

The fiber choice is done by setting $f_{out} = f_{in} + 1$ ($f_{out} = 0$ at the source output). The bundle choice, instead, depends on the $X$ string associated to the considered channel. In particular, $b_{out}$ is set to $x_c$, $c$ being the $c$th digit of $X$ and $c = f_{out} = f_{in} + 1$. Consequently, $b_{in}$ is not utilized for routing.

Since there is a correspondence between $X$ string and wavelength of a given channel, so that different strings are associated to different channels, it can be demonstrated that two different end-to-end paths never share the same fiber [26]. Therefore, an intermediate node, on the basis of the wavelength and of the input fiber, can identify the source-destination pair and its relative routing choice.

Through suitable optimization of this strategy, it has been shown that the number of wavelengths needed to distinguish all the possible end-to-end paths is equal to

$$\Lambda_r = \Delta^{D-1} \tag{3.101}$$

### *3.5.4.1 Comparison with Twin Shuffle*

In this section we report a brief performance evaluation of MDB, using twin shuffle (TS) for comparison. The latter is obtained from SN, by considering two fibers per bundle connecting any pair of adjacent nodes. In this way, adopting the space diversity, we also double the connectivity.

Furthermore, we modify the physical topology of a single SN, having $F = P = 1$, $B = p$, and $C = p$, in two parallel SNs with $F = 2$, $P = 1$, $B = p$, and $C = 2p$. Generalizing this concept, we refer to TS when $P > 2$ as well.

A twin shuffle that is suitable to be compared with MDB (assuming constant values of $B$ in both the networks) can utilize simple wavelength-routing schemes and seems to allow the use of a lower number of wavelengths $\Lambda_r$ with respect to other proposed solutions [116]. It is given by

$$\Lambda_r = \frac{kN}{p} - \left\lceil \frac{k-1}{2} \right\rceil p^{k-1} \tag{3.102}$$

The diameter of TS equals again $2k - 1$ and, similarly, the average number of hops and the $h$-$g_h$ table are the same as those reported in (3.35) and Table 3.5, respectively.

In Table 3.15, the diameter $D$, the average number of hops $\langle h \rangle$, the number of wavelengths, and the normalized number of wavelengths, which is $\Lambda_{\text{norm}} = FP\Lambda_r$, are listed for different TS and DB networks. It has been assumed $B = 4$ for both the classes of networks and $p = \Delta$ for TSs.

**Table 3.15**
Comparison Between Twin Shuffle and Modified de Bruijn Network, Assuming $B = 4$

| *Topology* | | $N$ | $\Delta$ | $P$ | $D$ | $\langle h \rangle$ | $\Lambda_r$ | $\Lambda_{\text{norm}}$ |
|---|---|---|---|---|---|---|---|---|
| Twin shuffle | $k = 2$ | 32 | 4 | 4 | 3 | 2.258 | 3 | 24 |
| Twin shuffle | $k = 3$ | 192 | 4 | 4 | 5 | 3.691 | 32 | 256 |
| Twin shuffle | $k = 4$ | 1024 | 4 | 4 | 7 | 5.173 | 224 | 1792 |
| Twin shuffle | $k = 5$ | 5120 | 4 | 4 | 9 | 6.668 | 1472 | 11776 |
| Twin shuffle | $k = 6$ | 24576 | 4 | 4 | 11 | 8.197 | 8448 | 67584 |
| Modified de Bruijn | $D = 3$ | 64 | 4 | 4 | 3 | 3.000 | 4 | 48 |
| Modified de Bruijn | $D = 4$ | 256 | 4 | 2 | 4 | 4.000 | 32 | 256 |
| Modified de Bruijn | $D = 5$ | 1024 | 4 | 2 | 5 | 5.000 | 128 | 1280 |
| Modified de Bruijn | $D = 6$ | 4096 | 4 | 2 | 6 | 6.000 | 512 | 6144 |
| Modified de Bruijn | $D = 7$ | 16384 | 4 | 1 | 7 | 7.000 | 4096 | 28672 |
| Modified de Bruijn | $D = 8$ | 65536 | 4 | 1 | 8 | 8.000 | 16384 | 131072 |

*Abbreviations*: $N$, number of nodes; $P$, number of parallel topologies; $\Lambda_r$, number of wavelengths.

Since we are analyzing physical topologies, we can observe that $D$ and $\langle h \rangle$ can give an indication of the maximum and average power losses that can characterize an end-to-end path.

The remarkable aspects shown by the table relates to the network diameter. MDB is advantageous since in this kind of topology there is equivalence between the average number of hops and the network diameter. Consequently, even though MDB and TS are comparable as regards the average number of hops, MDB is preferable considering the diameter.

Moreover, MDB behaves better also taking into account the number of wavelengths, that is lower than the one required by TS.

In any case, MDB seems always to perform better than TS, and this advantage goes with the feasibility of the switching architecture with the present state of the art of the optical technology [118].

Finally, the constant path length reduces the signal dynamics at the receivers, so simplifying the node implementation.

### 3.5.5 de Bruijn Network Variants

Other topologies can be found in the literature, which are based on a DB graph, either as basic backbone structures, or as subnetwork architectures. For the sake of brevity, in this section we briefly describe two examples, whereas we suggest that readers utilize the references for further details.

#### *3.5.5.1 Bidirectional de Bruijn Network*

From the basic unidirectional topology of the DB graph, presented in Section 3.5.1, derives the so-called bidirectional de Bruijn (BDB) graph, according to the following simple rules:

- All the unidirectional links of the DB graph are bidirectional in BDB;
- The same addressing scheme adopted in the DB graph is also used in BDB.

In this way, a given node $A = (a_1, a_2, \ldots, a_D)$ is connected with node $B = (b_1, b_2, \ldots, b_D)$ if one of the following conditions are satisfied:

- $b_i = a_{i-1}$, where $b_i, a_i \in \{0, 1, 2, \ldots, \Delta - 1\}$ and $2 \leq i \leq D$;
- $b_i = a_{i+1}$, where $b_i, a_i \in \{0, 1, 2, \ldots, \Delta - 1\}$ and $1 \leq i \leq D - 1$.

The nodes adjacent to $A$ can be classified in two classes, according to the condition satisfied by their addresses. Since $A$ is connected with both the classes of neighbors, it follows that the maximum connection degree is now equal to $2\Delta$. Nevertheless, for some nodes, it can assume the minimum value of $2\Delta - 2$. On the other hand, the number of nodes and the network diameter do not change with respect to the DB graph, so that they are respectively equal to $\Delta^D$ and $D$.

The example of a BDB is drawn in Figure 3.56 for $\Delta = 2$ and $D = 4$, where every link has to be intended as bidirectional and the redundant links have been omitted. In fact, the self-loops of nodes (0000) and (1111), as well as the multiple connections existing between nodes (0101) and (1010), are not reported.

It is worth observing that the node addressing is not unique, but two different and symmetric strings can be used to label the same vertex, according to the unidirectional DB substructure on which the node numbering is based. In fact, BDB can be seen as the merging of two coincident and symmetric unidirectional DB graphs. Once one is chosen, the node addressing follows consequently. With reference to the (2, 4) DB network of Figure 3.56, for example, node (0001) can be equivalently indicated by the string (1000) if the reverse addressing is used.

BDB have been considered in [119], where their performance has been compared with those offered by BS. Further analysis can be found in [120–123].

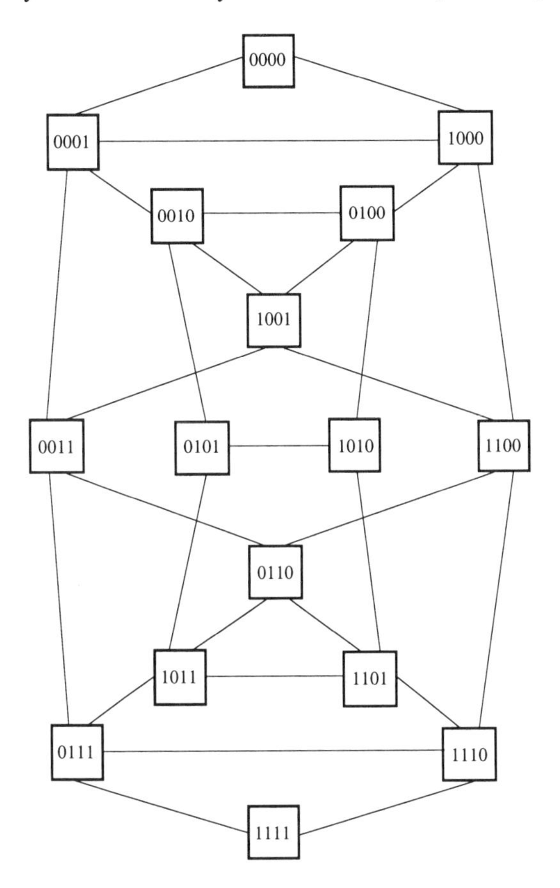


**Figure 3.56** Virtual topology of a (2, 4) bidirectional de Bruijn network.

### *3.5.5.2 Hyper de Bruijn Network*

To present the hyper de Bruijn (HDB) network, we need to introduce some new definitions. By $Z_n$ we denote the set $\{0, 1, \ldots, n-1\}$ for every integer $n$. Moreover, for any set $S$ and positive integer $k$, $S^k$ indicates the set of all the strings of length $k$ composed by the element of $S$.

If $G = (V_G, E_G)$ and $L = (V_L, E_L)$ are undirected graphs, being $V_G$, $V_L$ and $E_G$, $E_L$ the sets of vertexes and edges of graph $G$ and $L$, respectively, the direct product graph $G \times L$ has the node set $V_G \times V_L$. It can be seen as a graph $L$ where the nodes consist of graphs $G$ or vice versa.

Now, let us define the order-$n$ binary DB graph, denoted by $DB(n)$ as a DB graph composed by the set of nodes $Z_2{}^n$. On the contrary, the so-called order-$m$ hypercube (HC) graph, denoted by $H(m)$, consists of a set of node $Z_2{}^m$ and its edges connect two vertexes if and only if their addresses differ by only 1 bit [124,125].

From the above definitions, the product graph $HDB(m, n) = H(m) \times DB(n)$ is a multilevel graph where the upper layer can be organized as a DB (or HC) graph and the supernodes consist, at the lower layer, of HC (or DB) subgraphs. The network based on $HDB(m, n)$ is called a hyper de Bruijn network. It has been analyzed in [124] with special regard to its broadcasting properties.

***Some keypoints.*** The de Bruijn graphs, which can be used to represent the state transition diagram of the Δ-ary shift registers, have been proposed also as either virtual or physical topologies for WDM multihop lightwave networks. They exhibit good properties in terms of number of supported nodes, which approaches the theoretical optimal upper bound, and also in terms of number of utilized wavelength channels. Using suitable routing protocols, the source-destination distance can be made constant for every node pair. This opportunity reduces the signal dynamics at the receivers and produces a good packet flow distribution in the network. Finally, in this section some different versions of the basic de Bruijn graph are also presented, such as the modified de Bruijn network, the bidirectional de Bruijn network, and the hyper de Bruijn network.

## 3.6 MATRIX TOPOLOGY

### 3.6.1 Space Diversity to Avoid WDM Conversion

The basic characteristic of multihop networks is that, generally speaking, they force transmitted packets to pass through some intermediate nodes before reaching their final destination. This fact reduces the network throughput and its efficiency. In fact, as stated above, the number of hops is usually a random variable, which depends on the load conditions and on network architecture, and efficiency depends on the inverse of its mean value. Moreover, if the variance of such a number is considerable, there could be significant difficulties in managing real-time services into the network. In fact, if the packet interarrival time is longer than the sampling period, the signal cannot be rebuilt correctly. Therefore, we can say that the best operative conditions are reached when a multihop network is characterized by minimum average value and dispersion of the number of hops.

This can be optimally obtained by single-hop networks, that provide point-to-point and multicast communications through direct connections between source and destination nodes. Unfortunately, single-hop networks often imply the adoption of tunable lasers and filters, which can increase the cost and complexity of the node hardware.

In order to overcome these problems, the Multiwavelength All-optical TRansparent Information eXchange (MATRIX) topology has been proposed [126–128]. Its description is reported here as an example of an intermediate and hybrid solution between single and multihop techniques. In fact, it connects remote nodes by a number of hops never higher than two, using a wavelength allocation in multifiber cables, similar to the one utilized in some single-hop structures. This can be obtained without wavelength switching, so that point-to-point connections are performed using a single wavelength from source to destination along the entire path. This feature eliminates the necessity of utilizing tunable optical devices and simplifies the node structure, at the expense of an increased amount of installed fibers.

MATRIX is organized on a grid topology, as shown in Figure 3.57. Every node is directly connected through a first multifiber cable, composed by $m$ optical fibers, to the other $(m - 1)$ nodes of its same row and through a second multifiber cable, composed by $n$ optical fibers, to the other $(n - 1)$ nodes of its same column. No direct link is available for interconnecting it with the remaining nodes.

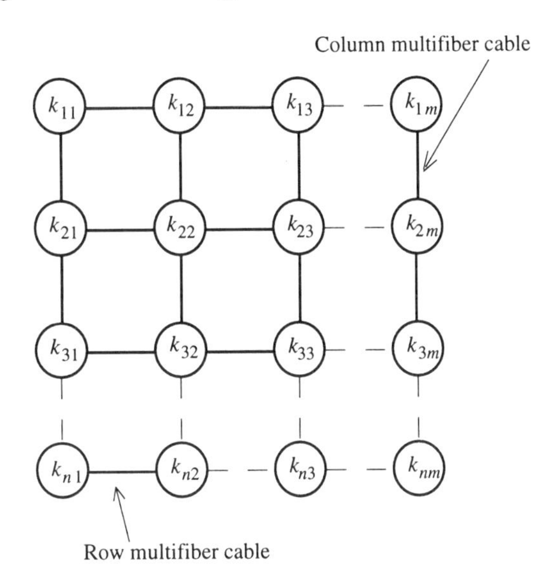


**Figure 3.57** Grid structure of a $n \times m$ MATRIX network.

As stated above, in MATRIX networks the number of hops $h$ is upper bounded by 2, thanks to the adoption of the following routing rules:

- If source and destination belong to the same row (column) the packet is sent along the row (column) itself;
- If source and destination do not belong to the same row (column) the packet is sent along the row (column) itself until it reaches the node that is placed on the same column (row) of destination, then the latter transmits it along the column (row).

The symmetry of the problem is evident. In both cases the propagation is not longer than two hops. In fact no intermediate node is crossed by the packets during their propagation along a row or a column. This aspect will be clarified below, when the node architecture will be described. For the time being, let us observe that the multiple access to the transmission medium is based on WDM technique with wavelength reuse. Considering once more the scheme of Figure 3.57, the wavelength assignment is reported in Figure 3.58 in the case $n = m = 4$. From this scheme it follows that the total number of available wavelengths $\lambda_N$ is given by $\lambda_N = \max(n, m)$. In each node, we have indicated by $\lambda_i$, the node identification wavelength (NIW) of the node itself, which is the only wavelength that it uses to receive the packets addressed to it by whichever node on its same row, or the only wavelength that it utilizes to send packets to every other node on its same column.

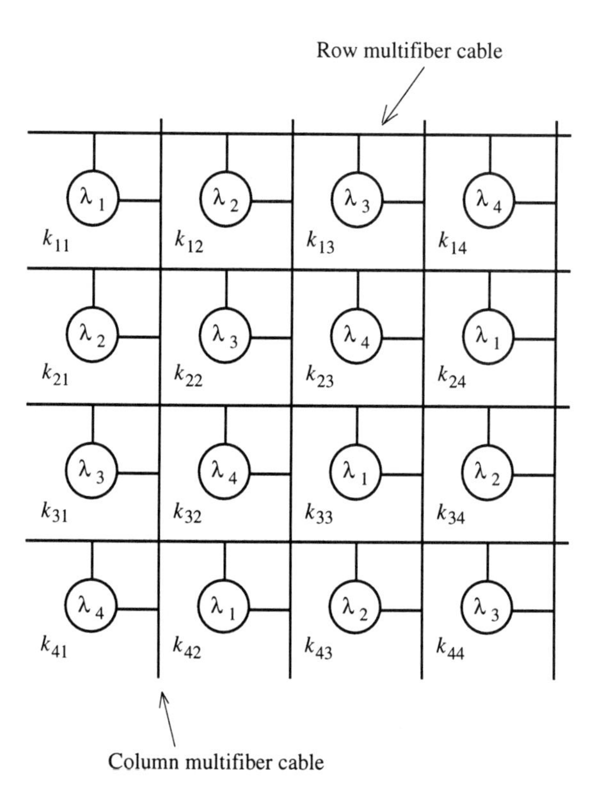


**Figure 3.58** Wavelength reuse in a 4×4 MATRIX network.

The possibility of having a complete end-to-end connection, without any wavelength conversion, results from the node architecture depicted in Figure 3.59, with reference to the example of Figure 3.58. It is basically composed by three elements:

1. A set of $(n - 1)$ optical filters working at a fixed wavelength, which is the one assigned to the given node, faced to the row multifiber cable;
2. A $n \times m$ space division switching fabric;

3. An optical transceiver, whose receiver section is interfaced with the column cable, whereas the transmitter section is linked with the row cable.

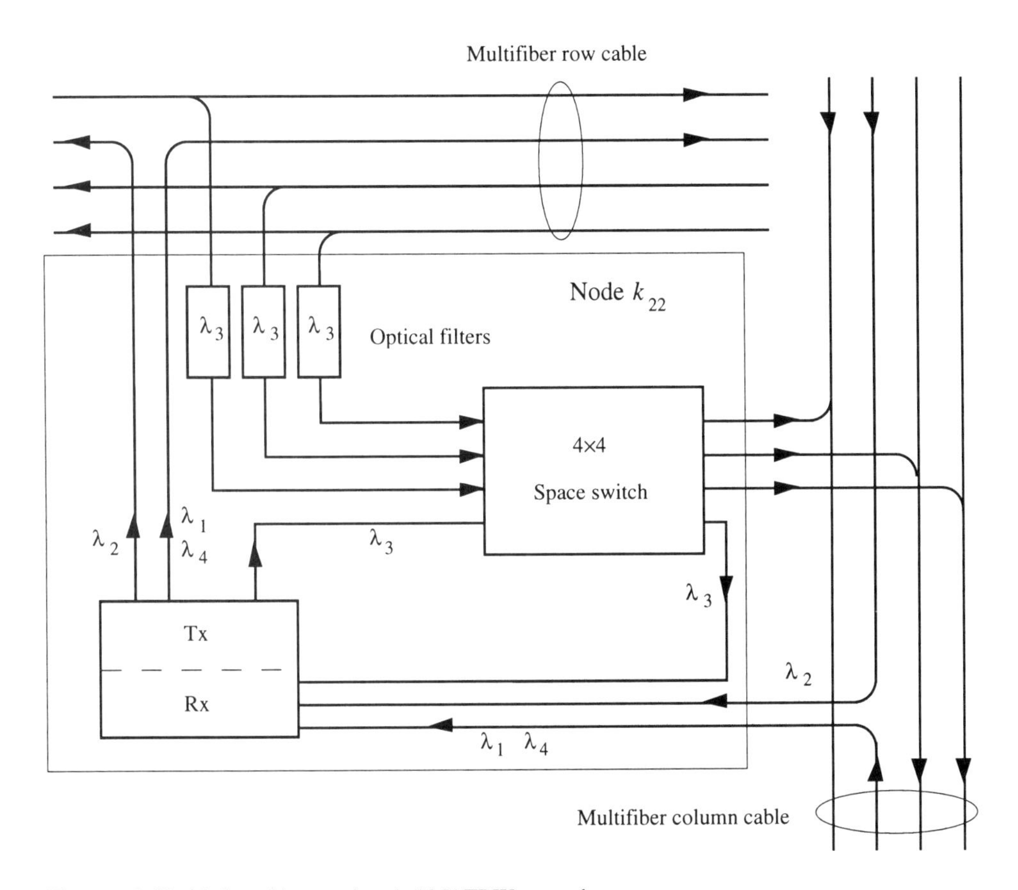


**Figure 3.59** Node architecture in a 4×4 MATRIX network.

Looking at Figure 3.59, it is possible to observe that every node placed on the second row having a packet to transmit to the considered node $k_{22}$ inserts it in the proper fiber at wavelength $\lambda_3$. It follows that messages destined to $k_{22}$ will flow along three different fibers but at the same wavelength. Node $k_{22}$ selects the packets addressed to itself, and traveling along the second row, by means of optical filters centered at its own NIW. On the other hand every message emitted by $k_{22}$, and destined to another node of the same row, will be transmitted at the NIW of the final destination, using only one fiber of the row cable. This means that, in the examined case, three external fibers are connected to the space switch. In other words, a given source node, to connect itself to other stations on its row, allocates packets on different wavelengths in a single fiber; whereas it receives messages emitted by those stations from different fibers but at the same wavelength, determined by its own NIW.

No collision due to multiple access to the transmission medium occurs, but packet losses can happen within the switching fabric because of packet contention [128]. To solve this problem, optical buffers have to be available in the nodes, such as those based on recirculating optical delay lines [129], so that suitable queueing discipline can be adopted to store the cells that otherwise would be lost.

The connection among nodes located along the same column is realized in a symmetric way, so that packets emitted by $k_{22}$ and addressed toward different nodes are allocated on different fibers with NIW = $\lambda_3$, whereas messages destined to $k_{22}$ are allocated on the same fiber but at different wavelengths. Once again, collision and loss of messages are impossible.

It is now evident that NIW characterizes the receiving wavelength of a node along the row and the transmitting one of the node itself along a column. In this way, messages utilizing a given node as an intermediate transit station do not require any wavelength conversion, independently of their origin and destination. Let us consider for example the point-to-point connection between nodes $k_{21}$ and $k_{42}$, using $k_{22}$ as intermediate node. This situation is shown in Figure 3.60: $k_{21}$ transmits its cells at the NIW of $k_{22}$, that is, NIW = $\lambda_3$, in its own output fiber. Then node $k_{22}$, working at its own NIW, switches the cells on the fiber connected with the receiver of $k_{42}$, so leaving the initial wavelength allocation unchanged. Finally, $k_{42}$ receives the cells. The signaling information, that the space switch needs to route the cells correctly, must be stored in the header of each packet.

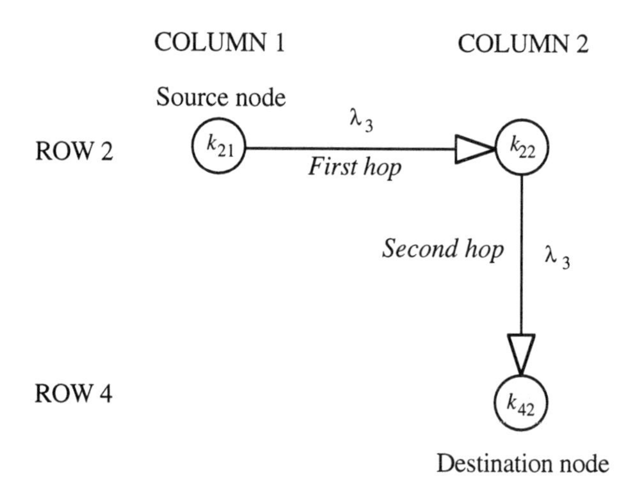


**Figure 3.60** Packet switching without wavelength conversion in a two-hop connection.

### 3.6.2 Network Parameters

In order to quantify the performance offered by MATRIX, let us consider some parameters that play an important role with regard to the behavior of the whole network.

From a given station, in a $n \times m$ MATRIX network, $(n + m - 2)$ nodes can be reached in a first hop and all the others in the second one. Therefore, the average number of hops is given by

$$\langle h \rangle = \frac{2nm - n - m}{nm - 1} \quad (3.103)$$

with $1 < \langle h \rangle < 2$. The connectivity degree $c$, defined as the number of channels leading to or originating from a node, equals $c = n + m - 2$. An example of connection diagram is shown in Figure 3.61, considering a 4×4 MATRIX and indicating also the proper NIWs.

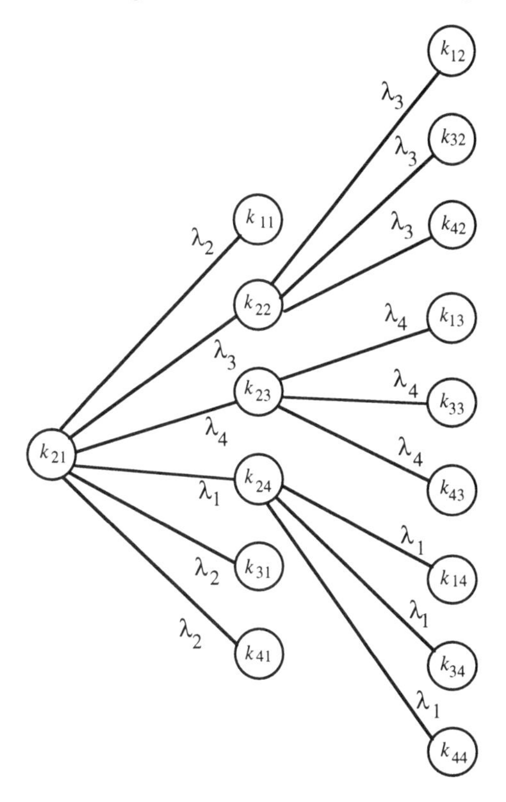


**Figure 3.61** Connection diagram relative to node $k_{21}$ in a 4×4 MATRIX.

The total number of channels in the whole network is then equal to $W = cN$, where $N = nm$. Now, let us define the parameter $b$ through the following function:

$$b = \begin{Bmatrix} \dfrac{n}{m} & \text{if } m \geq n \\ \dfrac{m}{n} & \text{if } n > m \end{Bmatrix} = \frac{\min(n, m)}{\max(n, m)} \qquad b \leq 1 \quad (3.104)$$

which is the ratio between the number of rows and columns. In Figure 3.62, $W$ is plotted as a function of $N$ for different values of parameter $b$. The best condition is reached for $b = 1$, that is, $n = m$, whereas the worst case is represented by a network where all the nodes are placed along a single row (or column). In general $W$ increases with $b$ decreasing.

The mean channel efficiency can be expressed as [130]

$$\eta = \frac{nm-1}{\max(n, m)\cdot(n+m-2)} \tag{3.105}$$

From the previous equation the throughput per user follows, that is,

$$\gamma = c\cdot\eta = \frac{nm-1}{\max(n, m)} \tag{3.106}$$

Finally, the network capacity under uniform traffic conditions is given by [127]

$$C = \gamma\cdot N = \eta\cdot W = \min(n, m)\cdot(nm-1) \tag{3.107}$$

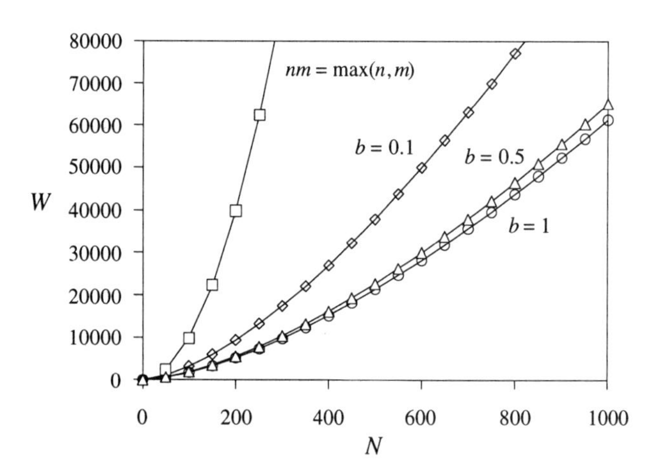


**Figure 3.62** Total number of channels versus the total number of nodes, for different values of the ratio $b$ between the number of rows and columns.

The capacity of MATRIX networks can reach considerable values. In a 30×30 configuration, assuming a data rate of 1 Gbps, the aggregate throughput is almost equal to 25 Tbps. Considering an average data rate of 10 Mbps for each subscriber, 3,000 users per node can be supported by the network.

Through some suitable routing rules, activated when necessary, MATRIX can also overcome problems due to link failures. In fact, the network manager has the possibility of deflecting point-to-point paths when one or many malfunctions are worsening the quality-of-service on the considered connection. This can be done by increasing the path length, passing to an adjacent row or column before the interruption and going back to the original cable after it.

An example of the considered situation is given in Figure 3.63, assuming node $k_{13}$ as a source. The occasional variation of $h$, experienced in some communications, is compensated by the augmented robustness and flexibility of the whole network.

In conclusion, MATRIX topology offers the possibility of realizing end-to-end connections without any WDM conversion. This opportunity is paid in terms of increased transmission medium cost.

***Some keypoints.*** The MATRIX network represents a particular multihop architecture whose diameter, independently of the net size, is always equal to two. Furthermore, it exploits the availability of multifiber cables to avoid the necessity of wavelength conversion in the intermediate nodes, so that an entire end-to-end path is allocated on the same WDM channel. Space diversity compensates the absence of wavelength switching and, combined with wavelength reuse, makes the number of channels reasonably low.

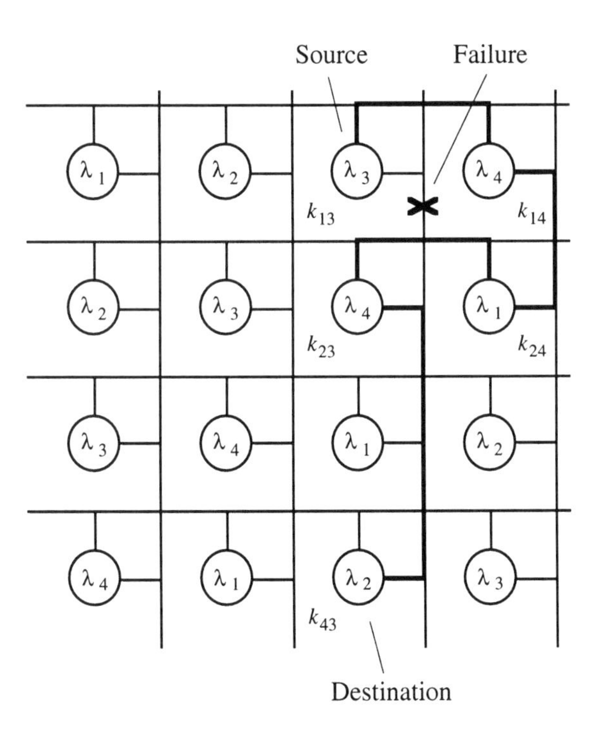


**Figure 3.63** Example of routing rearrangement in case of link failure.

## 3.7 SWIFT ARCHITECTURE

### 3.7.1 The SWIFT Approach

In this section we describe a possible approach to the problem of traffic managing in high-speed networks [131]. It is based on a dynamic sharing of communication resources in TDM systems, in order to avoid two fundamental problems that affect the network based on a fixed allocation of channels and time slots:

- Inability to provide on-demand variable bandwidth allocation, according to the users' necessities;
- Delay depending on TDM cycle length, rather than traffic load.

This new approach is called Store and forWard with Integrated Frequency and Time switching (SWIFT). It adapts the fixed multichannel allocation to the dynamically changing node requirements by the addition of a third layer (at the network level of the open systems interface (OSI) [1] reference model), while preserving the simple switching and routing procedures, implied by the adoption of fixed channel allocation schemes [132].

This method is considered in this chapter since it is based on a distributed routing algorithm that is applied in high-speed multihop packet switching networks. It is worth noting that SWIFT does not require a particular network topology but, on the contrary, it can work in different multihop networks, like those examined in the previous sections.

The basic feature of SWIFT scheme is that it does not use information relative to the topological characteristics of the network only, in order to identify the shortest path between any pair of nodes, but needs some knowledge of the load conditions on the whole communication systems too. This is done to avoid traffic bottlenecks that can be present along selected paths. In fact, the SWIFT goal is the selection of a multihop path covering underutilized channels and buffers, which leads to a service delay shorter than the one implied by direct transmission. As will be shown below, the use of the considered path can be advantageous in terms of throughput increase too, even though the packet propagation is not optimized from a topological point of view.

It must be taken into account that this solution needs the adoption of a suitable signaling system, able to update the traffic information relative to the network stored in nodes' database continuously. Moreover, the nodes have to be provided with a sufficient computing capability, to allow them to handle the routing procedures locally. As a counterpart, the role of the centralized network manager can be reduced or eliminated, at least in normal conditions, and, similarly, the source node is not responsible any more of the path identification for its transmissions.

The SWIFT routing procedure is based, at every intermediate node of a multihop connection, on the following principles:

- The packet has to be addressed toward a channel and a buffer not used by another packet in a given slot;
- The packet has to be forwarded along a route that implies a delay shorter than that of direct transmission from the current node to destination.

The first rule implies that the routing schemes are not handled by the source node only, since it is not able to predict the buffer and channel occupancy in advance. Moreover, the second rule guarantees that the global end-to-end path does not exceed the delay caused by direct transmissions.

### 3.7.2 The Data Link Layer

We have stated that a node cannot send a packet to another node having a full transmission buffer. If any alternatives are not available, the packet has to be stored in the considered node, until one of the adjacent nodes has emptied its buffer. This can be done if the subsequent node can inform its predecessor (where the packet is waiting for transmission) about the buffer state. For this reason SWIFT procedure needs a suitable protocol at the data link level, able to realize the flow control [1].

Different solutions can be used to this purpose, according to the network topology (bus, ring, or other) [133]. Here we will not present specific examples since, as stated above, SWIFT is suitable for application in multihop networks independently of their structure, whereas we report a very general method based on a source-destination allocation algorithm with immediate acknowledgment capability.

Let $N$ be the number of stations in the network, connected by $b$ subchannels having equal bandwidth, with $1 \leq b \leq N$. In every time slot, to simplify hardware implementation, each node can transmit and receive on a single subchannel [134]. By $A$ a $N \times T$ matrix is denoted, where $T = \lceil (N-1)N / b \rceil$ is the TDM allocation cycle length. We have that the generic element $A(s, t)$ of this matrix is alternatively equal to 0 or $d$, with $1 \leq d \leq N$ and $d \neq s$.

Since $s$ and $d$ are the source and destination nodes, respectively, and $t$ is the time slot, it follows that the rows of $A$ represents the sources, its columns are the time slots and the contents of the elements are the relative destinations. Finally, we define the logic function $\delta(x)$ that is equal to 1 if $x$ is true, whereas it equals 0 otherwise.

To solve the multichannel allocation problem, the following constraints have to be observed:

1. Two sources having a packet directed to the same destination cannot send it in the same time slot; so that

$$\forall A(s, t) \neq 0 \text{ then } A(s, t) \neq A(k, t) \ \forall s, k, t \text{ and } s \neq k \tag{3.108}$$

2. Every source can transmit a packet to every destination during a given TDM cycle; this can be expressed through the following notations

$$\forall A(s, t) \neq 0 \text{ then } A(s, t) \neq A(s, k) \ \forall s, k, t \text{ and } t \neq k \tag{3.109}$$

$$\sum_{t=1}^{T} \delta\big(A(s, t) \neq 0\big) = N - 1 \qquad \forall s \tag{3.110}$$

3. If a node receives a packet in a given time slot, it transmits an acknowledgment in the following slot; that is,

$$\text{if } A(i, t) \neq 0 \text{ then } A(A(i, t), (t + 1) \bmod T) \neq 0 \qquad \forall t \tag{3.111}$$

4. All the channels have to be utilized in each time slot; thus

$$\sum_{s=1}^{N} \delta\big(A(s, t) \neq 0\big) = b \qquad \forall t \tag{3.112}$$

Condition (1) derives from the principles indicated in Section 3.7.1; condition (2) allows a source to completely access the network users during the same TDM cycle; condition (3) forces the immediate acknowledgment; finally condition (4) guarantees an exhaustive channel assignment in every slot. The algorithm that implements the source-destination allocation with immediate acknowledgment capability and satisfies the above requirements is presented and discussed in [133]. In Table 3.16 an example of its application in a multihop network is reported, giving the correspondent matrix $A$.

**Table 3.16**
Matrix $A$ for the Allocation Algorithm Applied in a Network with $N = 8$, $b = 4$, and $T = 14$

| *Node* \ *Slot* | **1** | **2** | **3** | **4** | **5** | **6** | **7** | **8** | **9** | **10** | **11** | **12** | **13** | **14** |
|---|---|---|---|---|---|---|---|---|---|---|---|---|---|---|
| **1** | 2 | | 3 | 4 | | | 5 | 6 | | | 7 | 8 | | |
| **2** | | 3 | | | 4 | 5 | | | 6 | 7 | | | 8 | 1 |
| **3** | 4 | | 5 | 6 | | | 7 | 8 | | | 1 | 2 | | |
| **4** | | 5 | | | 6 | 7 | | | 8 | 1 | | | 2 | 3 |
| **5** | 6 | | 7 | 8 | | | 1 | 2 | | | 3 | 4 | | |
| **6** | | 7 | | | 8 | 1 | | | 2 | 3 | | | 4 | 5 |
| **7** | 8 | | 1 | 2 | | | 3 | 4 | | | 5 | 6 | | |
| **8** | | 1 | | | 2 | 3 | | | 4 | 5 | | | 6 | 7 |

About the algorithm considered above, we can develop some observations [133]. In every row of $A$ there are always $N - 1$ non-null elements. The correspondent values can be indicated by $r_j$ with $1 \leq j \leq N - 1$, beginning the numbering from the first nonnull element. We note that

$$r_{j+1} = r_j \bmod N + 1 \tag{3.113}$$

For the sake of clarity, looking at the third row of $A$ in Table 3.16, we see that starting with $r_1 = 4$ it follows $r_2 = 4 \bmod 8 + 1 = 5$, and so on. Equation (3.113) analytically describes the possibility of installing and maintaining a end-to-end communication with all the other nodes during the same cycle. Then, the consecutive elements $c_i$ on a given column of $A$, with $1 \le i \le b$, satisfy the following identity:

$$c_{i+1} = (c_i + y - 1) \bmod N + 1 \tag{3.114}$$

with $y$ being the quotient of the division $n/b$. This result derives from the exhaustive channel assignment condition.

Finally, it is evident that every node being destination in a given slot $t - 1$, for which we have $A(i, t - 1) \neq 0$ and $1 \le t \le T$, becomes a source at slot $t$.

### 3.7.3 The Routing Layer

The core of the SWIFT technique can be found at the routing layer where it acts like an adaptive control on a communication system that, because of the characteristics of the allocation protocol described in the previous section, can be seen as a multihop network in the time domain. SWIFT routing utilizes the delivery of a packet emitted by node $i$ and destined to node $j$ through an intermediate node $k$, at the place of the direct connection, if the transmission slot $t_{kj}$ occurs prior to $t_{ij}$. Theoretically, a continuous and updated knowledge of the state of each buffer should be available to optimize this method, as well as a predictive algorithm able to estimate the future buffer occupancy conditions, taking into account the statistical properties of traffic flows. In fact, the source, basing the routing choices on the present network situation, cannot optimize the path at the time slot in which the packet will be located in a given intermediate node. Moreover, a complete knowledge of the network state would require a significant amount of bandwidth devoted to signalling.

For these reasons, a suboptimal method has been proposed [133]. It distributes the routing procedures at the node level, working on local information. In this way, to know the state of the buffers of the adjacent nodes at the subsequent slot, we can exploit the immediate acknowledgment capability made available by the application of the data link protocol described in the previous section, expressed through (3.111). In fact, we have said that every node being destination in a given slot becomes a source at the next one.

To explain the SWIFT routing operation, let us consider a clique, that is to say a mesh network where for every node pair a direct connection exists. A link is characterized by a time-varying cost $l(i, j)$, where $i$ and $j$ are the ends of the considered edge. This cost depends on the time slot $t$ in which the transmission on the link is requested and on the fixed data link allocation. More precisely, the link cost represents the distance, measured in time slots, from a reference slot $t$ to the slot where the packet sent by $i$ to $j$ is allocated; it is determined by the allocation matrix $A$. In particular, the reference slot is chosen as the slot in which the routing choice has to be taken.

Furthermore, the weight of the same link is given by

$$d_{ij}(t) = (t_{ij} - t + T) \bmod T \quad (3.115)$$

with $t_{ij}$ being the slot used to transmit a packet from $i$ to $j$. In Figure 3.64 a clique with $N = 4$ and $b = 2$ is represented at slot $t = 1$ together with the relative link weights. The corresponding allocation matrix is shown in Table 3.17.

Now let us explain how it is possible to evaluate the link weights using the allocation matrix. For example, looking at Table 3.17, we calculate the weight of the link connecting node 3 to node 4, assuming the slot $t = 1$ as a reference. Applying (3.115), it turns out to be $d_{3,4}(t) = (1 - 1 + 6) \bmod 6 = 0$, as indicated in Figure 3.64.

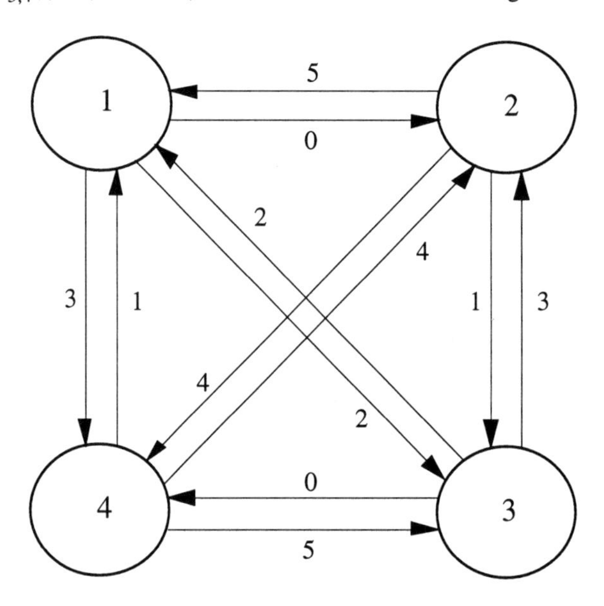


**Figure 3.64** Clique graph (fully meshed) for $N = 4$ and $b = 2$ at time slot $t$ with weighted links.

**Table 3.17**
Allocation Matrix $A$ for the Network Shown in Figure 3.64

| *Node* \ *Slot* | **1** | **2** | **3** | **4** | **5** | **6** |
|---|---|---|---|---|---|---|
| **1** | 2 | | 3 | 4 | | |
| **2** | | 3 | | | 4 | 1 |
| **3** | 4 | | 1 | 2 | | |
| **4** | | 1 | | | 2 | 3 |

The so-called routing table $R(t)$, for $1 \leq t \leq T$, is built in each node at slot $t$ to implement the routing algorithm. It is a $N \times N$ matrix whose generic element $r_{id}(t)$ is given by

$$r_{id}(t) = \begin{cases} k & (1 \leq k \leq N) \text{ node } i \text{ transmits to node } k \text{ a packet destined to } d \\ 0 & \text{node } i \text{ does not transmit a packet destined to } d \end{cases} \quad (3.116)$$

where $k$ is the destination of the direct transmission as determined by the allocation algorithm.

The $R(t)$ table is created using the algorithm reported here by means of the following pseudocode sequence [133]:

```
for each slot t (1 ≤ t ≤ T)
    for each node i (1 ≤ i ≤ N) {forwarding node}
        for each node d (1 ≤ d ≤ N, d ≠ i) {final destination node}
            for each node k (1 ≤ k ≤ N, k ≠ i) and (d_ik(t) = 0)
            {next forwarding node in slot t}
                if d_kd(t + 1) < d_id(t) then r_id(t) = k
            end
        end
    end
end
```

The structure of the algorithm is intuitive. If the weight of the $k$-$d$ connection at the next slot is lower than the one of the $i$-$d$ connection at the present slot, the packet emitted by node $i$ is sent to $k$, since it is the node with an empty buffer closer to $d$. This is done, at slot $t$, by forcing $r_{id}(t) = k$ in the routing table $R(t)$. Nevertheless, it is not sure that the considered packet will reach node $d$ at the subsequent slot because, having stored the packet in the buffer of $k$, the latter will result as a forwarding node at the $t + 1$ slot. Therefore, another node with empty buffer and closer to $d$ can be found at the slot $t + 2$. In this way, hop after hop, the algorithm constructs a path in the network that is not optimal but is characterized, however, by a delay never greater than the one relative to the direct connection.

It can be observed that every node needs to know the allocation matrix $A$, but it is not requested to build the whole table $R(t)$ to perform routing. In fact, a node only needs to identify an intermediate station having an empty buffer and whose direct path delay to the final destination is shorter than that of the node itself.

### 3.7.4 SWIFT Performance

The performance evaluation of a network managed by SWIFT is realized considering either the throughput or the average delays relative to end-to-end connections.

We base this analysis on the following assumptions [133]:

- Every node is provided with a couple of buffers, placed at the transmitter and at the receiver sections, respectively;
- An idle node has an empty output buffer;
- A backlogged node has a packet stored in the output buffer, queued for transmission;
- Packet arrivals occur only in idle nodes;
- At an idle node $i$, the arrival of a packet destined to node $j$ occurs at the beginning of slot, with probability $p_{ij}$.

The system can be modeled by a discrete Markov chain. The state of the system at the beginning of the $t$th slot is represented by the $(X, t)$ $N \times N$ matrix, with $1 \le t \le T$, whose generic element is given by

$$x_{id} = \begin{cases} s & (1 \le s \le N) \text{ if a packet with source } s \text{ exists at node } i \text{ with destination } d \\ 0 & \text{otherwise} \end{cases} \tag{3.117}$$

Denoting by $\pi_t(X)$ the steady-state probability of $X$ at slot $t$, the throughput $S_{sd}$ is the number of packets originated by source $s$ and transmitted by any node (source node or intermediate nodes) via direct transmission to their final destination $d$ [133]

$$S_{sd} = \frac{1}{T}\sum_{t=1}^{T}\sum_{X}\pi_t(X)\cdot\sum_{\substack{k=1 \\ k \neq d}}^{N}\delta(x_{kd} = s)\cdot\delta(r_{kd} = d) \tag{3.118}$$

Now, we can finally calculate the average network throughput as

$$S = \sum_{s}\sum_{d} S_{sd} \tag{3.119}$$

To obtain the average packet delay $\tau$ relative to the whole network, first we need to define the average packet delay $\tau_{sd}$, that is relative to the $s$-$d$ connection only. It is defined as the average time between the packet arrival at source node $s$ and its reception at destination node $d$. Using the Little's result [135], it can be written as

$$\tau_{sd} = \frac{Q_{sd}}{S_{sd}} \tag{3.120}$$

where $Q_{sd}$, the average backlog relative to the $s$-$d$ connection, is calculated through the following equation:

$$Q_{sd} = \frac{1}{T}\sum_{t=1}^{T}\pi_t(X)\cdot\sum_{\substack{k=1 \\ k \neq d}}^{N}\delta(x_{kd} = s) \tag{3.121}$$

The second sum in (3.121) is equal to the number of packets emitted by source $s$ and destined to destination $d$ that are present in the network at the considered slot. Finally, the average packet delay $\tau$, measured in slots, is equal to

$$\tau = \sum_{s} \sum_{d} \frac{S_{sd}}{S} \tau_{sd} \tag{3.122}$$

To complete the model, the steady-state probability $\Pi(X) = (\pi_1(X), \pi_2(X), \ldots, \pi_T(X))$ has to be known, since it is necessary for the calculus of the above-indicated parameters. This subject has been considered in [133]; it has not been reported here for the sake of brevity. However, the analysis here presented is sufficient to compare the SWIFT performance with the one of fixed allocation systems. For example, in Figures 3.65 and 3.66 the average queueing delay $\tau$ is plotted versus the average network throughput $S$ with regard to networks having $N = 16$ nodes with $b = 16$ subchannels and $N = 32$ nodes with $b = 32$ subchannels, respectively.

In both cases, all the nodes, provided by single buffers, are characterized by homogeneous arrival rates, that is, $p = p_{ij}, \forall i, j$. The advantage that is obtained using SWIFT instead of fixed time allocation is evident for low or medium loads, whereas it tends to become negligible for high values of $S$, until the two techniques reach convergence. Nevertheless this is not a drawback, since fixed allocation behaves optimally for high loads. On the other hand, the improvement produced by SWIFT results to be amplified as the number of nodes increases. In fact, in this condition the efficiency of fixed allocation systems worsens, since the higher $N$ the longer the TDM cycle length.

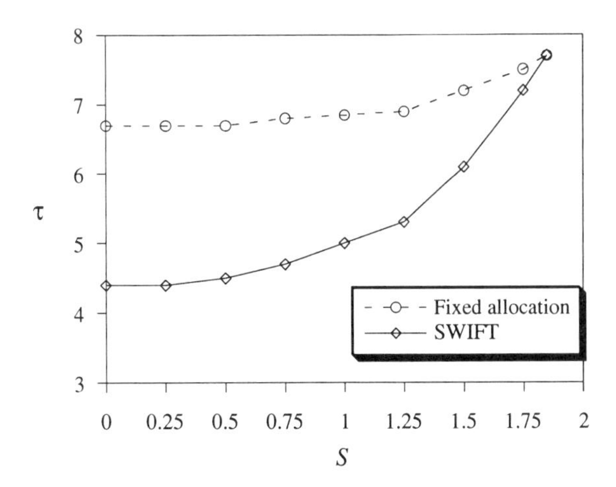


**Figure 3.65** Average queueing delay versus throughput for homogeneous arrival rates at the nodes of a network having $N = 16$ and $b = 16$. (*Source*: [133]. © 1990 IEEE.)

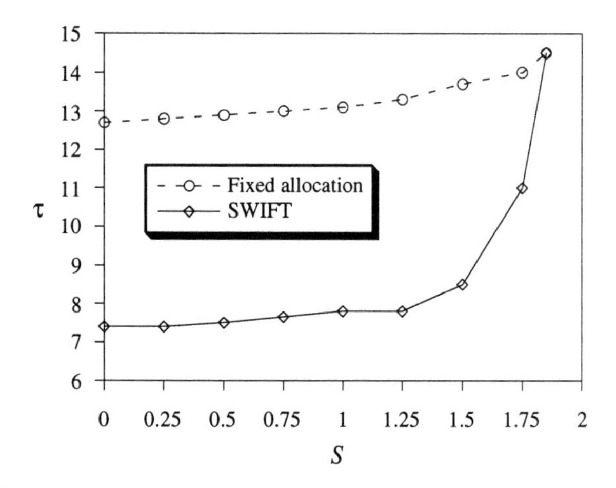


**Figure 3.66** Average queueing delay versus throughput for homogeneous arrival rates at the nodes of a network having $N = 32$ and $b = 32$.

A significant improvement can also be observed taking into account the throughput levels with respect to the arrival rates at the nodes, as indicated in Figure 3.67. This occurs since, as said before, SWIFT constructs paths corresponding to delays that are never higher than the direct connection delay. Therefore, using SWIFT, the buffers are emptied earlier than under fixed allocation schemes.

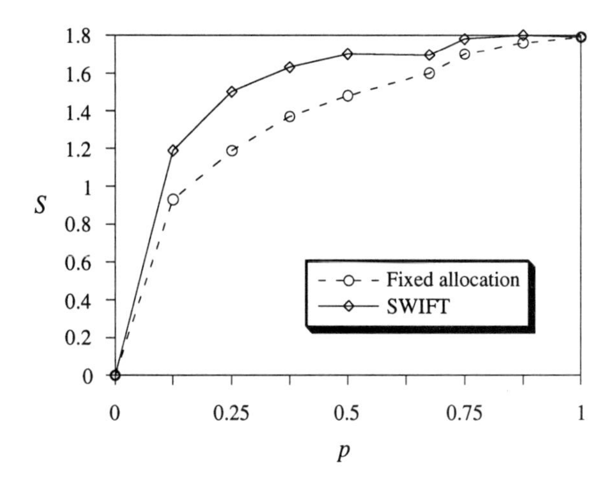


**Figure 3.67** Throughput for homogeneous arrival rates at the nodes of a network having $N = 16$ and $b = 16$.

Abandoning the hypothesis of homogeneous arrival rates, let us examine what happens when the nodes are differently loaded. This aspect is considered in Figures 3.68 and 3.69, with reference to a situation where $N = 16$ and half the nodes have an arrival probability equal to the 10% of the corresponding one in the remaining nodes.

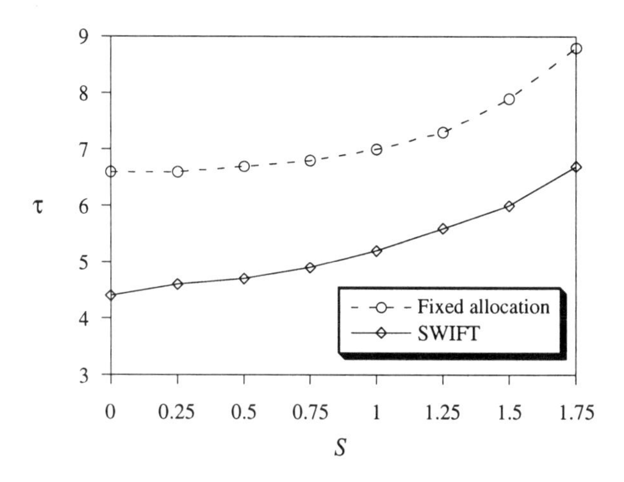


**Figure 3.68** Average queueing delay versus throughput for heterogeneous arrival rates at the nodes of a network having $N = 16$ (eight nodes with arrival rate equal to $p$, eight nodes with arrival rate equal to $0.1p$).

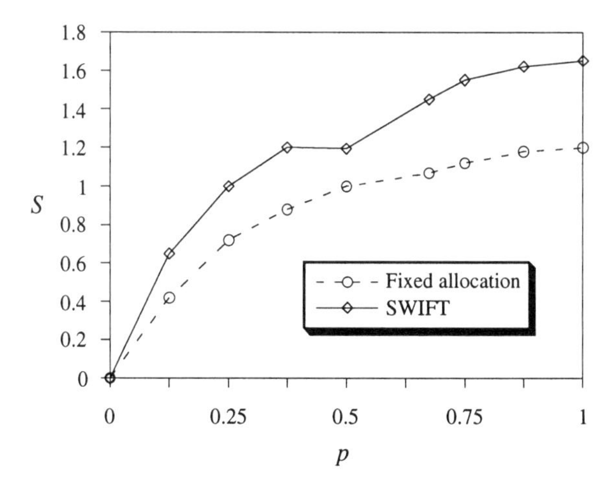


**Figure 3.69** Throughput for heterogeneous arrival rates at the nodes of a network having $N = 16$ (eight nodes with arrival rate equal to $p$, eight nodes with arrival rate equal to $0.1p$).

We see that, considering either the average queueing delay or the throughput, SWIFT operates always better than fixed allocation, and this is also the case when the load is high. In conclusion, SWIFT seems to offer the following interesting features:

- Improvement of network performance for low to medium traffic levels;
- Effective exploitation of network resources;
- Possibility of removing the problems due to time-varying traffic demands from users, experienced by fixed allocation schemes.

***Some keypoints.*** The SWIFT architecture, unlike previous considered networks, does not imply a particular virtual or physical topology since, in essence, the goal of this technique is the sharing of network resources in a dynamic fashion, carried out at data link and network layers, to improve the effectiveness and the performance of a given TDM communication system. Nevertheless, it has been studied in the context of multihop networks, because it adopts a routing strategy that takes advantage of the availability of multihop paths, crossing intermediate nodes, to reduce the source-destination packet delay. The performance of the SWIFT architecture is in general better than that achievable using fixed allocation schemes.

## 3.8 STARNET ARCHITECTURE

### 3.8.1 Starnet Basic Characteristics

In this last section we present the starnet [136] architecture, already realized and tested experimentally [137], conceived to implement both a packet-switched network and a broadband circuit interconnect on the same communication system. This structure is interesting since it seems to be able to exploit the bandwidth made available by the optical medium effectively, with an acceptable hardware and software complexity, and because it tries to fit the network resource availability to the real necessities implied by the different kinds of traffic demands. In fact, according to their characteristics, we can distinguish three different categories of traffic circulating in a broadband network:

- Low speed, bursty, or continuous (including telephony), that can be handled adequately by a packet-switched network;
- High-speed, continuous, and call-oriented (file transfers, video conferencing, etc.), that can be handled by a circuit-switched network, since usually it does not imply short switching times;
- High-speed, bursty.

The last class of traffic, typically related to supercomputer interconnection, could be managed either by circuit-switched or packet-switched networks, according to the necessities of point-to-point or multicast communications.

To satisfy these different requirements, the starnet approach includes both the switching techniques (circuit switching and packet switching), which operate on the same network simultaneously and independently. In this way, thanks to the presence of an alternative, the system is not requested to offer extreme performance on each of the two subnetworks, so relaxing the global building constraints.

Moreover, the coexistence of the two switching techniques can be utilized to simplify the installation of communication sessions, since the packet switching network is asked to support the signaling relative to all the end-to-end connections to be handled by the circuit switching network. This solution makes starnet particularly competitive with respect to other networks, for example, RAINBOW (see Section 2.3.2), that do not adopt packet switching and are forced to explore the network continuously, looking for call requests emitted by nodes waiting for transmissions. This procedure limits in practice the maximum number of connectable nodes, whereas starnet does not.

In order to explain the basic features of starnet, in the next section we will examine the node structure, with reference to the technical solutions applied in a first experimental prototype [138].

### 3.8.2 Node Structure

Every node in a starnet topology is linked with all the other ones by means of a passive star. It uses two fibers, connected, respectively, to the transmitter and the receiver sections, where the latter is composed by two independent receiving chains. According to the basic node configuration 1 (BNC1) architecture, schematically depicted in Figure 3.70 and, with more details in Figure 3.71, a node is provided with a transmitter that emits two independent streams of data: stream P for packet switching subnetwork and stream C for circuit switching subnetwork. The BNC2 architecture instead admits the handling of two streams P, using again only one transmitter and a couple of receivers. A simple multiplexing method is based on a suitable time sharing, organized to realize an interleaving of stream C and stream P data.

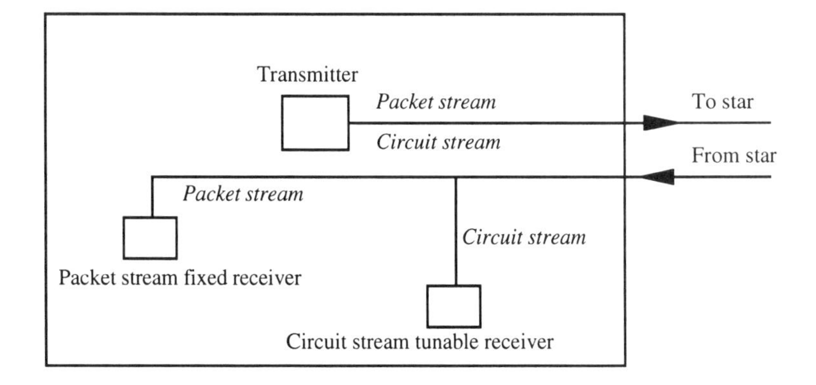


**Figure 3.70** Basic starnet node configuration (BNC1).

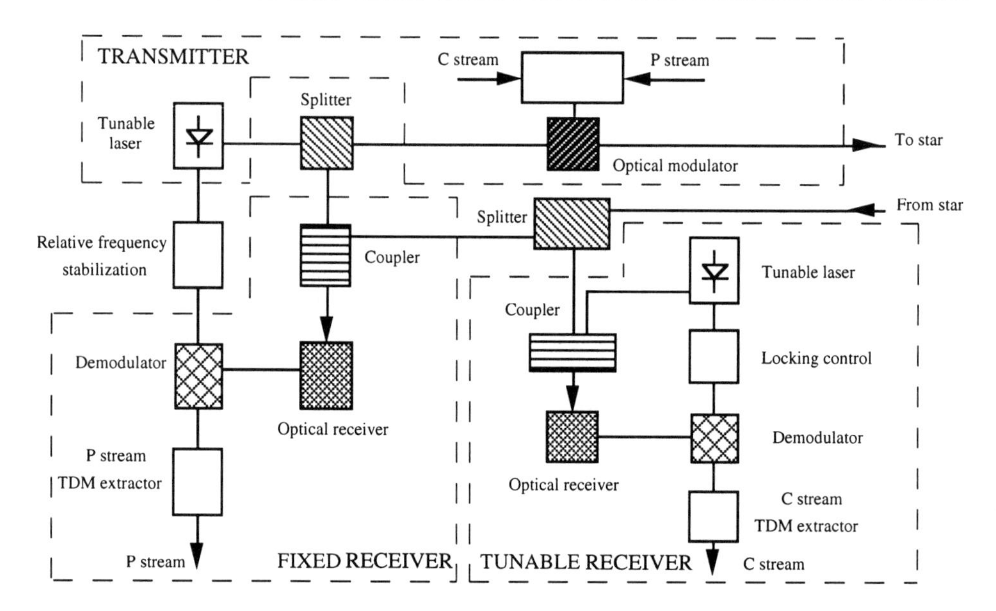


**Figure 3.71** Architecture of a BNC1 starnet node. (*Source*: [139]. © 1993 IEEE.)

Both BNC1 and BNC2 features have been studied in [139], whereas the experimental implementation of a starnet node has been presented and discussed in [34].

A tunable laser is utilized to guarantee network flexibility and good fault-tolerant properties, but not for realizing WDM switching operation. In fact, the transmitter of each node is tuned at a different fixed wavelength, in a way that a group of optical carriers is multiplexed on the fiber. The optical separation between channels can be reduced to increase the bandwidth efficiency, without causing interference on the signal reception [140].

As stated above, a pair of receivers compose the receiver section; the tunable receiver decodes the stream C only and is used to implement a circuit interconnection among nodes. On the other hand, the fixed receiver is permanently tuned at the transmitting wavelength of the previous node, along the wavelength comb, to intercept its stream P data. The latter, once received, is then forwarded to the node which is the subsequent one along the considered wavelength sequence.

In this way, a sort of store-and-forward chain is organized among nodes in the packet switching subnetwork. It is closed by setting the fixed receiver of the first node, that is, the node that is tuned at the first wavelength of the comb, at the transmitting wavelength of the last one. An example is given in Figure 3.72 for a starnet having $N$ nodes, whereas the corresponding logical unidirectional ring is shown in Figure 3.73.

In the BNC2 configuration, as stated above, three data streams can be multiplexed by the same transmitter, one stream C and two distinct streams P. In such a case, this second node arrangement permits the implementation of a bidirectional virtual ring, simply using two counter-oriented streams P and extending in the two opposite directions the same tuning techniques above described.

Incidentally, we note that BNC1 and BNC2 are suitable for application in other existing networks, using optical fibers as transmission medium, such as ANSI (American National Standards Institute) X3T9 fiber distributed data interface (FDDI) networks [141,142] or IEEE 802.6 distributed queue dual bus (DQDB) networks [143,144], respectively. A detailed description of both node configurations is reported in [139], with special attention to the problems of frequency stabilization, coherent modulation, and data multiplexing.

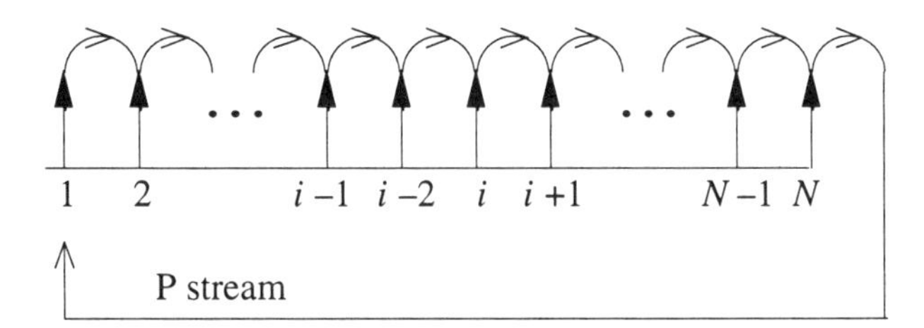


**Figure 3.72** Wavelengths comb for stream P transmission.

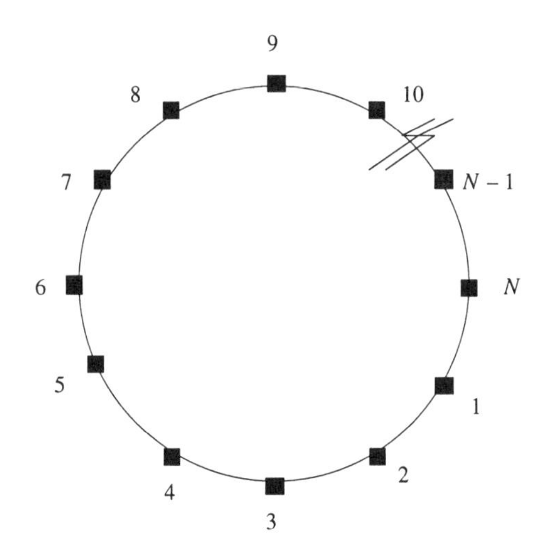


**Figure 3.73** Virtual ring overlapping the passive optical star in a $N$ node starnet.

We wish to underline that the above-mentioned configurations are those originally proposed for starnet, but other structures have been also considered and implemented. In particular, in [137] two packet-switched subnetworks have been used: a reconfigurable subnetwork working at high speed and a fixed-tuned subnetwork, operating at a lower speed. Both of them handle data traffic, in addition a low-speed subnetwork is also used to support signaling and network control functionalities.

We do not discuss here the various possible implementations at the hardware level, since this is not the goal of this book. Nevertheless we are going to give further information about the multiplexing techniques that are very important to understanding fundamental starnet operation.

In fact, the basic idea of starnet architecture consists in the multiplexing of two distinct classes of traffic in the same optical medium, having very different characteristics and speeds.

Until now, we have assumed TDM multiplexing of C and P streams, even though this is not the unique possible solution. In the sequel, other techniques are briefly analyzed and compared in terms of performance and implementation costs.

In particular, we consider the following alternative solutions:

- Time division multiplexing (TDM);
- Combined modulation formats;
- Subcarrier multiplexing;
- Multilevel modulation.

To realize a comparison between these solutions, the definition of power penalty $\Delta$ due to multiplexing is useful. It represents the ratio between the power needed to transmit both C and P streams and the power needed to transmit stream C only, at the same fixed error probability [139].

### *3.8.2.1 Time Division Multiplexing*

Using TDM multiplexing the time allocation of stream P requires that the transmission speed of stream C is increased. The relation existing between the corresponding bit rates can be given through the bit rate ratio $\rho$ that is equal to

$$\rho = \frac{B_C}{B_P} \tag{3.123}$$

where $B_C$ and $B_P$ are the bit rate of stream C and P, respectively. Now, it is necessary to calculate the power penalty $\Delta$, that can be given as [139]

$$\Delta = 10 \cdot \log_{10}\left(1 + \frac{1}{\rho}\right) \tag{3.124}$$

As will be shown below, this is the technique that guarantees the best performance in terms of power penalty, but it requires that both receivers operate at the bit rate of the aggregate stream resulting from multiplexing.

### *3.8.2.2 Combined Modulation*

Combined modulation is realized when different modulation formats are simultaneously used on the same carrier such as, for example, low-index amplitude shift keying (ASK) with phase shift keying (PSK), differential PSK (DPSK), frequency shift keying (FSK) [145], and others.

We can consider the ASK-PSK modulation, assuming that ASK is adjusted to make the error probability on its stream equal to the error probability relative to PSK stream. In particular, the P stream is ASK encoded whereas the C stream is PSK encoded. The bit rate ratio is that indicated in (3.123) whereas the power penalty turns out to be [139]

$$\Delta = 20 \cdot \log_{10}(1-\delta) \tag{3.125}$$

where $\delta$ determines the depth of the ASK modulation.

The implementation of combined modulation could be simpler than TDM, since no buffering or synchronization devices are needed to multiplex the P and C streams. In addition, the transmission speed of these streams is not increased like in the TDM case.

A BNC1 prototype using a ASK-PSK transmitter based on this technique has been realized and tested [138]. It has been demonstrated to be able to handle a 2.488-Gbps data stream in the circuit switching subnetwork and a 125-Mbps data stream in the packet switching subnetwork.

### 3.8.2.3 *Subcarrier Multiplexing*

According to this solution, channel amplitude, instead of channel power, is shared among multiplexed streams. Assuming the same modulation format for both the multiplexed streams C and P and the usual expression for the bit rate ratio, it follows that in this case the power penalty is equal to [139]

$$\Delta = 20 \cdot \log_{10}\left(1 + \frac{1}{\sqrt{\rho}}\right) \tag{3.126}$$

Subcarrier multiplexing offers better performance than combined modulation but at the expense of a greater bandwidth occupancy. Moreover, its implementation becomes more difficult considering channel tuning and frequency stabilization, with respect to TDM or combined modulation techniques.

On the contrary, a limited application of the solution considered seems to be convenient in BNC2 for the multiplexing of two P streams, using the combined modulation to multiplex the resulting signal with stream C.

### 3.8.2.4 *Multilevel Modulation*

In multilevel communication systems more than 1 information bit can be carried in a single symbol. If $n$ is the number of bits per symbol, starnet can assign $k$ of them to P transmissions and, consequently, $n - k$ bits to C transmissions.

This solution, which does not require either TDM multiplexing-demultiplexing devices, or special combined formats modulators, could be useful in WDM networks having limited bandwidth resources. Unfortunately, even though coherent multilevel optical transmissions are theoretically feasible [146], the present state of the art of the lightwave technology is not still consolidated to be at an experimental level [147].

#### *3.8.2.5 Performance Comparison*

A comparison among the multiplexing techniques considered so far can be performed, with regard to power penalty. However, we do not extend this compared analysis to multilevel modulation because, as stated above, its practical feasibility has still not been demonstrated. For this reason, in Figure 3.74, the power penalty relative to the remaining three cases is plotted versus the bit rate ratio. For all the curves, a bit error probability equal to $10^{-9}$ is assumed.

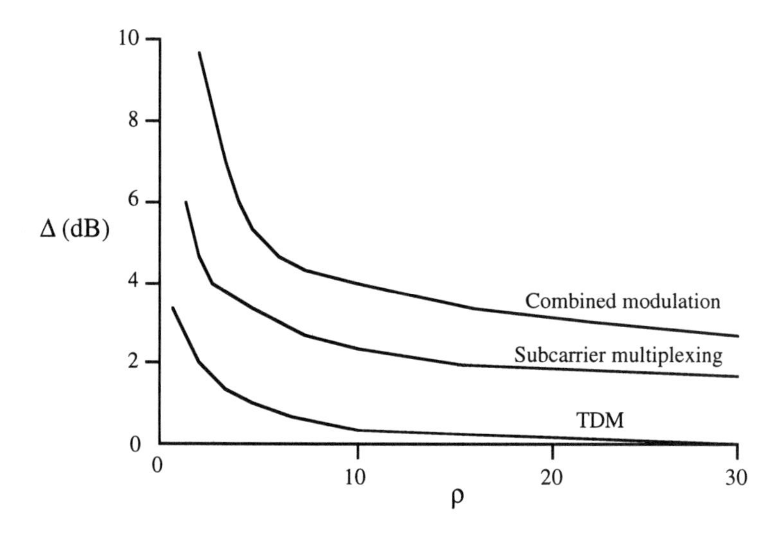


**Figure 3.74** Power penalty versus bit rate ratio for different multiplexing techniques, assuming an error probability equal to $10^{-9}$. (*Source*: [139]. © 1993 IEEE.)

As indicated above, TDM exhibits the best performance, followed by subcarrier multiplexing, whereas combined modulation performs worst. However, taking into account other performance parameters, we have to remind the reader that the adoption of subcarrier multiplexing, even though suitable for application in BNC2 nodes, implies a too-high bandwidth wasting.

Therefore, confining the comparison to TDM and combined modulation, we observe that the latter can compensate the worsening in the power penalty, that in any case tends to be reduced as ρ increases, with a simpler hardware implementation. Moreover, TDM forces the receivers to operate at the aggregate bit rate, higher than those of the two streams, whereas combined modulation does not.

The detailed analysis of all these aspects would require a wide discussion of topics that cannot be properly included in this book. Nevertheless, readers interested in having further technical information on these subjects can find it in [139] and in the more recent [34].

### 3.8.3 The Circuit Switching and Packet Switching Subnetworks

As seen in the previous section, the coexistence of two independent subnetworks, based on circuit and packet switching, does not imply hardware implementation that is too complex; furthermore the increased node costs are compensated by the achievement of a significant network flexibility. The latter is obtained, in particular, thanks to the possibility of using the packet switching subnetwork for delivering the signaling information relative to the circuit interconnect. This fact also simplifies network management procedures.

The maximum number of nodes that can be supported by the circuit switching subnetwork has been calculated in [139] as a function of many parameters relative to the network hardware architecture, the modulation characteristics, the power budget, and so on. It is essentially upper-bounded by three fundamental characteristics of the network, that are listed here in order of importance:

- Laser tunability;
- Power budget;
- Fiber bandwidth.

On the other hand, as regards the packet switching subnetwork, some choices can be made related to its logical architecture. A first topological choice can be based on a ring network where the access to the common channel is shared among nodes using the IEEE 802.5 token ring protocol [148]. According to this solution, the overall number of nodes is inversely proportional to the capacity per node and proportional to the propagation delay. With reference to the network capacity, a certain improvement could be achieved by a data transmission speed-up. Unfortunately, we cannot reduce the propagation delays in this manner, since they depend on the physical star topology and its extent on a geographical basis. In fact, independently of the considered pair of nodes that are to be connected and their relative distance, every transmitted packet has to cross the center of the star before reaching its destination. Similarly, the performance of the starnet is not significantly improved even if a dual-bus topology is adopted, which is allowable by BNC2 nodes utilization.

Even though the packet switching subnetwork is not necessarily requested to offer an optimal behavior in call-oriented end-to-end connections (they can be effectively managed by the circuit interconnect), its poor performance, exhibited when unidirectional or bidirectional ring topology is used at the virtual level, suggests the adoption of different network configurations, accepting some complexity and cost increments. In fact, we cannot forget that an excessive worsening in the performance of the packet switching subnetwork can make the operation of the circuit interconnect signaling protocol impossible, thus affecting the behavior of this second subnetwork as well.

A most interesting opportunity consists in the realization of a multilevel network, or network of networks, following a procedure that is briefly discussed here and then applied in a starnet environment, since it will be widely described in Chapter 4. A certain number of independent packet switching subnetworks, organized according to unidirectional ring or dual-bus structures, are allocated in adjacent wavelength bandwidth. In every subnetwork a special node plays the role of bridge; all the bridge nodes are then connected through a backbone ring. The latter can operate at a speed higher than the one of subrings, if a part of the capacity of the circuit switching network is devoted to supporting the backbone. An example of this multilevel topology is proposed in Figure 3.75.

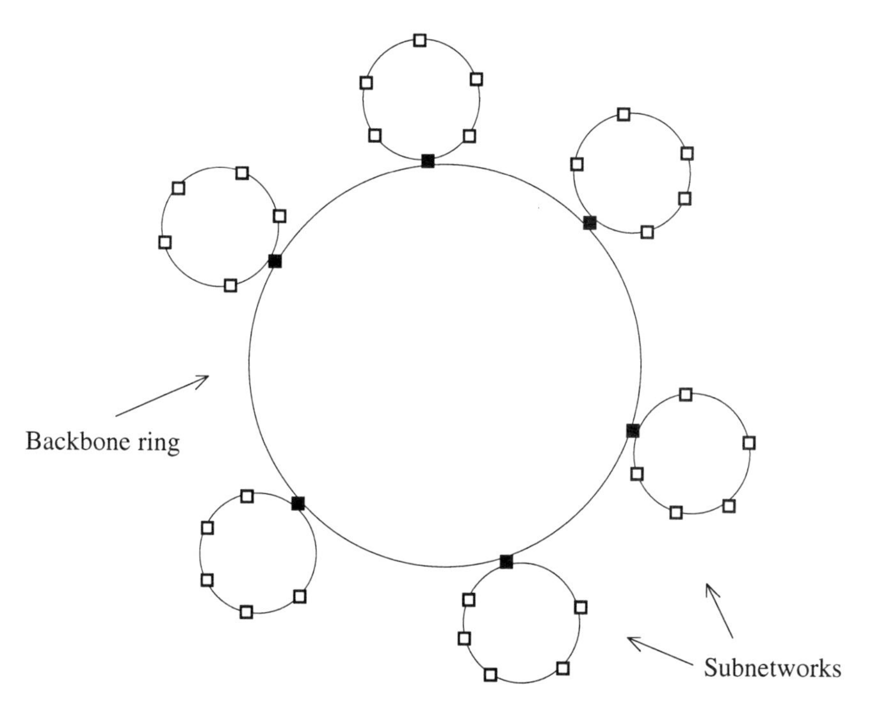


**Figure 3.75** Virtual topology of a starnet ring of rings (white squares = nodes; black squares = bridges).

All the transmissions among nodes belonging to the same subring are realized without involving the backbone. On the contrary, when a packet is destined to a node belonging to a subring different from the one of its source, the latter sends it to its proper bridge. Then the bridge retransmits the packet along the backbone ring, so that the bridge node corresponding to the destination subring can receive it. The packet is finally injected in that ring to be delivered to its destination. This procedure, considered at a virtual level, is realized in practice by bridge nodes switching the packet from P streams to C streams and vice versa. This means that this virtual partitioning of the network does not necessarily need a corresponding multilevel physical architecture.

It can be analytically demonstrated [139] that the network has to be organized in a way that the nodes to be grouped in the same subring have an high probability of packet exchange. In other words, network partitioning becomes useful when the backbone is moderately used by nodes, since in general their messages are delivered inside the subrings. The cluster of nodes composing each subring is realized by assigning adjacent wavelengths to nodes belonging to the considered partition.

Even though the analytical determination of a good network subdivision is not simple, it must be noted that the number of clusters should be chosen accurately, since excessive partitioning does not lead to optimal network behavior. Further details about this aspect can be found in [139], with reference to specific starnet applications, and in Chapter 4 as regards the general approach to multilevel network analysis.

For the time being, we can say that an alternative solution with respect to the backbone network can be found, by simplifying the bridges interconnection. A first trivial possibility, conceived to increase the network reliability when bridge nodes are dispersed in a wide geographical area, consists in a complete mesh network that contains all the end-to-end links between every pair of bridges. However this solution, which substitutes the implementation of an optical star, is not cost-effective at all.

A second very promising solution is based on the idea of concentrating the bridges in the same location, where they can be gathered and interfaced for example through an asynchronous transfer mode (ATM) [149] switch. Therefore, through bridge concentration high aggregate throughputs can also be achieved. In case of bridge failure, the corresponding subring can be merged with another one, with a simple and fast recovery procedure. Similarly, if all the bridges fail together, starnet architecture can be reset to the original single-ring topology or rearranged in a new two-level structure, using other nodes as bridges.

### 3.8.4 Multihop Networks Supported by Starnet

From the above descriptions, the possibility of realizing a multihop network on a starnet background is evident. In this section, we will show the implementation of a multihop network with out- and in-degree equal to 2, using the standard configurations BNC1 and BNC2. We explain this opportunity by means of an example, considering the case of a 4×4 MSN (Section 3.2.1). Every node is provided with a couple of receivers and transmitters. One receiver is tunable, whereas the other operates at a fixed wavelength. The latter can extract either low-speed packet information or broadband packet information from one of the two adjacent nodes.

Inside the network we can draw an Hamiltonian ring, that is, a closed path that crosses all the nodes of the network only once, and then comes back to the source. It can be represented as a sequence of nodes, beginning and ending at the source node, where each node appears only once. Because of its topology, the Hamiltonian ring can be realized, using fixed receivers, in a way similar to the case represented in Figure 3.73. The application of this concept to the multihop network considered is depicted in Figure 3.76, where all the unidirectional links are not oriented to simplify the representation.

Exploiting the flexibility of starnet, the remaining links, which are not included in the Hamiltonian path, are arranged in subrings using the tunable receivers to complete the topology. The 4×4 MSN is therefore organized on four rings, which are reported in Figure 3.77. A similar approach is used in [141], applied to a shufflenet (Section 3.3.3).

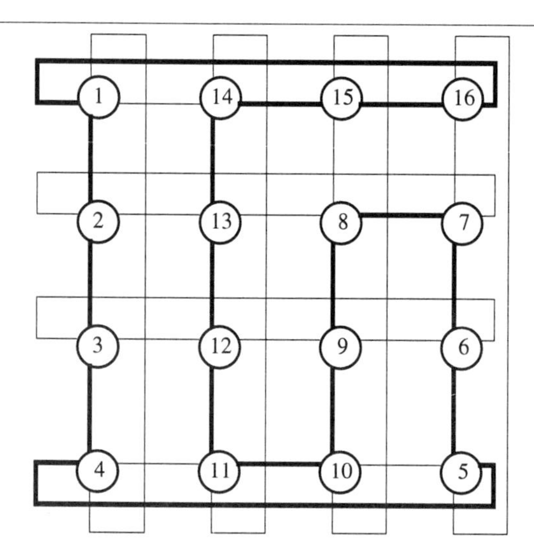


**Figure 3.76** Hamiltonian ring (bold links) in a 4×4 Manhattan street network.

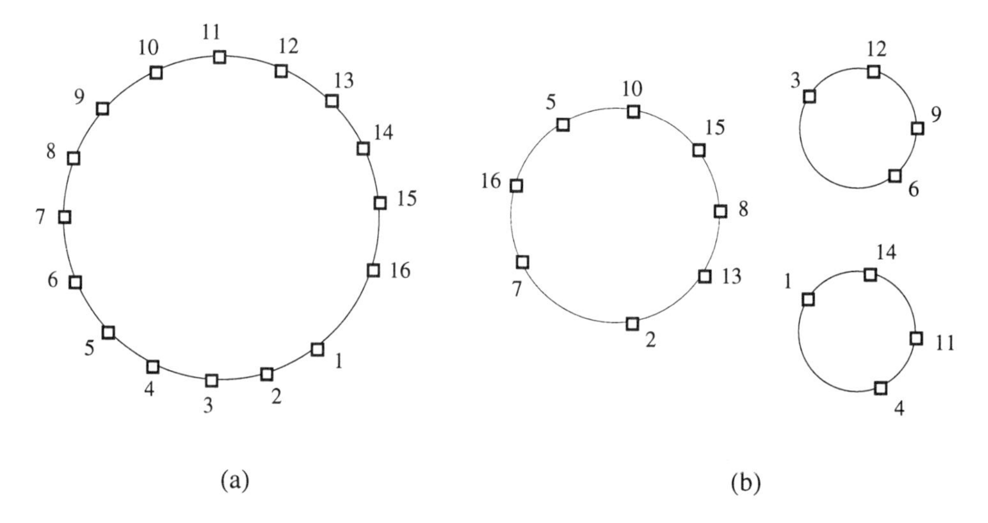


**Figure 3.77** Node arrangement for the 4×4 Manhattan street network of Figure 3.76 in a starnet architecture: (a) Hamiltonian ring composed by nodes using fixed receivers, (b) other rings composed of nodes using tunable receivers.

The MSN node operation is not affected by the utilization of this particular configuration, since the network topology has not been changed at the virtual level. Furthermore, the hardware architecture of the nodes can be implemented on BNC1 or BNC2 basis, without requiring any additional optic devices. This is true for all but one node, which is requested to close the Hamiltonian ring. In this node we need the presence of two tunable receivers. Alternatively, two directly connected distinct nodes can be used, located in the same place and adopting a standard configuration.

The resulting multihop network has all the properties of a starnet network, as regards robustness, flexibility, and reconfigurability. In addition, this solution allows two access modes at different speeds, so that starnet does not force all the nodes to operate at high transmission rates. Therefore, only users requiring broadband services need upgraded transceivers, whereas the others (nonmultihop nodes) can be implemented with a low cost architecture. This fact makes the network expansion gradual, reducing the initial investment costs and fitting the resources to the real necessities.

Higher-order multihop networks can be implemented with similar procedures, adopting suitable node configurations. Finally, in a starnet architecture two or more independent multihop networks can coexist without interfering, and multilevel multihop networks can be realized too. As stated above, the techniques to be used for this kind of segmented networks are the subject of Chapter 4.

***Some keypoints.*** Starnet is a broadband backbone optical WDM network based on a passive star topology. It is, in turn, composed by two subnetworks, working at different speeds to support different classes of traffic. In their basic configuration, one subnetwork utilizes packet switching, whereas the other one is circuit switched, but both of them share the same optical transmission medium, thanks to the adoption of suitable multiplexing techniques. Through starnet architecture, common fully-connected multihop networks, as well as segmented (multilevel) multihop networks, can be implemented.

## REFERENCES


[1] Tanenbaum, A. S., *Computer Networks*, Englewood Cliffs, NJ, Prentice Hall, 1986.

[2] Brackett, C. A., A. S. Acampora, J. Sweitzer, G. Tangonan, M. T. Smith, W. Lennon, K.-C. Wang, and R. H. Hobbs, "A scalable multiwavelength multihop optical network: A proposal for research on all-optical networks," *IEEE/OSA J. Lightwave Technol.*, Vol. 11, No. 5/6, May/June 1993, pp. 736–753.

[3] Mukherjee, B., "WDM-based local lightwave networks – Part II: multihop systems," *IEEE Network Mag.*, Vol. 6, No. 4, July 1992, pp. 20–32.

[4] Walters, R., *Computer Telephone Integration*, Norwood, MA, Artech House, 1993.

[5] Maxemchuk, N. F., "Regular mesh topologies in local and metropolitan area networks," *AT&T Tech. J.*, Vol. 64, No. 7, Sept. 1985, pp. 1659–1685.

[6] Maxemchuk, N. F., "Routing in the manhattan street network," *IEEE Trans. Commun.*, Vol. COM-35, No. 5, May 1987, pp. 503–512.

[7] White, H. E., and N. F. Maxemchuk, "An experimental TDM data loop exchange," *Proc. ICC '74.*

[8] Brassil, J., A. K. Choudhury, and N. F. Maxemchuk, "The manhattan street network: A high performance, highly reliable metropolitan area network," *Computer Networks and ISDN Systems*, Vol. 26, March 1994, pp. 841–858.

[9] Cheng, Z., and T. Berger, "Reliability and availability analysis of manhattan street networks," *IEEE Trans. Commun.*, Vol. 42, No. 2/3/4, Feb./Mar./April 1994, pp. 511–522.

[10] Chung, T. Y., and D. P. Agrawal, "On network characterization and optimal broadcasting for the MSN," *Proc. INFOCOM '90*, San Francisco, CA, June 5-7, 1990, pp. 465–472.

[11] Chung, T. Y., N. Sharma, and D. P. Agrawal, "Cost-performance trade-offs in manhattan street network versus 2-D torus," *IEEE Trans. Computers*, Vol. 43, No. 2, Feb. 1994, pp. 240–243.

[12] Prosser, R. T., "Routing procedures in communication networks-part I: random procedures," *IRE Trans. Commun. Systems*, Dec. 1962, pp. 322–329.

[13] Maxemchuk, N. F., "Comparison of deflection and store-and-forward techniques in the manhattan street network and shuffle-exchange networks," *Proc. INFOCOM '89*, Ottawa, Canada, April 1989, pp. 800–809.

[14] Baran, P., "On distributed communication networks," *IEEE Trans. Commun. Sys.*, Vol. CS-12, Mar. 1964, pp. 1–9.

[15] Acampora, A. S., and S. I. A. Shah, "Multihop lightwave networks: A comparison of store-and-forward and hot-potato routing," *IEEE Trans. Commun.*, Vol. 40, No. 6, June 1992, pp. 1082–1090.

[16] Fultz, G. L., and L. Kleinrock, "Adaptive routing techniques for store-and-forward computer-communications networks," *Proc. ICC '71*, Montreal, Canada, June 1971, pp. 39.1–8.

[17] Choudhury, A. K., and V. O. K. Li, "Performance analysis of deflection routing in the manhattan Street network," *Proc. ICC '91*, Denver, CO, June 23-26, 1991, Vol. 3, pp. 1659–1665.

[18] Goodman, J., and A. G. Greenberg, "Sharp approximate analysis of adaptive routing in mesh networks," *Proc. International Seminar, Teletraffic Analysis and Computer Performance Evaluation*, Amsterdam, The Netherlands, June 2-6, 1986, pp. 255–270.

[19] Robertazzi, T., and A. A. Lazar, "Deflection strategies for the manhattan street network," *Proc. ICC '91*, Denver, CO, June 23-26, 1991, Vol. 3, pp. 1652–1658.

[20] Ayanoglu, E.,"Signal flow graphs for pattern enumeration and deflection routing analysis in multihop metworks," *Proc. GLOBECOM '89*, Nov. 1989, pp. 1022–1029.

[21] Iyengar, A., and M. El Zarki, "Switching prioritized packets," *Proc. GLOBECOM '89*, Nov. 1989, pp. 1181–1186.

[22] Hajek, B., and A. Krishna, "Performance of shuffle-like switching networks with deflection," *Proc. INFOCOM '90*, Vol. 2, San Francisco, CA, June 1990, pp. 473–480.

[23] Cotter, D., and M. C. Tatham, "Dead reckoning – a primitive and efficient self-routing protocol for ultrafast mesh networks," *IEE Proc. Commun.*, Vol. 144, No. 3, June 1997, pp. 135–142.

[24] Cotter, D., and S. C. Cotter, "Algorithm for binary word recognition suited for ultrafast nonlinear optics", *Electron. Lett.*, Vol. 29, 1993, pp. 945–946.

[25] Cotter, D., J. K. Lucek, M. Shabeer, K. Smith, D. C. Rogers, D. Nesset, and P. Gunning, "Self-routing of 100 Gbit/s packets using 6-bits keyword address recognition," *Electron. Lett.*, Vol. 31, 1995, pp. 2201–2202.

[26] Marsan, M. A., A. Bianchi, E. Leonardi, and F. Neri, "Topologies for wavelength-routing all-optical networks," *IEEE/ACM Trans. Networking*, Vol. 1, No. 5, 1993, pp. 534–546.

[27] Ozeki, T., "Ultra optoelectronic devices for photonic ATM switching system with tera-bits/sec throughput," *IEICE Trans. Commun.*, Vol. E77-B, No. 2, Feb. 1994, pp. 100–109.

[28] Masetti, F., "System functionalities and architectures in photonic packet switching," in G. Prati (ed.), *Photonic Networks*, Springer-Verlag, London, England, 1997, pp. 331–348.

[29] Yamanaka, Y., T. Numai, K. Kasahara, and K. Kubota, "Optical fiber loop memory using vertical to surface transmission electro-photonic device," *IEEE/OSA J. Lightwave Technol.*, Vol. 11, No. 12, Dec. 1993, pp. 2140–2144.

[30] Huffaker, D. L., W. D. Lee, D. G. Deppe, C. Lei, T. J. Rogers, J. C. Campbell, and B. G. Streetman, "Optical memory using a vertical-cavity surface-emitting laser," *IEEE Photon. Technol. Lett.*, Vol. 3, No. 12, Dec. 1991, pp. 1064–1066.

[31] Nakajima, K., H. Kan, and Y. Mizushima, "Power-speed product of an optical flip-flop memory with optical feedback," *IEEE J. Solid-State Circuits*, Vol. 26, No. 1, Jan. 1991, pp. 75–76.

[32] Oudar, J. L., R. Kuszelewicz, and J. L. Panel, "Optical bistable arrays: Prospects for ultimate performances," *OSA Proc. on Photonic Switching*, Vol. 8, H. S. Hinton and J. W. Goodman (eds.), 1991, pp. 239–243.

[33] Kurihara, K., Y. Tashiro, I. Ogura, M. Sugimoto, and K. Kasahara, "32×32 two-dimensional array of vertical to surface transmission electrophotonic devices with a pnpn structure," *IEE Proc.*, Vol. 138, No. 2, Apr. 1991, pp. 161–163.

[34] Kazovsky, L., R. T. Hofmeister, and S. M. Gemelos, "From STARNET to CORD: Lessons learned from Stanford WDM project," in G. Prati (ed.), *Photonic Networks*, Springer-Verlag, London, England, 1997, pp. 300–330.

[35] Habara, K., T. Matsunaga, and K.Yukimatsu, "Large capacity WDM packet switching", in G. Prati (ed.), *Photonic Networks*, Springer-Verlag, London, England, 1997, pp. 285–299.

[36] Minami, T., K. Yamaguchi, H. Shimizu, N. Fujino, H. Hamano, N. Suyama, K. Iguchi, and I. Yamada, "A 200 Mbps synchronous TDM loop optical LAN suitable for multiservice integration," *IEEE J. Select. Areas Commun.*, Nov. 1985.

[37] Suzuki, S., T. Terakado, K. Komatsu, K. Nagashima, A. Suzuki, and M. Kondo, "An experiment of high-speed optical time-division switching," *IEEE/OSA J. Lightwave Technol.*, July 1986.

[38] Prucnal, P., M. Santoro, and S. Sehgal, "Ultrafast all-optical synchronous multiple access fiber networks," *IEEE J. Select. Areas Commun.*, Dec. 1986.

[39] Prucnal, P., M. Santoro, and T. R. Fan, "Spread spectrum fiber optical local area network using optical processing," *IEEE/OSA J. Lightwave Technol.*, May 1986.

[40] Davidson, A. C., and S. K. Chaudhuri, "A self-routing packet network via optical processing of the header," *Europ. Trans. Telecommun.*, Vol. 4, No. 2, Mar.-Apr. 1993, pp. 201–211.

[41] Kazovsky, L. G., "Multichannel coherent optical communication systems," *IEEE/OSA J. Lightwave Technol.*, Aug. 1987.

[42] Kaminow, I. P., P. Iannone, J. Stone, and W. Stulz, "FDMA-FSK star network with a tunable optical filter demultiplexer," *IEEE/OSA J. Lightwave Technol.*, Sept. 1988.

[43] Acampora, A. S., and N. J. Karol, "An overview of lightwave packet network," *IEEE Network*, Jan. 1989, pp. 29–41.

[44] Marsan, M. A., G. Albertengo, A. Francese, and F. Neri, "Manhattan topologies for all-optical networks," *Proc. EFOC/LAN 91*, London, England, June 19-21, 1991, pp. 144–150.

[45] Borgonovo, F., and E. Cadorin, "Routing in the bidirectional manhattan network," *Proc. Third International Conference on Data Communication Systems and Their Performance*, Rio de Janeiro, Brazil, June 1987, pp. 193–201.

[46] Borgonovo, F., and E. Cadorin, "Locally-optimal deflection routing in the bidirectional manhattan network," *Proc. INFOCOM '90*, Vol. 2, San Francisco, CA, June 1990, pp. 458–464.

[47] Borgonovo, F., and L. Fratta, "Deflection networks: Architectures for metropolitan and wide area networks," *Computer Networks and ISDN Systems*, Vol. 24, April 1992, pp. 171–183.

[48] Maxemchuk, N. F., "Distributed clocks in slotted networks," *Proc. INFOCOM '88*, New Orleans, LA, Mar. 1988.

[49] Marsan, M. A., G. Albertengo, A. Francese, E. Leonardi, and F. Neri, "All-optical bidirectional manhattan networks," *Proc. ICC '92*, Chicago, IL, June 14-18, 1992, pp. 1461–1467.

[50] Kovačevíc, M., A. S. Acampora, and G. Brown, "Analysis of bidirectional manhattan network with uplinks," *Proc. Third International Conference on Computer Communication and Networks*, San Francisco, CA, Sept. 11-14, 1994, pp. 302–308.

[51] Albertengo, G., P. L. Civera, R. L. Cigno, G. Piccinini, M. Zamboni, F. Borgonovo, L. Fratta, and G. Panizzardi, "Deflection networks: principles, implementation, services," *Eur. Trans. Telecommun.*, Vol. 3, No. 2, Mar.-April 1992, pp. 195–206.

[52] Albertengo, G., R. Lo Cigno, and G. Panizzardi, "Optimal routing algorithms for the bidirectional manhattan street network," *Proc. ICC '91*, Vol. 3, Denver, CO, June 23-26, 1991, pp. 1676–1680.

[53] Borgonovo, F., and E. Cadorin, "HR4-net: A hierarchical random-routing reliable and reconfigurable network for metropolitan area," *Proc. INFOCOM '87*, San Francisco, CA, Mar. 31-April 2, 1987, pp. 320–326.

[54] Borella, A., and F. Chiaraluce, "All-optical 16×16 random packet switching fabric," *Proc. NOC '96*, Heidelberg, Germany, June 25-28, 1996, pp. 176–183.

[55] Borella, A., and F. Chiaraluce, "Proposal of a packet switch based on purely random routing," *Int. J. Commun. Sys.*, Vol. 9, No. 3, May-June 1996, pp. 145–150.

[56] Bellezza, B., *Analytical model of nondeterministic routing techniques in bidirectional manhattan networks*, Degree Thesis, University of Ancona, Ancona, Italy, 1996 (in Italian).

[57] Albertengo, G., R. Lo Cigno, and G. Panizzardi, "Simplified routing algorithms for the bidirectional manhattan street network," *Proc. 3rd IFIP Conference on High Speed Networking*, Berlin, Germany, Mar. 1991.

[58] Albertengo, G., R. Lo Cigno, and G. Panizzardi, "The deflection network: A reliable high speed packet network for computer communication," *Proc. COMPEURO 91*, Bologna, Italy, May 1991.

[59] Todd, T. D., and A. M. Bignell, "Performance modeling of the SIGnet MAN backbone," *Proc. INFOCOM '90*, San Francisco, CA, June 5-7, 1990.

[60] Maxemchuk, N. F., "Problems arising from deflection routing: Live-lock, lockout, congestion and message reassembling," *Proc. NATO Advanced Research Workshop,* Sophia Antipolis, France, June 25-27, 1990.

[61] Borgonovo, F., and E. Cadorin, "Packet-switching network architectures for very high speed services," *Proc. 1990 Zurich Seminar on Digital Communications*, Zurich, Switzerland, Mar. 5-8, 1990.

[62] Borgonovo, F., L. Fratta, and F. Tonelli, "Circuit service in deflection networks," *Proc. INFOCOM '91*, Miami, FL, April 9-11, 1991, pp. 2B.4.1–2B.4.7.

[63] Acampora, A. S., "A multichannel multihop local lightwave network," *Proc. GLOBECOM '87*, Tokyo, Japan, 1987, pp. 37.5.1–37.5.9.

[64] Hluchyj, M. G., and M. J. Karol, "Shufflenet: an application of generalized perfect shuffles to multihop lightwave networks," *IEEE/OSA J. Lightwave Technol.*, Vol. 9, No. 10, Oct. 1991, pp. 1386–1397.

[65] Stone, H. S., "Parallel processing with the perfect shuffle," *IEEE Trans. Computers*, Vol. C-20, No. 2, Feb. 1971, pp. 153–161.

[66] Patel, J. H., "Performance of processor-memory interconnections for multiprocessors," *IEEE Trans. Computers*, Vol. C-30, No. 10, Oct. 1981, pp. 771–780.

[67] Chen, W.-K., *Applied Graph Theory–Graph and Electrical Networks*, New York, North Holland, 1976.

[68] Matera, F., E. Ripani, and M. Settembre, "Proposal of all optical soliton shuffle multihop network," *Electron. Lett.*, Vol. 28, No. 17, Aug. 1992, pp. 1570–1571.

[69] Matera, F., E. Ripani, M. Romagnoli, and M. Settembre, "Proposal of an all optical shuffle multihop network," *Europ. Trans. Telecommun.*, Vol. 4, No. 2, Mar.-Apr. 1993, pp. 213–219.

[70] Acampora, A. S., M. J. Karol, and M. G. Hluchyj, "Terabit lightwave networks: The multihop approach," *AT&T Tech. J.*, Vol. 66, No. 6, Nov.-Dec. 1987, pp. 21–34.

[71] To, P. P., T. S. P. Yum, and Y. W. Leung, "Multistar implementation of expandable shufflenets," *IEEE/ACM Trans. Networking*, Vol. 2, No. 4, Aug. 1994, pp. 345–351.

[72] Karol, M. J., "Optical interconnection using shufflenet multihop networks in multi-connected ring topologies," *Proc. ACM SIGCOMM '88 Symposium*, 1988, pp. 25–34.

[73] Brackett, C. A., "Dense wavelength division multiplexing network: principles and applications," *IEEE J. Select. Areas Commun.*, Vol. 8, No. 6, Aug. 1990, pp. 948–964.

[74] Birk, Y., M. E. Marhic, and F. A. Tobagi, "Selective broadcast interconnections (SBI) for wideband fiber-optic local area networks," *Proc. 2nd International Technical Symposium in the Conference of Broadband Networks*, Cannes, France, Nov. 1985, pp. 28–39.

[75] Y. Birk, "Fiber-optic bus-oriented single-hop interconnection among multi-transceiver stations," *IEEE/OSA J. Lightwave Technol.*, Vol. 9, No. 12, Dec. 1991, pp. 1657–1664.

[76] Y. Birk, "Power efficient lay-out of a fiber-optic multistar that permits $\log_2 N$ concurrent baseband transmissions among N stations," *IEEE/OSA J. Lightwave Technol.*, Vol. 11, No. 5/6, May/June 1993, pp. 908–913.

[77] Karol, M. J., and S. Shaikh, "A simple adaptive routing scheme for congestion control in shufflenet multihop lightwave networks," *IEEE J. Select. Areas Commun.*, Vol. 9, No. 7, Sept. 1991, pp. 1040–1051.

[78] Han, H.-K., and Y.-K. Jhee, "A WDM channel sharing scheme for multihop lightwave networks using logically bidirectional perfect shuffle interconnection pattern," *IEICE Trans. Commun.*, Vol. E77-B, No. 9, Sept. 1994, pp. 1152–1161.

[79] Hayes, J. F., *Modeling and Analysis of Computer Communications Networks*, New York, Plenum Press, 1984.

[80] Bannister, J. A., L. Fratta, and M. Gerla, "Topological design of the wavelength-division optical networks," *Proc. INFOCOM '90*, San Francisco, CA, June 5-7, 1990, pp. 1005–1013.

[81] Labourdette, J.-F. P., and A. S. Acampora, "Wavelength agility in multihop lightwave networks," *Proc. INFOCOM '90*, San Francisco, CA, June 5-7, 1990, pp. 1022–1029.

[82] Liew, S. C., "On the stability of shuffle-exchange and bidirectional shuffle-exchange deflection networks," *IEEE/ACM Trans. Networking*, Vol. 5, No. 1, Feb. 1997, pp. 87–94.

[83] Bharath-Kumar, K., and P. Kermani, "Analysis of a resequencing problem in communication networks," *Proc. INFOCOM '83*, San Diego, CA, April 1983.

[84] Yum, T. S. P., and T.-Y. Ngai, "Resequencing of messages in communication networks," *IEEE Trans. Commun.*, Vol. COM-34, No. 2, Feb. 1986, pp. 143–149.

[85] Eisenberg, M., and N. Meharavari, "Performance of the multichannel multihop lightwave network under nonuniform traffic," *IEEE J. Select. Areas Commun.*, Vol. 5, No. 7, Aug. 1988, pp. 1063–1078.

[86] Jeng, J.-W., and T. G. Robertazzi, "Duplex routing with reverse channels in WDM lightwave shufflenet," *Proc. Third International Conference on Computer Communications and Networks*, San Francisco, CA, Sept. 11-14, 1994, pp. 396–400.

[87] Pavan, A., S. Bhattacharya, and D. Du, "Reverse channel augmented multihop lightwave networks," *Proc. INFOCOM '93*, San Francisco, CA, Vol. 3, March 1993, pp. 1127–1134.

[88] Li, B., and A. Ganz, "Virtual topologies for WDM star LANs – the regular structure approach," *Proc. INFOCOM '92*, Florence, Italy, Vol. 3, May 1992, pp. 2134–2143.

[89] Jeng, J.-W., and T. G. Robertazzi, "Duplex routing with reverse channels in WDM lightwave shufflenet," SUNY at Stony Brook College of Engineering and Applied Science, Technical Report. 695, July, 1994.

[90] Iness, J., S. Banerjee, and B. Mukherjee, "GEMNET: A generalized, shuffle-exchange-based, regular, scalable, and modular multihop network based on WDM lightwave technology," Dept. of Computer Science, University of California, Davis, Technical Report CSE-94-8, June 1994.

[91] Ramaswami, R., "Multiwavelength lightwave networks for computer communications," *IEEE Commun. Mag.*, Vol. 31, No. 2, February 1993, pp. 78–88.

[92] Ramaswami, R., and K. Sivarajan, "A packet-switched multihop lightwave network using subcarrier and wavelength division multiplexing," *IEEE Trans. Commun.*, Vol. 42, No. 2/3/4, Feb./March/April 1994, pp. 1198–1211.

[93] Iness, J., S. Banerjee, and B. Mukherjee, "GEMNET: a generalized, shuffle-exchange-based, regular, scalable, modular, multihop, WDM lightwave network," *IEEE/ACM Trans. Networking*, Vol. 3, No. 4, August 1995, pp. 470–476.

[94] Tridandapani, S. B., and B. Mukherjee, "Channel sharing in multi-hop WDM lightwave networks: Realization and performance of multicast traffic," *IEEE J. Select. Areas Commun.*, Vol. 15, No. 3, April 1997, pp. 488–500.

[95] Borella, A., and F. Chiaraluce, "An enlarged shufflenet topology for optical networks," *Proc. ICAPT '96*, Montreal, Canada, July 29-August 1, 1996, pp. 16.3–16.4.

[96] Ragni, G., *Wavelength reuse and virtual topology improvement in WDM optical networks*, Degree Thesis, University of Ancona, Ancona, Italy, 1995 (in Italian).

[97] Mestdagh, D. J. G., *Fundamentals of Multiaccess Optical Fiber Networks*, Norwood, MA, Artech House, 1995.

[98] Borella, A., and F. Chiaraluce, "Modification of the Shufflenet connectivity graph for balancing the load in the case of uniform traffic," *IEICE Trans. Fundamentals*, Vol. E80-A, No. 2, Feb. 1997, pp. 423–426.

[99] Acampora, A. S., *An Introduction to Broadband Networks*, New York, Plenum Press, 1994.

[100] Labourdette, J.-F. P., and A. S. Acampora, "Logically rearrangeable multihop lightwave networks," *IEEE Trans. Commun.*, Vol. 39, No. 8, August 1991, pp. 1223–1230.

[101] Ayadi, F., J. F. Hayes, and M. Kaverhad, "Bilayered shufflenet: A new logical configuration for multihop lightwave networks," *Proc. GLOBECOM '93*, Houston, TX, Nov. 29-Dec. 2, 1993, pp. 1159–1163.

[102] Wu, C. L., and T. Y. Feng, "On a class of multistage interconnection networks," *IEEE Trans. Computers*, Vol. 29, No. 8, Aug. 1980, pp. 694–702.

[103] Shaikh, S. Z., M. Schwartz, and T. H. Szymanski, "A comparison of the shufflenet and the banyan topologies for broadband packet switches," *Proc. INFOCOM '90*, San Francisco, CA, June 5-7, 1990, pp. 1260–1267.

[104] Lawrie, D. H., "Access and alignment of data in an array processor," *IEEE Trans. Computers*, Vol. 24, 1975, pp. 1145–1155.

[105] Goke, L. R., and G. J. Lipovski, "Banyan networks for partitioning multiprocessors systems," *Proc. 1st Annual Computer Architecture Conference*, 1973, pp. 21–28.

[106] Tang, K. W., "BanyanNet – a bidirectional equivalent of shufflenet," *IEEE/OSA J. Lightwave Technol.*, Vol. 12, No. 11, Nov. 1994, pp. 2023–2031.

[107] Jenq, Y.-C., "Performance analysis of a packet switch based on single-buffered banyan network," *IEEE J. Select. Areas Commun.*, Vol. SAC-1, No. 6, Dec. 1983.

[108] Szymanski, T. H., and S. Z. Shaikh, "Markov chain analysis of packet-switched banyans with arbitrary switch sizes, queue sizes, link multiplicities and speedups," *Proc. INFOCOM '89*, Ottawa, Canada, April 1989, pp. 960–971.

[109] Youssef, A. S., and B. Narahari, "The banyan-hypercube networks," *IEEE Trans. Parallel and Distributed Systems*, Vol. 1, No. 2, April 1990, pp. 160–169.

[110] Tang, K. W., "BanyanNet: A multihop WDM-based lightwave network," *Proc. Second International Conference on Computer Communications and Networks*, San Diego, CA, June 28-30, 1993, pp. 208–212.

[111] Bollobas, B., *Extremal Graph Theory with Emphasis on Probabilistic Methods*, Providence, RI, American Mathematical Society, 1986.

[112] Golomb, S. W., *Shift Register Sequences*, Aegean Park Press, 1982.

[113] Sridhar, M. A., C. S. Raghavendra, "Fault-tolerant networks based on the de Bruijn graphs," *IEEE Trans. Computers*, Vol. 4, No. 10, Oct. 1991, pp. 1167–1174.

[114] Kar, B. K., and D. K. Pradhan, "Scalability of binary de Bruijn networks," *Proc. 5th IEEE Symposium on Parallel and Distributed Processing*, Dallas, TX, Dec. 11-14, 1993, pp. 796–799.

[115] Sivarajan, K., and R. Ramaswami, "Multihop lightwave networks based on de Bruijn graphs," *Proc. INFOCOM '91*, Vol. 3, Miami, FL, April 7-11, 1991, pp. 1001–1011.

[116] Sivarajan, K. N., and R. Ramaswami, "Lightwave networks based on de Bruijn graphs," *IEEE/ACM Trans. Networking*, Vol. 2, No. 1, Feb. 1994, pp. 70–79.

[117] Baumslag, M., "Fault-tolerance properties of de Bruijn and shuffle-exchange networks," *Proc. 5th IEEE Symposium on Parallel and Distributed Processing*, Dallas, TX, Dec. 11-14, 1993, pp. 556–563.

[118] Marsan, M. A., A. Bianco, E. Leonardi, and F. Neri, "De Bruijn topologies for self-routing all-optical networks," *Proc. SPIE International Symposium on Optical Applied Science and Engineering*, San Diego, CA, July 1993, pp. 99–111.

[119] Feng, Z., and O. W. Yang, "Routing algorithms in the bidirectional de Bruijn graph metropolitan area networks," *Proc. Milcom '94*, Fort Monmouth, NJ, Oct. 2-5, 1994, pp. 957–961.

[120] Hyatt, C., and D. P. Agrawal, "Bidirectional versus unidirectional networks: Cost/performance trade-offs," *Proc. Mascots '95*, Durham, NC, Jan. 18-20, 1995, pp. 128–133.

[121] Esfahanian, H., and S. L. Hakimi, "Fault-tolerant routing in de Bruijn communication networks," *IEEE Trans. Computers*, Vol. C-34, No. 9, Sept. 1985, pp. 777–788.

[122] Imase, M., T. Soneoka, and K. Okada, "Connectivity of regular directed graphs with small diameter," *IEEE Trans. Computers*, Vol. C-34, No. 3, Mar. 1985, pp. 267–273.

[123] Sridhar, M. A., "The unidirected de Bruijn graph: Fault tolerance and routing algorithms," *IEEE Trans. Circuits and Systems – I: Fundamental Theory and Applications*, Vol. 39, No. 1, Jan. 1992, pp. 45–48.

[124] Ganesan, E., and D. K. Pradhan, "Optimal broadcasting in binary de Bruijn networks and hyper-de Bruijn networks," *Proc. 7th International Parallel Processing Symposium*, Newport, CA, Apr. 13-16, 1993, pp. 655–660.

[125] Chou, C. H., and D. H. Du, "Uni-directional hypercubes," *Proc. Supercomputing '90*, Vol. 11, 1990, pp. 254–263.

[126] Gipser, T., and M. S. Kao, "MATRIX: an all-optical high speed network architecture," *Proc. ECOC '94*, Florence, Italy, Sept. 1994, pp. 499–502.

[127] Gipser, T., "Survivability and scalability aspects of the all-optical network MATRIX," *Proc. EFOC&N '95*, Brighton, England, June 27-30, 1995, pp. 196–199.

[128] Gipser, T., and M. S. Kao, "MATRIX: a new network for multiwavelength all-optical transparent information exchange," *Proc. EFOC&N '94*, Heidelberg, Germany, June 21-24, 1994, pp. 146–150.

[129] Spring, J., R. S. Tucker, "Photonic 2×2 packet switch with input buffers," *Electron. Lett.*, Vol. 29, No. 3, Feb. 1993, pp. 284–285.

[130] Gipser, T., and M. S. Kao, "An all-optical network architecture," *IEEE/OSA J. Lightwave Technol.*, Vol. 14, No. 5, May 1996, pp. 693–702.

[131] Limb, J. O., "Performance of local area networks at high speed," *IEEE Commun. Mag.*, Vol. 22, Aug. 1984, pp. 41–45.

[132] Chlamtac, I., and A. Ganz, "Toward alternative high-speed networks," *Proc. INFOCOM '87*, San Francisco, CA, Mar. 31-Apr. 2, 1987.

[133] Chlamtac, I., and A. Ganz, "Toward alternative high-speed network concepts: The SWIFT architecture," *IEEE Trans. Commun.*, Vol. 38, No. 4, Apr. 1990, pp. 431–439.

[134] Chlamtac, I., and A. Ganz, "Channel allocation protocols in frequency-time controlled high speed networks," *IEEE Trans. Commun.*, Vol. 36, No. 4, April 1988, pp. 430–440.

[135] Kleinrock, L., *Queueing Systems – Vol. 1 Theory*, New York, Wiley, 1975.

[136] Poggiolini, P., and L. G. Kazovsky, "Starnet: An integrated services broadband optical network with physical star topology," *Proc. SPIE Advanced Fiber Communication Technology*, Boston, MA, Sept. 1991, SPIE Vol. 1579, pp. 14–29.

[137] Chiang, T.-K., S. K. Agrawal, D. T. Mayweather, D. Sadot, C.F. Barry, M. Hickey, and L. G. Kazovsky, "Implementation of Starnet: A WDM computer communication network," *IEEE J. Select. Areas Commun.*, Vol. 14, No. 5, June 1996, pp. 824–839.

[138] Hickey, M., and L. G. Kazovsky, "The Starnet coherent WDM computer communication network: Experimental transceiver employing a novel modulation format," *IEEE/OSA J. Lightwave Technol.*, Vol. 12, No. 5, May 1994, pp. 876–884.

[139] Kazovsky, L. G., and P. Poggiolini, "Starnet: A multi-gigabit-per-second optical LAN utilizing a passive WDM star," *IEEE/OSA J. Lightwave Technol.*, Vol. 11, No. 5/6, May/June 1993, pp. 1009–1027.

[140] O'Byrne, V., "A method for reducing the channel spacing in a coherent optical heterodyne system," *IEEE Photon. Technol. Lett.*, Vol. 2, No. 7, July 1990.

[141] Karol, M. J., and R. D. Gitlin, "High-performance optical local and metropolitan area networks: Enhancements of FDDI and IEEE 802.6 DQDB," *IEEE J. Select. Areas Commun.*, Vol. 8, No. 8, Oct. 1990, pp. 1439–1448.

[142] ANSI X3T9.5, *FDDI specification*, Rev. 5.1, Sept. 1989.

[143] IEEE Std 802.6-1990, *IEEE Standards for Local and Metropolitan Area Networks: Distributed Queue Dual Bus (DQDB) Subnetwork of a Metropolitan Area Network (MAN)*, Piscataway, NJ, IEEE Standards Board, Dec. 1990.

[144] Borella, A., D. Broglio, G. Cancellieri, and F. Chiaraluce, "Performance of DQDB protocol for MANs with bandwidth balancing mechanism in a multi-priority traffic environment," *Proc. EFOC/LAN '92*, Paris, France, June 24-26, 1992, pp. 218–222.

[145] Kazovsky, L. G., and O. K. Tonguz, "ASK and FSK coherent lightwave systems: A simplified approximate analysis," *IEEE/OSA J. Lightwave Technol.*, Vol. 8, No. 3, Mar. 1990, pp. 338–352.

[146] Viterbi, A. J., *Principles of Coherent Communication*, New York, McGraw-Hill, 1966.

[147] Cvijetic, M., *Coherent and Nonlinear Lightwave Communications*, Norwood, MA, Artech House, 1996.

[148] ISO/IIEC 8802.5, IEEE Std 802.5, *Token Ring Access Method and Physical Layer Specifications*, 1st ed., June 1992.

[149] de Prycker, M., *Asynchronous Transfer Mode: Solution for Broadband ISDN*, Englewood Cliffs, NJ, Prentice Hall, 1990.

# Chapter 4

# Multilevel Optical Networks

## 4.1 NETWORKS OF NETWORKS

The realization of future national or international infrastructures involving all-optical architectures seems naturally related to the development of large scalable networks, born by the aggregation of many subnetworks hierarchically organized [1]. In practice, users will be first connected in local or metropolitan areas at low level and then aggregated in one or more upper levels, up to a backbone for complete connectivity. Each subnetwork will be itself an all-optical network, capable of autonomous operation.

Although geographical proximity probably will be the main criterion for creating subnetworks in the first phase, optimization of the design will take into account the traffic features, and privileging solutions that fit traffic localities, with their own requirements of bandwidth and type of service. This should guide the choice of the topology, either virtual or physical, with a special look at the cost and flexibility of the layout.

While the major benefits of the optical option will become more and more evident with the advent of new wideband services (such as video on demand) and the increasing thirst of information (for research, business, education, but also for entertainment and recreation), some priorities can be identified now and may be thought in terms of optical implementation. A significant example, in this sense, is given by the need for interconnecting large computer centers or supercomputer sites, in massively parallel systems able to carry a huge amount of data. Modular structures, such as those permitted by high-performance hierarchical optical networks, appear to be good candidates for these kinds of applications.

Among the wide series of proposals that can be conceived for interconnecting subnetworks and managing the traffic at the various levels, some will be presented in this chapter. Our goal is to give an idea of the methods of analysis, and of the level of performance achievable. In this sense, attention will be mainly focused on regular structures, since only such structures allow one to draw general conclusions.

## 4.2 STAR-OF-STARS NETWORK

A simple way to group single-hop broadcast-and-select networks consists in realizing a multiple WDM star [2] of the type depicted in Figure 4.1. In such a system there are $n_c$ local stars (clusters) and one remote star. The $i$th local star, with $i = 1, 2, \ldots, n_c$, is interconnected to $N_i$ disjoint nodes and makes the intracluster communications possible among them. All the nodes are also interconnected to the remote star which allows, instead, intercluster communications. In Figure 4.1, any arrow represents a bidirectional link realized, as usual, through two distinct fibers.

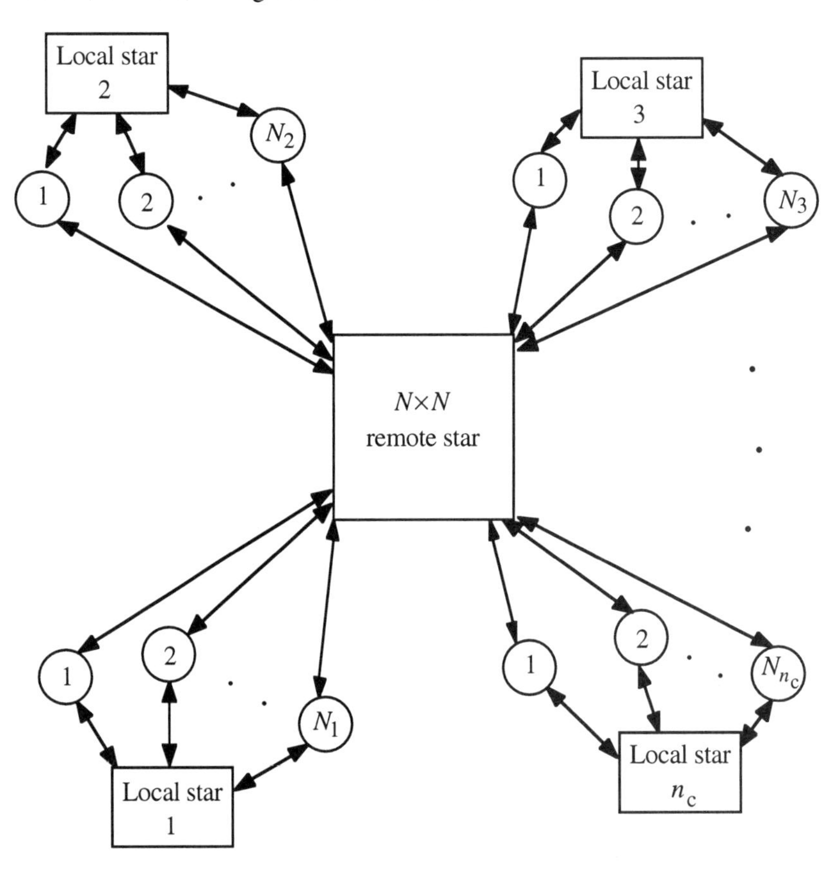


**Figure 4.1** Multiple WDM star. $N$ is the total number of nodes.

The total number of nodes connected by the system is $N = N_1 + N_2 + \cdots + N_{n_c}$. Each node has two transmitters and two receivers, both tunable over a range including $W$ different wavelengths, that can be shared by the local and remote networks. Since the local stars are physically distinct as well as separated by the remote star, up to a maximum of $(n_c + 1)W$ parallel transmissions can occur in the network. To prevent traffic conflicts, the source/destination allocation (S/DA) fixed assignment protocol described in Section 2.2.1, can be used.

As discussed in Section 1.8, the network traffic demand can be described by a traffic matrix $\boldsymbol{T}$. Once having numbered the nodes in such a way that the $j$th node of the $i$th cluster is $j + N_1 + N_2 + \cdots + N_{i-1}$, its element $T_{xy}$ gives the average load addressed from node $x$ to node $y$ . Actually, since S/DA is based on a combined WDM/TDM approach, $T_{xy}$ also represents the demand in time slot per frame, required for interconnection from node $x$ to node $y$. In this case, $\boldsymbol{T}$ is an integer matrix.

The object of time-wavelength scheduling is to minimize the overall transmission duration, thereby maximizing system resource utilization. Each frame is divided into a number of tuning periods during which the assignment of wavelengths to the system transmitters and receivers does not change. At the beginning of each tuning period, however, transmitters and receivers must be tuned to the correct wavelengths, and this leads to a penalty which increases the transmission time. So, if $\tau$ stands for the tuning time, $M_t$ for the number of tuning periods, $t$ the packet transmission time, and $M_s$ the number of time slots required for the transmission of the information packets only (i.e., with $\tau = 0$), then the quantity to minimize is $T = \tau \cdot M_t + t \cdot M_s$. This problem is generally very difficult to treat, even in the case of single stars [3]. A possibility consists in minimizing the value of $M_s$, subject to the constraint that $M_t$ is minimum. To this purpose, matrix $\boldsymbol{T}$ is decomposed into $M_s$ slot matrices (i.e., $\boldsymbol{T} = \boldsymbol{C}^{(1)} + \boldsymbol{C}^{(2)} + \cdots + \boldsymbol{C}^{(M_s)}$) and distributed among $M_t$ tuning periods. Each $\boldsymbol{C}^{(k)}$ is a 0-1 matrix, with $C_{xy}^{(k)} = 1$ if node $x$ is scheduled for transmission to node $y$ in slot $k$, or $C_{xy}^{(k)} = 0$ otherwise. Each tuning period is opened and identified by a 0-1 tuning matrix $\boldsymbol{S}^{(m)}$, $m = 1, 2, \ldots, M_t$, where $S_{xy}^{(m)} = 1$ if during the $m$th tuning period source node $x$ will transmit at the same wavelength on which destination $y$ will receive, or $S_{xy}^{(m)} = 0$ otherwise. Obviously, $S_{xy}^{(m)} = 1$ if at least one of the elements $C_{xy}^{(k)}$ included in the $m$th tuning period equals 1 in its turn.

The so-called minimum tuning duration algorithm (MTDA) [2,3] allows one to minimize the number of tuning matrices for each frame. In applying the algorithm, it is useful to denote by $\boldsymbol{T}^{(p,q)}$ the $N_p \times N_q$ submatrix of $\boldsymbol{T}$ which represents the traffic demand, in time slots per frame, required for the interconnection between the nodes connected directly to the $p$th local star and the nodes connected directly to the $q$th local star; $\boldsymbol{T}^{(p,p)}$ has the meaning of intracluster submatrix, while $\boldsymbol{T}^{(p,q)}$, with $p \neq q$, is an intercluster submatrix. So, taking into account the above system hardware constraints, in terms of the number of transmitters and receivers per node and the number of wavelengths, it is easy to verify that the minimum value of $M_t$ is equal to

$$M_{t_{\min}} = \max\left\{ \max_{1 \le i \le N} \left\{ \lceil RN_i / 2 \rceil, \lceil CN_i / 2 \rceil \right\}, \max_{1 \le p \le n_c} \left\{ \lceil LN_{pp} / W \rceil, \left\lceil \sum_{\substack{1 \le p,q \le n_c \\ p \neq q}} LN_{pq} / W \right\rceil \right\} \right\} \tag{4.1}$$

where $RN_i$ is the number of nonzero elements on row $i$ of $\boldsymbol{T}$, $CN_i$ is the number of nonzero elements on column $i$ of $\boldsymbol{T}$, and $LN_{pq}$ is the total number of nonzero elements in submatrix $\boldsymbol{T}^{(p,q)}$. Furthermore, in (4.1) $\max\{a,b\}$ gives the maximum between $a$ and $b$, while $\lceil a \rceil$ represents the smallest integer greater than $a$. Once having generated the tuning matrices (minimum in number), the algorithm goes on by creating the corresponding slot matrices. Omitting details for the sake of brevity (the procedures are described in [2]), we limit ourselves to discuss the simple example shown in Figure 4.2. The system consists of $N = 16$ nodes, grouped in four local stars of equal size; $W = 6$ wavelengths are available inside each cluster, as well as in the remote star for intercluster communications. The $\boldsymbol{T}^{(p,q)}$'s are 4×4 matrices, easily recognizable in the partition of matrix $\boldsymbol{T}$, evidenced in the figure. From (4.1), it is possible to prove that $M_{t_{min}} = 1$. Thereafter, using MTDA, only one tuning period is obtained, while the number of transmission slots is $M_s = 6$. The corresponding structure of $\boldsymbol{C}^{(k)}$, $k = 1, 2, \ldots, 6$ is also reported in Figure 4.2, together with that of $\boldsymbol{S}^{(1)}$.

The system of Figure 4.1 does not properly define a hierarchical structure. All the nodes inside a cluster have, in fact, the same functionalities and no single node is more important than the others. Moreover, each node can be connected directly, using single-hop communications, to all the others, via one of the local stars or the remote star. The basic idea of the proposal is, in fact, to use the same number of wavelengths ($W$) to reach a relatively small number of users (intracluster communications) as well as a larger number of users (intercluster communications). It is evident that the channel efficiency of the multiple-star solution is better than that of the single star when the traffic is more likely to occur within the groups. This is the case of the example shown in Figure 4.2, where $LN_{pp}$ is greater than $LN_{pq}$, for any $q \neq p$ and, more generally, as stressed in Section 4.1, this is the situation of any multilevel network. But the way in which better channel efficiency (or, equivalently, a higher transmission concurrency) is achieved can obviously be different.

In Figure 4.3, for instance, the same objective of realizing a star-of-stars network is achieved by resorting to another structure [4], which is more likely to be hierarchical, in the sense specified above. Only one node, in each cluster, is connected directly to the remote star, and all the traffic generated inside the cluster but directed toward nodes of other clusters must necessarily pass through it. Therefore, such a node, noted by $gw_i$, $i = 1, 2, \ldots, n_c$, in Figure 4.3, plays the role of a gateway (usually electronic) to separate the various clusters in the backbone constituted by the $n_c \times n_c$ remote star. Each node has a specific wavelength to transmit and a tunable filter to receive, that is, the network architecture can be described as FT-TR, according to the classification scheme of Chapter 2. The number of wavelengths inside each cluster, noted by $W_i$, equals the number of nodes, so that $W_i = N_i$, $i = 1, 2, \ldots, n_c$. Actually, since it is often required that the bandwidth per channel in the backbone be $R$ times greater than in the clusters, the gateway of the $i$th cluster must be equipped with $R$ transceiver modules and the number of ordinary end-nodes is therefore $Q_i = N_i - R$. Wavelength reuse, in different stars, is obviously possible. Intracluster communications are single-hop, while intercluster communications are multihop, since a packet exchange between nodes of different clusters, passes across two intermediate gateways (those of the involved local stars), with a maximum number of hops equal to three.

$$
T = \left[\begin{array}{cccc|cccc|cccc|cccc}
0 & 3 & 0 & 3 & 0 & 0 & 0 & 0 & 0 & 0 & 0 & 0 & 0 & 0 & 0 & 0 \\
0 & 0 & 6 & 0 & 0 & 0 & 0 & 0 & 0 & 0 & 0 & 0 & 0 & 0 & 0 & 0 \\
0 & 3 & 0 & 0 & 0 & 0 & 0 & 0 & 0 & 0 & 0 & 0 & 0 & 0 & 0 & 0 \\
6 & 0 & 0 & 0 & 0 & 0 & 0 & 0 & 0 & 0 & 0 & 0 & 0 & 0 & 0 & 0 \\
\hline
0 & 0 & 0 & 0 & 0 & 0 & 0 & 3 & 0 & 0 & 3 & 0 & 0 & 0 & 0 & 0 \\
0 & 0 & 0 & 0 & 0 & 0 & 6 & 0 & 0 & 0 & 0 & 0 & 0 & 0 & 0 & 0 \\
0 & 0 & 0 & 0 & 0 & 3 & 0 & 0 & 3 & 0 & 0 & 0 & 0 & 0 & 0 & 0 \\
0 & 0 & 0 & 0 & 6 & 0 & 0 & 0 & 0 & 0 & 0 & 0 & 0 & 0 & 0 & 0 \\
\hline
0 & 0 & 0 & 0 & 0 & 0 & 0 & 0 & 0 & 0 & 0 & 6 & 0 & 0 & 0 & 0 \\
0 & 0 & 0 & 0 & 0 & 0 & 0 & 3 & 0 & 0 & 3 & 0 & 0 & 0 & 0 & 0 \\
0 & 0 & 0 & 0 & 0 & 0 & 0 & 0 & 0 & 6 & 0 & 0 & 0 & 0 & 0 & 0 \\
0 & 0 & 0 & 0 & 0 & 3 & 0 & 0 & 3 & 0 & 0 & 0 & 0 & 0 & 0 & 0 \\
\hline
0 & 0 & 0 & 0 & 0 & 0 & 0 & 0 & 0 & 0 & 0 & 0 & 0 & 0 & 0 & 3 \\
0 & 0 & 0 & 0 & 0 & 0 & 0 & 0 & 0 & 0 & 0 & 0 & 0 & 0 & 6 & 0 \\
0 & 0 & 0 & 0 & 0 & 0 & 0 & 0 & 0 & 0 & 0 & 0 & 0 & 3 & 0 & 0 \\
0 & 0 & 0 & 0 & 0 & 0 & 0 & 0 & 0 & 0 & 0 & 0 & 6 & 0 & 0 & 0
\end{array}\right]
$$

$$
S^{(1)} = C^{(1)} = C^{(2)} = C^{(3)} = \left[\begin{array}{cccccccccccccccc}
0 & 1 & 0 & 1 & 0 & 0 & 0 & 0 & 0 & 0 & 0 & 0 & 0 & 0 & 0 & 0 \\
0 & 0 & 1 & 0 & 0 & 0 & 0 & 0 & 0 & 0 & 0 & 0 & 0 & 0 & 0 & 0 \\
0 & 1 & 0 & 0 & 0 & 0 & 0 & 0 & 0 & 0 & 0 & 0 & 0 & 0 & 0 & 0 \\
1 & 0 & 0 & 0 & 0 & 0 & 0 & 0 & 0 & 0 & 0 & 0 & 0 & 0 & 0 & 0 \\
0 & 0 & 0 & 0 & 0 & 0 & 0 & 1 & 0 & 0 & 1 & 0 & 0 & 0 & 0 & 0 \\
0 & 0 & 0 & 0 & 0 & 0 & 1 & 0 & 0 & 0 & 0 & 0 & 0 & 0 & 0 & 0 \\
0 & 0 & 0 & 0 & 0 & 1 & 0 & 0 & 1 & 0 & 0 & 0 & 0 & 0 & 0 & 0 \\
0 & 0 & 0 & 0 & 1 & 0 & 0 & 0 & 0 & 0 & 0 & 0 & 0 & 0 & 0 & 0 \\
0 & 0 & 0 & 0 & 0 & 0 & 0 & 0 & 0 & 0 & 0 & 1 & 0 & 0 & 0 & 0 \\
0 & 0 & 0 & 0 & 0 & 0 & 0 & 1 & 0 & 0 & 1 & 0 & 0 & 0 & 0 & 0 \\
0 & 0 & 0 & 0 & 0 & 0 & 0 & 0 & 0 & 1 & 0 & 0 & 0 & 0 & 0 & 0 \\
0 & 0 & 0 & 0 & 0 & 1 & 0 & 0 & 1 & 0 & 0 & 0 & 0 & 0 & 0 & 0 \\
0 & 0 & 0 & 0 & 0 & 0 & 0 & 0 & 0 & 0 & 0 & 0 & 0 & 0 & 0 & 1 \\
0 & 0 & 0 & 0 & 0 & 0 & 0 & 0 & 0 & 0 & 0 & 0 & 0 & 0 & 1 & 0 \\
0 & 0 & 0 & 0 & 0 & 0 & 0 & 0 & 0 & 0 & 0 & 0 & 0 & 1 & 0 & 0 \\
0 & 0 & 0 & 0 & 0 & 0 & 0 & 0 & 0 & 0 & 0 & 0 & 1 & 0 & 0 & 0
\end{array}\right]
$$

$$
C^{(4)} = C^{(5)} = C^{(6)} = \left[\begin{array}{cccccccccccccccc}
0 & 0 & 0 & 0 & 0 & 0 & 0 & 0 & 0 & 0 & 0 & 0 & 0 & 0 & 0 & 0 \\
0 & 0 & 1 & 0 & 0 & 0 & 0 & 0 & 0 & 0 & 0 & 0 & 0 & 0 & 0 & 0 \\
0 & 0 & 0 & 0 & 0 & 0 & 0 & 0 & 0 & 0 & 0 & 0 & 0 & 0 & 0 & 0 \\
1 & 0 & 0 & 0 & 0 & 0 & 0 & 0 & 0 & 0 & 0 & 0 & 0 & 0 & 0 & 0 \\
0 & 0 & 0 & 0 & 0 & 0 & 0 & 0 & 0 & 0 & 0 & 0 & 0 & 0 & 0 & 0 \\
0 & 0 & 0 & 0 & 0 & 0 & 1 & 0 & 0 & 0 & 0 & 0 & 0 & 0 & 0 & 0 \\
0 & 0 & 0 & 0 & 0 & 0 & 0 & 0 & 0 & 0 & 0 & 0 & 0 & 0 & 0 & 0 \\
0 & 0 & 0 & 0 & 1 & 0 & 0 & 0 & 0 & 0 & 0 & 0 & 0 & 0 & 0 & 0 \\
0 & 0 & 0 & 0 & 0 & 0 & 0 & 0 & 0 & 0 & 0 & 1 & 0 & 0 & 0 & 0 \\
0 & 0 & 0 & 0 & 0 & 0 & 0 & 0 & 0 & 0 & 0 & 0 & 0 & 0 & 0 & 0 \\
0 & 0 & 0 & 0 & 0 & 0 & 0 & 0 & 0 & 1 & 0 & 0 & 0 & 0 & 0 & 0 \\
0 & 0 & 0 & 0 & 0 & 0 & 0 & 0 & 0 & 0 & 0 & 0 & 0 & 0 & 0 & 0 \\
0 & 0 & 0 & 0 & 0 & 0 & 0 & 0 & 0 & 0 & 0 & 0 & 0 & 0 & 0 & 0 \\
0 & 0 & 0 & 0 & 0 & 0 & 0 & 0 & 0 & 0 & 0 & 0 & 0 & 0 & 1 & 0 \\
0 & 0 & 0 & 0 & 0 & 0 & 0 & 0 & 0 & 0 & 0 & 0 & 0 & 0 & 0 & 0 \\
0 & 0 & 0 & 0 & 0 & 0 & 0 & 0 & 0 & 0 & 0 & 0 & 1 & 0 & 0 & 0
\end{array}\right]
$$

**Figure 4.2** Example of application of the MTDA for a multiple WDM star network with six wavelengths per star and four local stars, each with four nodes. (*Source*: [2] © 1992 IEEE.)

Similarly to the previous solution (but with the important difference of fixed transmitters), a preallocation-based WDM/TDM protocol can be used to prevent conflicts, that here can appear only when two or more nodes have to transmit simultaneously to the same node. The time slots allocation is based on the bandwidth demands of the nodes, which are transmitted in advance on the same channels as the data and can vary, at least in principle, frame per frame. A number of variants can be conceived to make the scheduling protocol more efficient [4]. For example, looking at the intracluster communications, one can distinguish between a high-priority and a low-priority operation mode, reserve a subset of slots for real-time guarantee-seeking services, foresee a mechanism to release reserved slots that are unused in a frame, and so on. Generally, these facilities are managed through some control slots, $N_i$ in number (one per each node), that are embedded with the data slots in a frame.

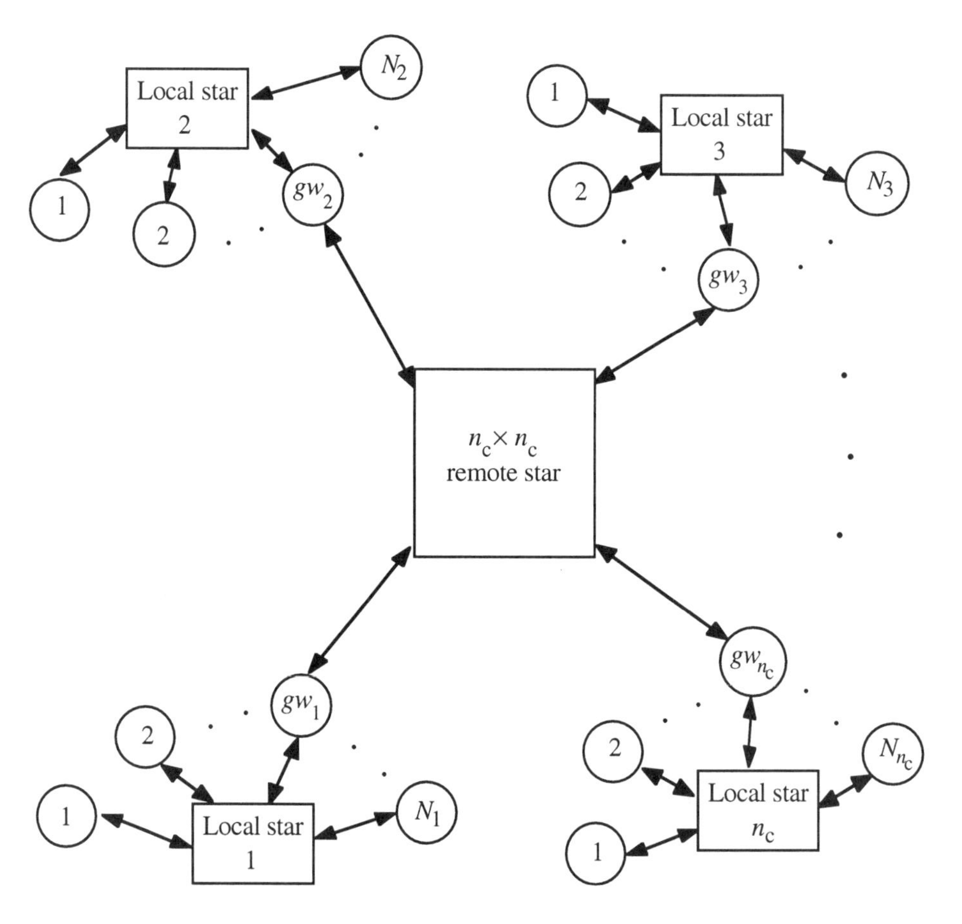


**Figure 4.3** Another realization of a multiple WDM star.

As regards the intercluster communications, the frame length in the backbone is the same as that in the clusters; this implies that the number of bits in a backbone slot is $R$ times greater than in a cluster slot. Each backbone slot is divided into $R$ subslots, all with the same pair of gateway nodes as source and destination, although generally with different pairs of end-nodes. The maximum throughput per node, normalized to the transmission rate, can be analytically determined under the hypothesis of full slot reservation or no slot reservation [4]. It has the meaning of minimum guaranteed bandwidth per source node, when the network is in overload conditions (i.e., each node has always at least one packet to transmit to any other). Hence, there are no unused slots and no action must be made to recover them. Furthermore, it is assumed that no slots are assigned a priori to guarantee-seeking services.

Let us suppose, for the sake of simplicity, that the clusters are all equal among them, each grouping $n_c$ nodes, and with a gateway having $R$ transceiver modules on the cluster side. In this way the total number of nodes is $N = n_c^2$, with a number of ordinary end-nodes $Q = n_c(n_c - R)$.

In the case of full slot reservation, a node has only one slot per frame to transmit in. So, since the total number of slots in a frame is assumed to be $S_F = n_c(n_c - 1) + n_c = n_c^2$ (the first term at the second side representing the number of data slots and the second term the number of control slots) the maximum normalized throughput achievable by considering the only transit from the source node to the gateway node of the source cluster is $C_1 = 1/n_c^2$. Actually, this number must be compared with the constraints imposed on the throughput by the message transit between the gateway nodes of the source cluster and the destination cluster, as well as those imposed by the transfer from the gateway node to the end-node, in the destination cluster. As regards the second transit, that is, through the backbone, the maximum throughput can be determined by considering that the gateway of the source node must multiplex all the slots coming from the $Q_i = n_c - R$ ordinary nodes of the cluster itself. Since the gateway has $R$ receivers, $Q_i/R$ frames are necessary, on average, to satisfy these requirements. Therefore, the maximum normalized throughput, as it would be imposed by the second transfer, is $C_2 = R/(Q_i n_c^2)$. Since it is reasonable to assume $Q_i > R$, we have $C_2 < C_1$. Finally, the gateway node of the destination cluster must multiplex slots from all $Q_i(n_c - 1)$ ordinary nodes outside the cluster. Using its $R$ transmitters on the cluster side, the gateway needs for $Q_i(n_c - 1)/R$ frames, on average, and the maximum throughput, as it would be imposed by the third transit, is $C_3 = R/[Q_i(n_c - 1)n_c^2] < C_2$. In conclusion, we can say that the last transit is the bottleneck, and the maximum normalized throughput per node, $C_u$, taking into account the three steps of the transmission, is fixed by the transfer in the destination cluster, that is, $C_u = C_3$.

In the case of no slot reservation, a node potentially has $n_c - 1$ slots per frame to transmit in. The previous analysis remains valid, but $C_1$, $C_2$, and $C_3$ must be multiplied by $n_c - 1$. In conclusion, we have $C_u = R/(Q_i n_c^2)$, which is higher, as qualitatively expected, than the value obtained in the case of full slot reservation.

The maximum normalized throughputs per node so determined are plotted in Figure 4.4, as a function of the total number of ordinary nodes in the network, and assuming $R = 2$. For increasing $R$, at a parity of $n_c$ (i.e., the network size) slightly higher curves can be found. In any case, the condition $Q_i > R$ always has to be verified.

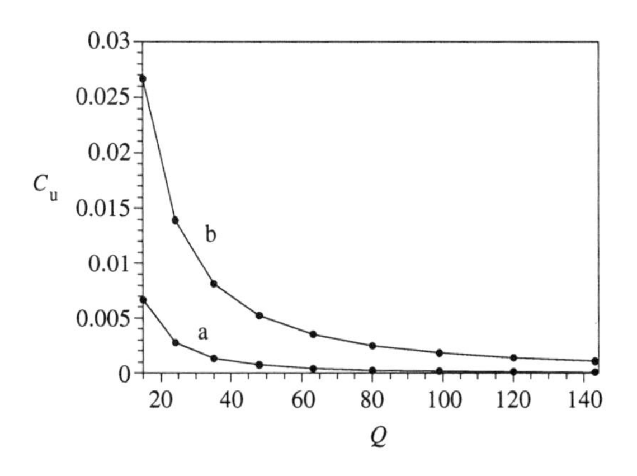


**Figure 4.4** Maximum normalized throughput per node in the case of: (a) full slot reservation and (b) no slot reservation. $Q$ is the total number of ordinary end-nodes in the network.

## 4.3 HIERARCHICAL LLN

The possibility of realizing single-hop hierarchical structures is not peculiar of broadcast-and-select networks. In Figure 4.5, for example, a hierarchical linear lightwave network (LLN) is shown, which is based on a recursive repetition of the Petersen graph (see Section 2.4.3.1). So it defines a highly ordered physical topology, simple to analyze by extending discussion of Section 2.4.

Beginning with a basic 10-node network, a second level of the hierarchy is created by adding subnetworks of the same form to one or more of the original nodes. This is the case, in Figure 4.5, of nodes C, D, and G, to which subnetworks $N_C$, $N_D$, and $N_G$ are added, respectively. One or more nodes of these subnetworks can be, in their turn, the root of other subnetworks, so creating a third level of the hierarchy. Once again looking at the specific example of Figure 4.5, this is the case of nodes DC and DE of $N_D$, to which subnetworks $N_{DC}$ and $N_{DE}$ are added, respectively. For simplifying and making notation univocal, we have denoted each (sub)network with the overall number of nodes it includes, and used the same local node labeling of Figure 2.50.

Figure 4.5 stops at the third hierarchical level, but it is quite evident that the growing procedure could continue, so achieving larger and larger network size. It is easy to prove that a complete network with $\psi$ levels, formed by growing all possible subnetworks at each level, contains a number of nodes equal to

$$N_t = \frac{5}{4}\left(9^{\Psi} - 1\right) \tag{4.2}$$

Therefore, the number of connectable users soon becomes impressively high.

Once having established the topology, in analyzing the network it is usual (see also Section 4.4) to number the levels in reverse order with respect to the growing procedure. In this way, the original 10-node network is at level $\psi$, while the last layer of subnetworks added is at level 1. This convention corresponds to the intuitive idea that a high hierarchical level should permit a degree of connectivity greater than that of a low level. Actually, in Figure 4.5 (where the network is not complete), the connection between nodes CB and CE only uses subnetwork $N_C$, while connection between nodes CB and DB must pass through the original network, which can therefore connect a larger number of users.

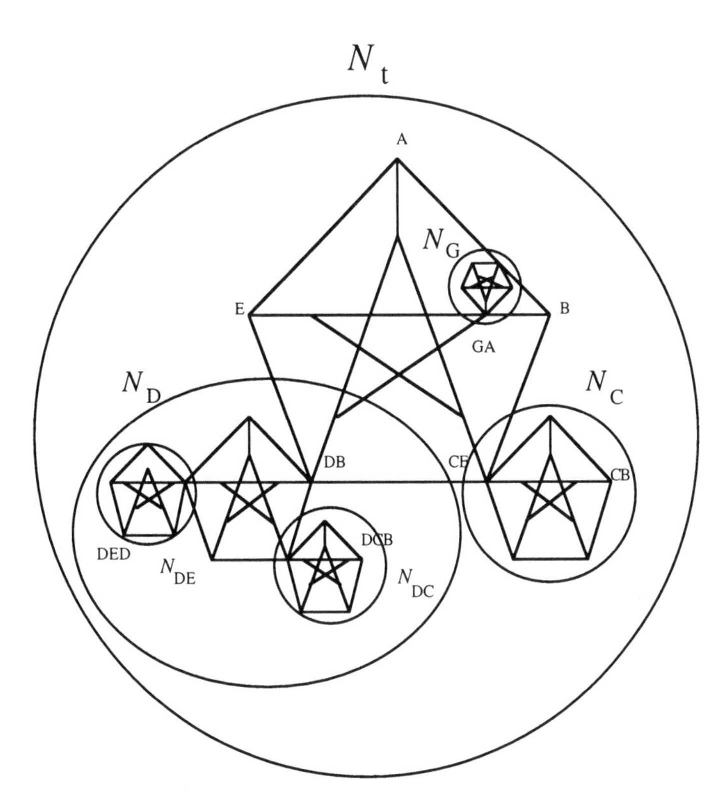


**Figure 4.5** Example of hierarchical LLN.

With this convention in mind, we can suppose that $n_{sc}$ stations are attached to each node of level 1 only, while the other nodes are used to transfer traffic from one level to another. If the network is complete, the number of users served is

$$N = n_{sc} \cdot \left[ N_t - \frac{5}{4} \sum_{i=2}^{\psi-1} \left(9^i - 1\right) \right] \tag{4.3}$$

Just with $\psi = 3$ and $n_{sc} = 20$, we find $N = 16{,}200$.

The hierarchical structure suggests a natural way for routing, according to the criteria discussed in Section 2.4 and based on the subdivision of the available spectrum into wavebands. In the most economical solution, one can use as many wavebands as the number of levels. Thus, in the example of Figure 4.5, we can split the spectrum into $K_b = \psi = 3$ wavebands, each of them transporting traffic at a given level of the hierarchy. Precisely, the waveband $WB_1$ can be used to support intracluster traffic (local traffic) within each subnetwork at level 1 (e.g., $N_{DC}$ and $N_{DE}$); the waveband $WB_2$ can be used to carry intercluster traffic limited to level 2 (long distance traffic, e.g., within $N_C$, $N_D$ and $N_G$); finally, the waveband $WB_3$ can be used for all other intercluster traffic involving also level 3 (very long distance traffic), that is, the original 10-node network (e.g., between $N_C$ and $N_D$). It should be noted that the terminology used to define the various kinds of traffic [5] implicitly assumes that the hierarchical organization reflects the geographical distance between users. As stressed above, this is undoubtedly the most frequent criterion for grouping stations but it is not the only one. In fixing the subnetwork organization, in fact, what it is really important is the existence (and exploitation) of a "community of interest" or "traffic locality," by virtue of which it is expected that demand decreases with the number of levels to cross.

With the goal of giving an idea of the performance achievable, we can refer to the complete three-level network mentioned after (4.3), and connecting $N$ = 16,200 users. Because of its symmetry, it is simpler to analyze than the incomplete structure of Figure 4.5. Referring to the notation introduced in Section 2.4.3, we limit discussion to the case of static routing, with two fiber-pairs per link. Furthermore, in agreement with the "community of interest" principle, it is assumed, as an example, that the total very long distance offered load is 400 Erlangs, the long distance load is 2,500 Erlangs, and the local load is 6,000 Erlangs [5].

In order to increase the efficiency, the number of wavebands can be significantly higher than the minimum value (for the strategy chosen) mentioned above; for example, one can set $K_b = 25$. For the same reason, ignoring possible limitations due to technology (calculus has explicative purposes), the number of channels is assumed rather high as well, for example, $W = 176$ [5]. Such resources are distributed as follows: 12 wavebands with eight channels at level 3, 12 wavebands with six channels at level 2, and 1 waveband with eight channels at level 1. This apparent disuniformity can be explained by considering that, because of the hierarchical structure, the longer distance traffic tends to concentrate at the "top" of the network; therefore, to avoid bottlenecks, more network resources must be assigned to these levels of traffic.

Let us consider the very long distance traffic: it must be routed upward through two levels of the hierarchy, entering level 3 via the three links attached to anyone of its nodes. A simple way to manage this interlevel routing is through a single spanning tree (see scheme (1) in Section 2.4.3.1). At level 3, instead, the situation is exactly that considered in scheme (5) in Section 2.4.3.1. Using the corresponding routing strategy, from Table 2.12 we see that the maximum load compatible with a blocking probability $p_B = 0.04$ and $12 \times 8 = 96$ local channels is 414 Erlangs, slightly larger than the designed value.

Long distance traffic can be routed in a similar way. In this case, however, level 3 is not involved by the transmission, and a single spanning tree is used only to convey this traffic to the three links of a node at level 2. Routing scheme (5) in Section 2.4.3.1 can be used again inside each subnetwork at level 2 and, repeating the calculus of Table 2.12 for the case of 12×6 = 72 local channels (maintaining $p_B = 0.04$), one finds 252 Erlangs as the maximum load. Since the intermediate level consists of 10 subnetworks, this yields a total of 2,520 Erlangs for the long distance traffic, larger in turn than the designed value.

Finally, local traffic can be routed using scheme (3) in Section 2.4.3.1: with eight channels, one finds 69 Erlangs of maximum load, which multiplied by 90 (the number of subnetworks at level 1) yields 6,210 Erlangs, that fully satisfy the designed requirements.

In conclusion, the considered complete hierarchical LLN, on three levels and $N$ = 16,200 stations, supports 9,144 Erlangs, with a blocking probability of 0.04. Assuming that each connection runs at 1 Gbps, the system ensures a throughput of more than 9 Tbps. It should be noted that multiplying the number of channels by the bit rate we obtain the maximum capacity of an equivalent star, as equal to 176 Gbps. This value is more than 50 times smaller than that permitted by the hierarchical LLN. Such a remarkable improvement is obviously made possible by the manifold spectrum reuse permitted by the topology. As an example, the single waveband allocated to local traffic can be used simultaneously within all the subnetworks at level 1, which means that it is reused 90 times.

## 4.4 COMBINATION OF SINGLE-HOP AND MULTIHOP CONNECTION MODES IN MONs

The analysis developed in Chapters 2 and 3 has shown the merits but also emphasized some limits of single-hop and multihop WDM networks. Generally speaking, we can say that single-hop networks are well suited to manage real time traffic and allow an easy implementation of multicast and broadcast transmissions. On the other hand, they cannot interconnect several stations without severely limiting the throughput of each station. Multihop networks, in turn, exhibit high throughputs and allow a large number of users to be connected. As a counterpart, however, they do not face efficiently the problem of broadcast transmission. In fact, the procedures for switching and routing data through these networks cannot be managed easily, while the hardware and software requirements often impose heavy node architecture. Moreover, the multihop strategy implies differences in the propagation delays, which may limit the implementation of real-time services, especially if a deflection routing approach is adopted.

Such considerations have led to the belief that, instead of using these architectures separately, one can combine them, thus compensating the disadvantages while enjoying the benefits of each scheme [6–8]. The resulting structure is a multilevel hybrid system, where either single-hop or multihop methods are used.

Following [6], in some previous papers we called these networks MONETs. The acronym MONET, however, is currently used to denote a very important research program [9], born from the cooperation of a number of partners (AT&T, Bellcore, Lucent Technologies, Bell Atlantic, BellSouth, Pacific Telesis, and SBC, with participation by the USA National Security Agency and the Naval Research Laboratory), with the primary aim of demonstrating the commercial viability of the WDM networking technology. While we have designated the acronym MONET to mean Multilevel Optical NETwork, the alternative meaning is Multiwavelength Optical NETworking.

Although the two research projects certainly have interesting synergies (in particular, the MONET program also assumes a hierarchical network architecture in order to achieve scalability to national size), it is evident that they have developed in quite different contexts. Therefore, in order to avoid confusion and false expectations, we decided to use the shorter acronym MON to denote the family of multilevel hybrid optical networks we have considered in this book.

MONs will be described in this section, with special emphasis on the presentation of an analytical model for their performance evaluation. Other details about the implementative aspects can be found in [6].

### 4.4.1 Basic Concepts for the Two-Level Case

A basic MON consists of two hierarchical levels: level 1 is a set of single-hop networks (clusters) interconnected by a multihop network, which operates at the uppermost level (level 2).

A simple way to introduce the proposed structure consists in regarding it as an evolution of the multiple WDM star shown in Figure 4.3. Similarly to that solution, stations are partitioned into single-hop clusters and, inside each cluster, a station acts as gateway, so permitting to interface, through the backbone, all the clusters together. In Figure 4.3, however, communications between different gateways were single-hop as well, while in the present solution they occur through a multihop network. The virtual circuit of this two-level MON has therefore the structure pictorially represented in Figure 4.6. Many different multihop solutions could be obviously used to realize the backbone. Among them, for the subsequent quantitative analysis, we have decided to choose the shufflenet. It is simple and significant enough to permit us to draw some remarkable conclusions.

As regards the physical topology which implements the MON, the optical star could continue to be used. Nevertheless, a tree topology seems more suitable to exploit the benefits and motivation of the combined single-hop and multihop connection. So, we can refer, as an example, to the structure shown in Figure 4.7. The virtual subnetworks are embedded in the physical topology and the stations of each cluster do not belong necessarily to the same physical subtree. In the example of the figure, a cluster can be constituted by the shadowed stations only. In other words, this means that geographical vicinity is a possible criterion (undoubtedly important) to create the clusters, but it is largely overcome by the identification of the "traffic localities" existing inside the network. If there is strong locality in the system and it is reflected in the hierarchical structure, the network performance can be optimized. In practice, it is reasonable to group in the same cluster stations which demonstrate particular needs to communicate to each other. This way, the requirement of bandwidth, per user pair, at level 2 is smaller than at level 1. On the other hand, the number of potential pairs communicating with each other through the uppermost level is higher. These two effects tend to compensate, and in the ideal case the product of the bandwidth per user and the number of user pairs should be constant for each level. Actually, these are the basic ideas that guide the optimization problem faced in the following sections.

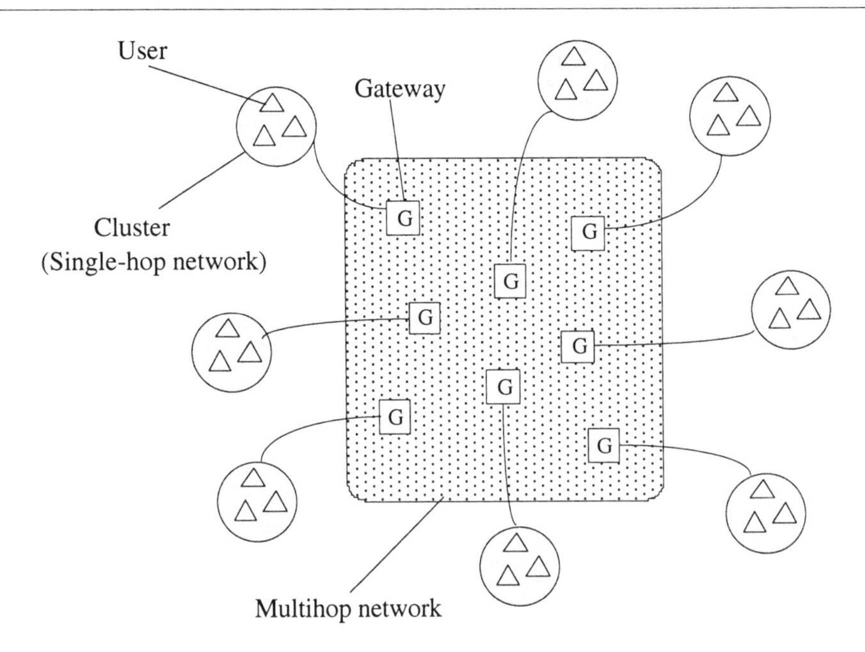


**Figure 4.6** Schematic representation of the virtual topology for a two-level MON.

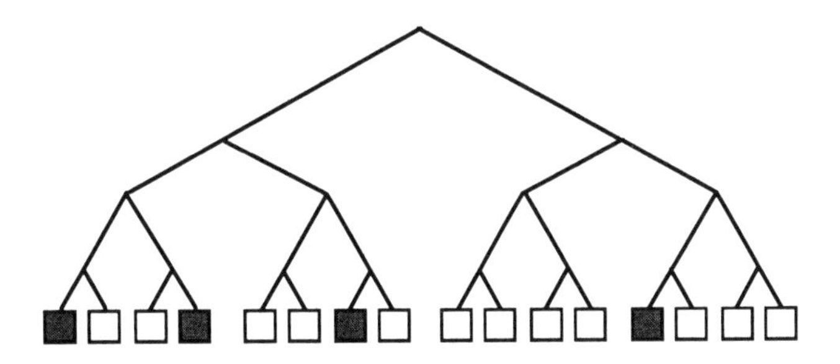

**Figure 4.7** Example of MON physical topology. Shaded stations belong to the same cluster.

From a different point of view, we can say that, in a MON, line cost is determined exclusively by the physical topology and is therefore totally independent of the arrangement chosen for the clusters. A separate comment, however, must be made about the possibility of implementing the clusters in disjoint subtrees, thus permitting reuse of the wavelengths. This solution seems natural when a large geographical area must be covered, in which case each physical subtree will serve a limited area, connecting neighboring stations.

Once having made clusters arrangement clear, we may state that one of the stations in each cluster plays the role of gateway, in the virtual topology, for intercluster communications. With the choice of the shufflenet, each gateway has $p$ transmitters and $p$

receivers on the backbone side. The number of clusters must equal the number of nodes in the shufflenet, that is, $n_c = kp^k$. An example is shown in Figure 4.8, again in the case of a tree with 16 nodes, which can be arranged in four clusters ($k = p = 2$) each with four stations. Unlike Figure 4.7, to simplify the graph, in the example, we have supposed that each cluster is associated with a physical subtree, and its stations have been enclosed into an ellipse. Figure 4.8 has the same meaning of Figure 4.6, but it refers to a specific physical topology (a tree) and a specific multihop virtual topology (a shufflenet).

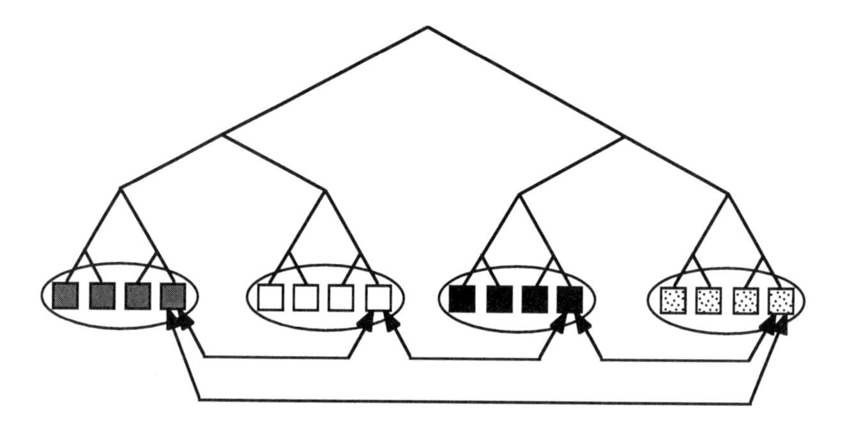

**Figure 4.8** Example of MON virtual topology.

Finally, it is worth mentioning that the above-described partitioning of the network is not necessarily static but can adapt (dynamically) to possible changes in the current traffic conditions. To obtain such a flexible network, however, a network manager is, on the one hand, required to monitor the variable traffic localities while, on the other hand, it is desirable that each station be equipped with the hardware and software typical of a gateway. In particular conditions, in fact, any station could be asked to control not only its own incoming traffic, but also the traffic exchanged between adjacent subnetworks and the subnetwork it belongs to.

### 4.4.2 Analysis and Optimization of MONs

Let us suppose, for simplicity, that each cluster has the same number of stations ($n_{sc} + 1$) and the same number of available channels for intracluster communications, denoted by $n_\lambda < n_{sc}$. As one of the stations in each cluster is designated as the gateway, the number of ordinary end-nodes is equal to $n_{sc}$, and the total number of users in the network results in

$$N = n_{sc} kp^k \qquad (4.4)$$

As regards the total number of wavelengths $W$, instead, we suppose that the physical topology is fully broadcast (as in Figures 4.7 and 4.8) so that wavelength reuse is not permitted. Under this condition, it is easy to verify that

$$W = kp^k (p + n_\lambda) \tag{4.5}$$

In the special case of one wavelength per cluster ($n_\lambda = 1$) and two wavelengths assigned to each gateway for its transmission toward the shufflenet ($p = 2$), (4.5) gives $W = 3n_c$.

From the expressions above, one can understand that the number of connectable users is very large, but that the required number of wavelengths becomes very high as well when a large number of clusters is configured in the MON. As mentioned in Section 4.4.1, to alleviate this problem, clusters could be mapped into geographically separate virtual subtrees, so that wavelengths could be reused [6]. With a solution that permits to use the same $n_\lambda$ wavelengths within each cluster, the total number of wavelengths lowers up to $W = kp^{k+1} + n_\lambda$, which is only slightly greater than that of the shufflenet. Keeping this possibility in mind, in the subsequent numerical analysis, however, we will refer to (4.5), so achieving a pessimistic evaluation of the network cost. Apart from the simplification in the model and the network description, it should be noted that the choice of a fully broadcast physical topology yields a structure that is easy to reconfigure dynamically, and whose performance is directly comparable with that of a pure shufflenet.

Similarly to the choice made for the star-of-stars (see Section 4.2) the transmission protocol (WDM/TDM) is based on a fixed bandwidth assignment rule [10], which means that, at any level, the channels are provided to the various source/destination pairs in such a way as to prevent any destination conflict and collision. As widely discussed in Chapter 2, the assumption of a fixed assignment protocol is optimal in the case of heavy loads but it becomes not very efficient at light loads, where many slots may be unused and the packet delay too high. The choice is justified, however, for MONs, since we expect that, through the hierarchical organization, any level is designed in such a way as to operate in conditions near saturation, thus fully exploiting the resources available. When this is not the case, other assignment protocols could obviously be considered, with suitable modifications in the analytical model.

The object of the analysis is to find the throughput of the network, and its related quantities. To this purpose, two different load conditions will be examined below [11]:

1. The traffic generated at each end-station is uniformly distributed among all the other users of the network;
2. The traffic generated at each end-station is nonuniformly distributed among all the other users of the network.

Case (1) is not very realistic, since it does not correspond to the motivation of the multilevel approach. Nevertheless, it is useful to introduce the fundamentals of the model. Case (2) is more practical and, although here limited to some specific situation, it seems relevant for further discussion.

#### *4.4.2.1 Uniform Traffic Distribution*

Though the traffic is uniformly distributed, the utilization of the shufflenet and of the single-hop networks are generally different from each other. In more explicit terms, it is possible that, when the single-hop network reaches its maximum throughput, the multihop network is still able to locate traffic and vice versa. Optimum operation conditions can be obviously reached when no bottleneck effect is present and the networks at different hierarchical levels operate in reciprocal agreement.

To understand the concept better, let us consider the following expression for the normalized throughput per user:

$$C_{\mathrm{u}} = \frac{N-1}{\max\left[T_{\mathrm{s}}^{\mathrm{sh}}, T_{\mathrm{s}}^{\mathrm{mh}}\right]} \tag{4.6}$$

Function max[$x$,$y$] gives the maximum value between $x$ and $y$, while $T_{\mathrm{s}}^{\mathrm{sh}}$ and $T_{\mathrm{s}}^{\mathrm{mh}}$ represent the number of time slots involved in the transmission of a packet (one time slot long) from each node of the network toward all the others in the single-hop and the multihop networks, respectively. In other words, when averaged on a sufficiently wide time interval, max[$T_{\mathrm{s}}^{\mathrm{sh}}$, $T_{\mathrm{s}}^{\mathrm{mh}}$] is the number of time slots that ensures that each user has transmitted and received exactly $N-1$ packets. We call this period "transmission cycle."

Two very simple examples are shown in Figure 4.9. They are of limited practical interest, and are only useful to illustrate (and emphasize) the problem. The architecture is basically that of Figure 4.9(a), consisting of two clusters with $n_{\mathrm{sc}} = 2$, at level 1, and a shufflenet with $k = 2$ and $p = 1$, at level 2. Actually, because of the assumption of $p = 1$, the shufflenet reduces itself to a unidirectional ring. The gateways have been denoted by $a$ and $b$; they belong to the single-hop clusters (being nodes but not users) as well as to the multihop backbone. So, for the sake of clarity, they have been extracted from the clusters. In Figure 4.9(b) a possible assignment of the time slots and wavelengths is shown, in the case of $n_\lambda = 1$. The channels of the single-hop subnetworks have been denoted by $\lambda_{\mathrm{sh1}}$ (in cluster $c_1$) and by $\lambda_{\mathrm{sh2}}$ (in $c_2$), while the channels of the shufflenet are $\lambda_{\mathrm{mh1}}$ and $\lambda_{\mathrm{mh2}}$. $x$-$y$ denotes a packet emitted by $x$ (source) and addressed to $y$ (destination); the unassigned slots, in the $i$th transmission cycle, are unused.

The choice made in Figure 4.9(b) is obviously not unique: different correspondences between packets and slots could be selected without altering the network performance. In any case, the single-hop networks need a number of slots, to complete the transfers, greater than that required by the multihop network. This means that the single-hop networks act as a bottleneck and, in steady-state conditions, the shufflenet wastes part of the available slots, thereby causing a reduction of the channel utilization with respect to that theoretically achievable with a better resource arrangement.

On the opposite side, Figure 4.9(c) shows a possible assignment in the case of $n_\lambda = 5$. This is a redundant situation that is useful, however, for reinforcing the concepts presented above. Wavelengths $\lambda_{\mathrm{sh}j}$, with $j = 1, \ldots, 5$, are available in cluster $c_1$, whereas wavelengths $\lambda_{\mathrm{sh}j}$, with $j = 6, \ldots, 10$, are available in cluster $c_2$. Bottleneck is due here to the multihop network, and slots are wasted in the single-hop ones.

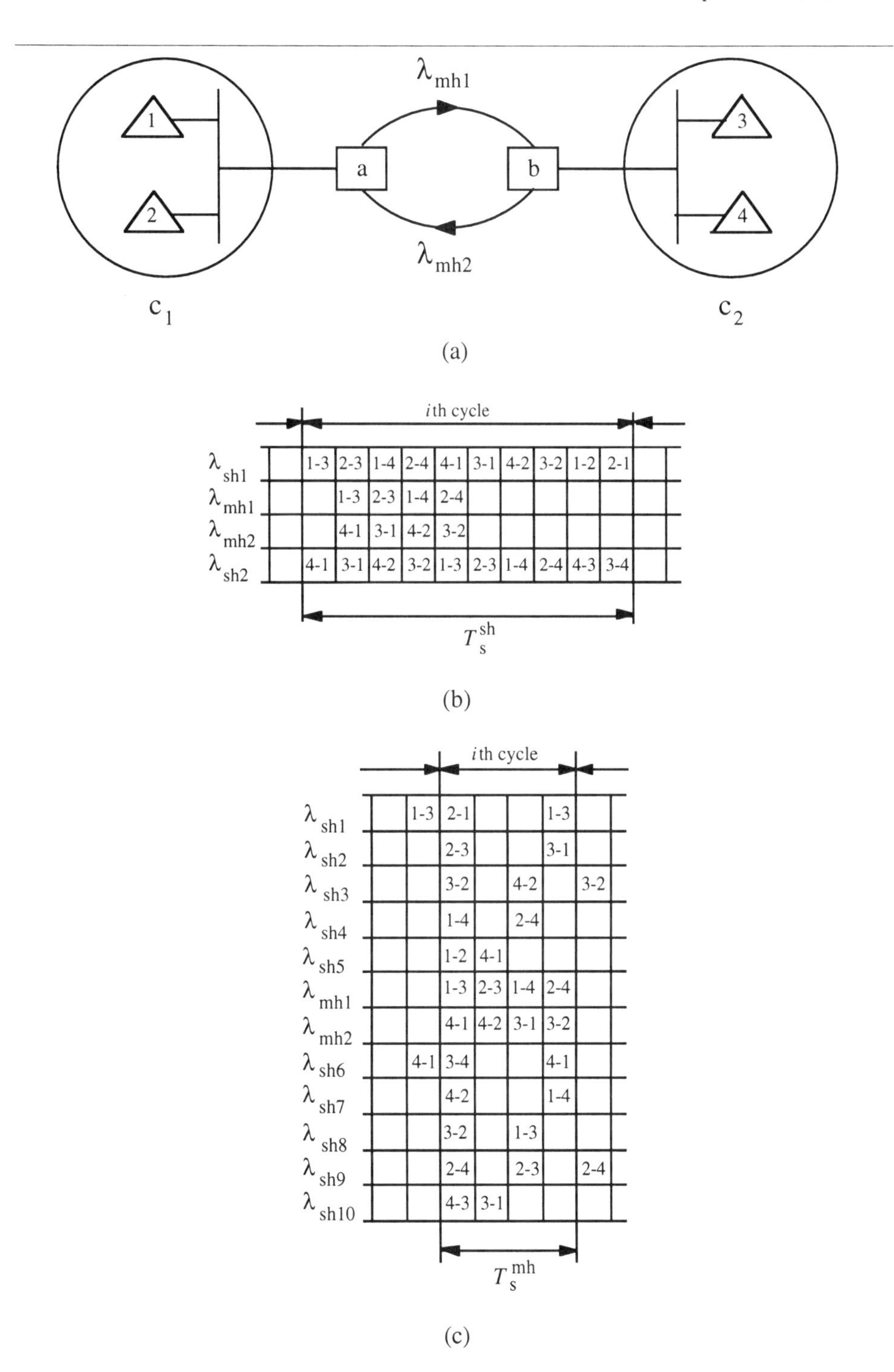


**Figure 4.9** Examples of wavelength and time slot assignment, with "bottleneck" effect.

In general, the values of $T_s^{sh}$ and $T_s^{mh}$ can be derived on the basis of some simple considerations concerning the network behavior. In particular, in each cluster $n_{sc}(n_{sc} - 1)$ packets are exchanged among its users (that do not enter the multihop network), $n_{sc}(N - n_{sc})$ packets outgoing the cluster through its gateway and addressed toward the external users and, finally, $n_{sc}(N - n_{sc})$ packets coming in from the other clusters.

Some algebraic manipulations, reminding one that the number of channels per cluster is equal to $n_\lambda$, allow us to determine

$$T_s^{sh} = \frac{n_{sc}^2\left(2kp^k - 1\right) - n_{sc}}{n_\lambda} \tag{4.7}$$

Similarly, as regards the multihop network, the overall number of packets passing through it, during a transmission cycle, is given by

$$M_s^{mh} = n_{sc}^2 kp^k \sum_{\text{shufflenet}} hg_h = n_{sc}^2 n_c \sum_{\text{shufflenet}} hg_h \tag{4.8}$$

where $g_h$ is the number of gateways (i.e., clusters, of the shufflenet $h$ hops far from any considered gateway). The sum in (4.8) is therefore extended over all the nodes of the shufflenet. In practice, taking into account the multihop mechanism, we can say that every end-station produces $M_s^{mh} / (n_{sc} \cdot n_c)$ packet generations in the shufflenet (the packet relative to a given source/destination pair can appear more than one time in the table of the time slot and channel assignment). It is important to stress that this argument holds only because shufflenet has the isotropic property which ensures that routing trees look the same regardless of the originating node, that is, we can speak of a single value of $g_h$ for all originating nodes.

As the multihop network has, globally, $kp^{k+1}$ channels, assuming the availability of a routing algorithm that balances the load on all these channels, we obtain

$$T_s^{mh} = \frac{n_{sc}^2}{p} \sum_{\text{shufflenet}} hg_h \tag{4.9}$$

Combining (4.6), (4.7), and (4.9), the following expressions result:

$$C_u = \frac{N-1}{n_{sc}^2\left(2kp^k - 1\right) - n_{sc}} n_\lambda \qquad \text{if } T_s^{mh} \le T_s^{sh} \tag{4.10a}$$

$$C_u = \frac{N-1}{n_{sc}^2 \sum_{\text{shufflenet}} hg_h} p \qquad \text{if } T_s^{mh} \ge T_s^{sh} \tag{4.10b}$$

According to the above considerations, best performance occurs when $T_s^{mh} = T_s^{sh}$; this equality, in turn, can be used as a direct design condition. For example, one may be interested in the evaluation of the number of channels per each cluster, having fixed all the other parameters. Equating (4.10a) and (4.10b), we have

$$n_{\lambda 0} = \frac{n_{sc}\left(2kp^k - 1\right) - 1}{n_{sc} \sum_{\text{shufflenet}} hg_h} p \tag{4.11}$$

which gives the optimum value of $n_\lambda$: for $n_\lambda > n_{\lambda 0}$ the single-hop throughput would increase but the throughput per user of the entire MON is limited by the multihop network; for $n_\lambda < n_{\lambda 0}$, instead, the MON would be underused since, also considering overloaded clusters, the potentiality of the multihop network is not completely exploited. In conclusion, it is not fruitful to assume a number of wavelengths inside each cluster that exceeds the smallest integer greater than (4.11); this would lead to higher costs, while the normalized throughput cannot be greater than (4.10b). The simple analysis here reported [11], considering $n_\lambda$ as a free design parameter, could be obviously repeated by fixing attention on other features of the network (i.e., the number of stations per cluster, and the values of $p$ or $k$). In the most general case, we have a multivariate optimization problem.

The dependence of $C_u$ on $n_\lambda$, discussed above, is confirmed by the example plotted in Figure 4.10, which refers to the case of $k = 2$ and $p = 4$, with different values of $n_{sc}$. In the first zone of each curve, the throughput increases linearly with $n_\lambda$; correspondingly, in fact, $C_u$ is upper bounded by the single-hop network bandwidth, as given in (4.10a). Once having reached the optimum $n_\lambda$, the throughput remains constant, since the multihop network acts as a bottleneck.

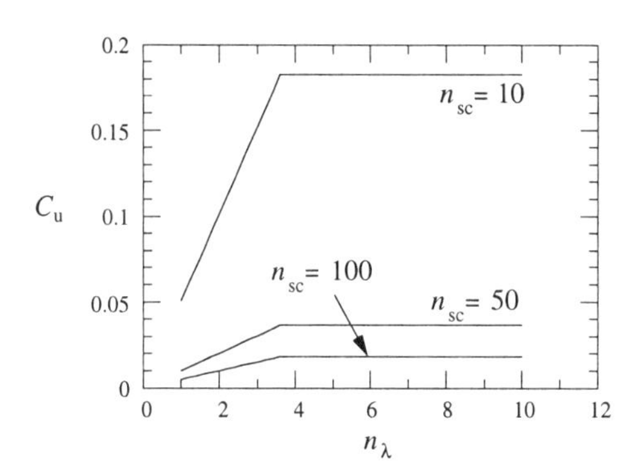


**Figure 4.10** Normalized throughput per user as a function of $n_\lambda$ under uniform traffic distribution.

The figure highlights a result that is not so obvious from the previous relationships. Precisely, from (4.11) one could conclude that $n_{\lambda 0}$ depends on the number of users per cluster $n_{sc}$. Actually, in Figure 4.10, all curves saturate in correspondence of nearly the same value of $n_\lambda$. This conclusion can be analytically justified by reminding that (from Table 3.5)

$$\sum_{\text{shufflenet}} hg_h = k\left[\frac{p^k(3k-1)}{2} - \frac{p^k - 1}{p-1}\right] \tag{4.12}$$

The latter, replaced into (4.11) allows one to approximate $n_{\lambda 0}$ as follows:

$$n_{\lambda 0} \approx \frac{2p(p-1)\left(2kp^k - 1\right)}{kp^k(3k-1)(p-1) - 2k\left(p^k - 1\right)} \tag{4.13}$$

which is independent of $n_{sc}$. Equation (4.13) holds if $N >> 1$, but it is clear that such a position is always verified in practice. On the other hand, it is possible to demonstrate that the maximum of $C_u$, reached when $n_\lambda \geq n_{\lambda 0}$, depends on $1/n_{sc}$. Assuming, in fact, that in (4.10a) (considered for $T_s^{mh} = T_s^{sh}$) $N - 1 \approx N$, we have

$$\left(C_u\right)_{max} \approx n_{\lambda 0} \frac{kp^k}{n_{sc}\left(2kp^k - 1\right)} \tag{4.14}$$

Instead of reasoning in terms of throughput per user, one can consider the normalized total network throughput, defined as

$$C_{tot} = N \cdot C_u \tag{4.15}$$

A graph of $C_{tot}$, for the same situations discussed in Figure 4.10, is plotted in Figure 4.11. The curves are practically independent of $n_{sc}$, as it is also easy to verify analytically. In particular, using (4.14), the maximum value results in

$$\left(C_{tot}\right)_{max} \approx n_{\lambda 0} \frac{k^2 p^{2k}}{\left(2kp^k - 1\right)} \tag{4.16}$$

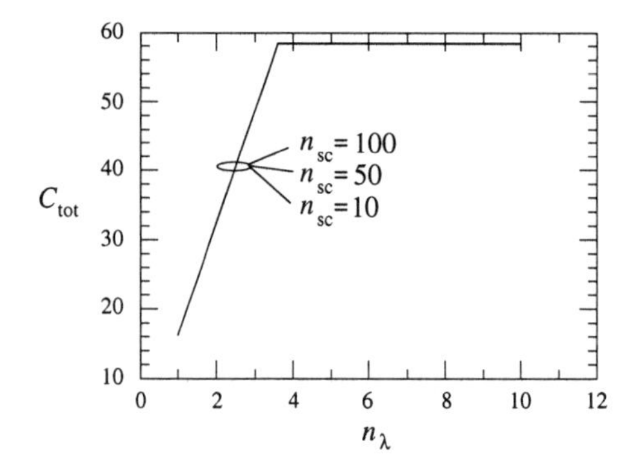


**Figure 4.11** Normalized total throughput as a function of $n_\lambda$ under uniform traffic distribution.

The latter result implies that, once given the topology of the multihop network, which also fixes the number of channels assigned to it, the total network throughput depends only on the number of the channels in each single-hop network. In conclusion, we can say that, for networks of not negligible size, the number of users per cluster modifies the throughput per user but does not affect the overall network performance.

There is an alternative, and more intuitive, way to derive the expression of the throughput when $T_s^{mh} = T_s^{sh}$. The optimum condition, in the sense specified above, is in fact equivalent to dividing the total number of time slots in a transmission cycle by the overall number of channels in the MON. Such a position implicitly assumes that no slot is wasted, and the throughput is consequently maximized. Thus, the following expression appears:

$$C_u = \frac{N-1}{N \sum_{\text{MON}} h u_h} W \qquad \text{if } n_\lambda = n_{\lambda 0} \tag{4.17}$$

where $u_h$ is the number of users $h$ hops far from any considered source in the MON. The value of the sum appearing in (4.17) can be determined on the basis of Table 4.1 [12]. The latter represents the extension of Table 3.5 to the enlarged multilevel architecture.

Moreover, similarly to (3.35), the expected number of hops between two randomly selected users of the MON can be easily computed as

$$\langle h \rangle = \frac{1}{N-1} \sum_{\text{MON}} h u_h = \frac{1}{N-1} \left\{ n_{sc} k \left[ \frac{3}{2} p^k (k+1) - \frac{p^k - 1}{p-1} \right] - (n_{sc}+1) \right\} \tag{4.18}$$

Adding the further hypothesis of balanced load among all the channels (verified, however, in optimum conditions), the reciprocal of $\langle h \rangle$ expresses the efficiency $\eta$ so that, keeping in mind (4.5), the expression of $C_u$, for $n_\lambda = n_{\lambda 0}$, can be rewritten as follows:

$$(C_u)_{max} = \eta \frac{n_{\lambda 0} + p}{n_{sc}} \tag{4.19}$$

Actually, as stated above, (4.19) holds even for $n_\lambda > n_{\lambda 0}$, and gives the maximum throughput more accurately than (4.14).

It is interesting to investigate the behavior of $\eta = 1/\langle h \rangle$ as a function of $n_{sc}$, $k$ and $p$, for various combinations of the other parameters [12]. As shown in Figure 4.12, the efficiency exhibits a substantial independence of the number of users per cluster, while it increases for smaller and smaller values of the number of clusters ($n_c = kp^k$). Figures 4.13 and 4.14 show the effect of the multihop backbone, for a fixed value of $n_{sc} = 10$. Particularly, in Figure 4.13, the channel efficiency has been plotted, as a function of $k$, for two values of $p$. The effect of the number of channels per gateway in the shufflenet is appreciable only for low $k$; when the latter parameter becomes large, the efficiency decreases, tending asymptotically to zero, regardless of the $p$ value. The independence of $p$ is made explicit in Figure 4.14, where three values of $k$ have been considered. With the exception of the region where $p$ is on the order of few units, the curves are practically constant and $\eta \approx 2/[3(k+1)]$.

**Table 4.1**
Distance, in Hops, from a Generic Source in the MON and Corresponding Number of Connectable Users

| $h$ | $u_h$ |
|---|---|
| 1 | $n_{sc} - 1$ |
| 2 | 0 |
| 3 | $n_{sc}p$ |
| 4 | $n_{sc}p^2$ |
| 5 | $n_{sc}p^3$ |
| . | . |
| . | . |
| . | . |
| $k$ | $n_{sc}p^{k-2}$ |
| $k + 1$ | $n_{sc}p^{k-1}$ |
| $k + 2$ | $n_{sc}(p^k - 1)$ |
| $k + 3$ | $n_{sc}(p^k - p)$ |
| $k + 4$ | $n_{sc}(p^k - p^2)$ |
| . | . |
| . | . |
| . | . |
| $2k + 1$ | $n_{sc}(p^k - p^{k-1})$ |

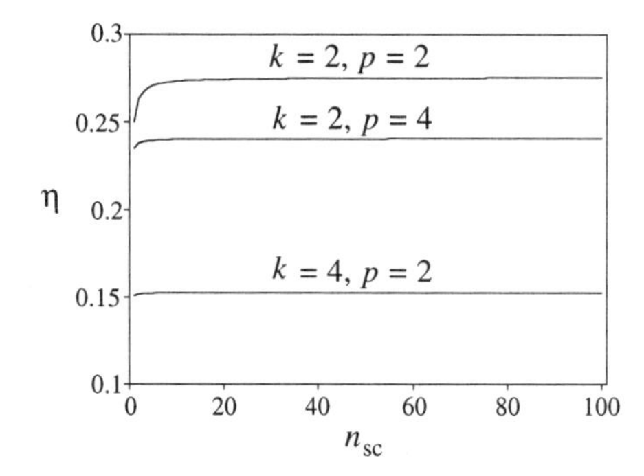


**Figure 4.12** Channel efficiency of the MON, under uniform traffic distribution, as a function of the number of users per cluster, and for different structures of the shufflenet.

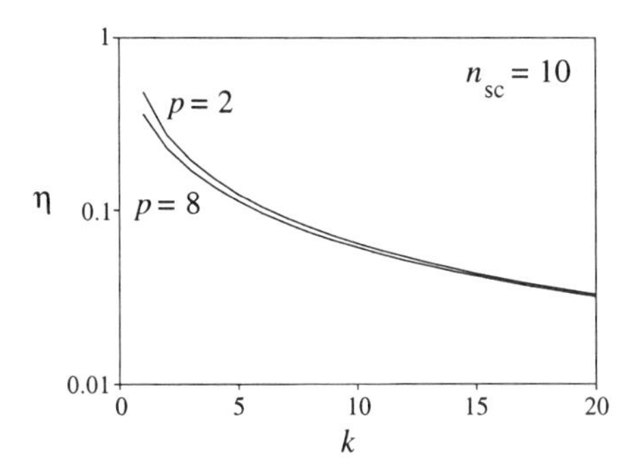


**Figure 4.13** Efficiency of the MON, under uniform traffic distribution, as a function of the number of columns in the shufflenet.

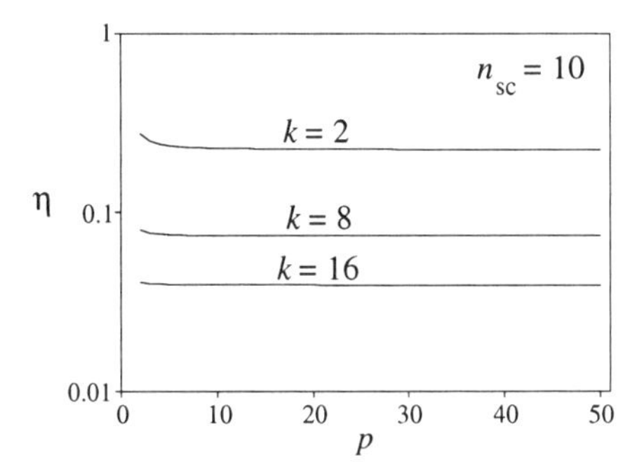


**Figure 4.14** Efficiency of the MON, under uniform traffic distribution, as a function of the number of channels of each gateway, for intercluster communications.

The asymptotic behaviors mentioned above can be analytically derived from (4.18); just the idea of evidencing them justifies the resort to so high values for $k$ and $p$, which obviously correspond to extremely large network sizes. On the other hand, combining the results of Figures 4.13 and 4.14, we can conclude that values of $\eta$ that are satisfactorily high can be reached only assuming $p$ much greater than $k$. Such information will be used also in the numerical examples of the next section.

Coming back, for a moment, to Figure 4.12, it is interesting to note that the dependence of $\eta$ on the pairs $(k, p)$, for a given value of $n_{sc}$, is not necessarily the same of the maximum throughput. This is shown in Figure 4.15: basically as a consequence of the increase in the factor $(n_{\lambda 0} + p)$, the value of $\left(C_u\right)_{max}$ for $k = 2$ and $p = 4$, though related to a smaller efficiency, is greater everywhere than that for $k = p = 2$.

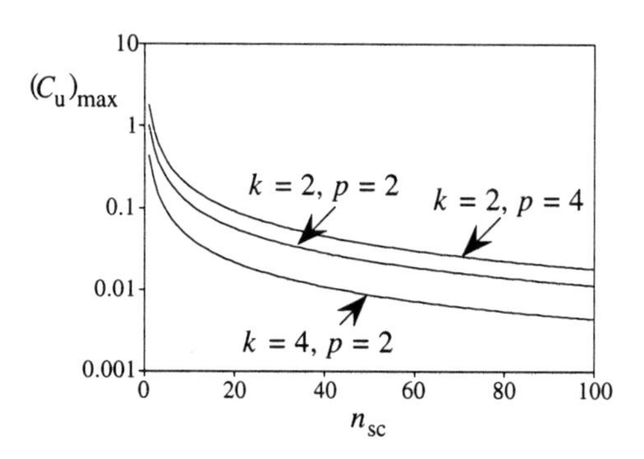


**Figure 4.15** Maximum normalized throughput per user under uniform traffic distribution.

Another parameter is worth being introduced to complete the analysis. It is defined as

$$\chi = \frac{W}{C_{tot}} \tag{4.20}$$

and can be seen as a measure of the network cost. To understand this statement, one can consider that the higher the number of channels is, the more expensive the network results, since more and more optical fibers and components (like lasers, couplers, and photodiodes) are necessary for its implementation. Similarly, increasing the total network throughput, the network cost can be made lower since more and more services are offered to the users. For these reasons, minimization of (4.20) seems a crucial point in the MON design optimization problem. It should be noted that, on the basis of (4.5), (4.15), and (4.19), when $n_\lambda = n_{\lambda 0}$ (i.e., in optimum conditions) we have $\chi = 1/\eta = \langle h \rangle$.

As an example, the behavior of $\chi$ for a MON with $k = 2$ and $p = 4$, $n_\lambda$ variable and $n_{sc}$ as a parameter is shown in Figure 4.16. Independently of the value of $n_{sc}$, this "cost figure" reaches its minimum in correspondence of $n_{\lambda 0}$, and then increases monotonically on the right as well as on the left.

#### 4.4.2.2 *Nonuniform Traffic Distribution*

Now the hypothesis of traffic uniformly distributed among all the destination nodes of the network will be removed [11]. This seems necessary in order to make more realistic, and then significant, the analysis as well as the design considerations. When the number of messages emitted by a generic end-station is not equally addressed toward all the other users, the organization of the nodes in clusters is efficiently made on the basis of information on the traffic distribution rather than simply following their geographical location. More precisely, each cluster includes all the users that have a higher probability of linking one another.

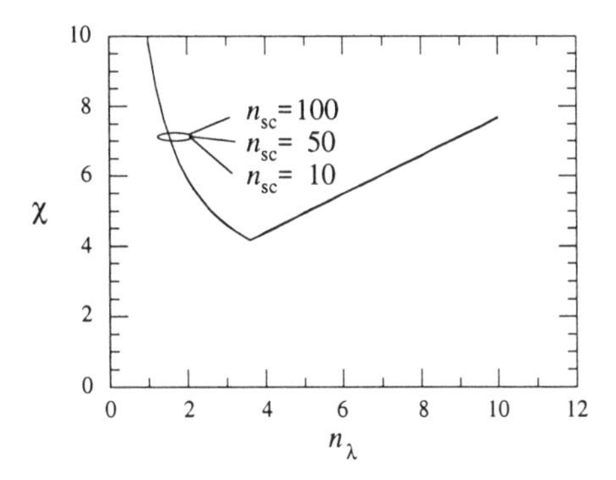


**Figure 4.16** Cost figure as a function of $n_\lambda$ under uniform traffic distribution.

The connection probability between users $i$ and $j$ can be seen as the element $T_{ij}$ of a real traffic matrix $\boldsymbol{T}$. Contrary to the example of Figure 4.2, the requirement of bandwidth for the various connections is therefore not expressed by integer numbers (the time slots needed in any transmission cycle), but this simply corresponds to the introduction of a sort of scale factor between the two representations. The meaning of $\boldsymbol{T}$ remains, in fact, the same.

For the sake of simplicity, we continue to assume that the clusters are all equal among them, and that each user has the same connection probability with the other end-nodes of its own cluster, different from that with the end-nodes of other clusters. More precisely, noting by $P_c$ the common rate of intracluster connections required by a generic user, the probability that two end-nodes inside the same cluster are connected is given by $P_c/(n_{sc} - 1)$. The probability of connection for two end-nodes belonging to different clusters, instead, results in $(1 - P_c)/(N - n_{sc})$. A correct cluster organization implies that

$$\frac{P_c}{n_{sc} - 1} \geq \frac{1 - P_c}{N - n_{sc}} \tag{4.21}$$

which, in turn, leads to

$$\frac{n_{sc} - 1}{N - 1} \leq P_c \leq 1 \tag{4.22}$$

The lower bound in (4.22), from now on denoted by $P_u$, returns the case of uniform traffic distribution. Expression (4.6) still holds, but its numerator has to be replaced by the number $L$ of packets emitted, on average, by each user in a transmission cycle (in general, $L \neq N - 1$); so we have

$$C_u = \frac{L}{\max\left[T_s^{sh}, T_s^{mh}\right]} \tag{4.23}$$

As regards the evaluation of $T_s^{sh}$ and $T_s^{mh}$, the procedure is conceptually similar to that described in Section 4.4.2.1, but it must take into account the implications of the nonuniform traffic distribution. More specifically, the traffic inside each cluster can be split up among three different parts: the first part is the traffic generated inside the cluster and addressed toward the other end-nodes of the same cluster; the second part is the traffic generated inside the cluster but addressed toward end-nodes belonging to different clusters; finally, the third part is the traffic coming from the multihop network. The traffic inside the cluster can be measured by the number of packets passing through it in a given time period. Adding the above three contributions, it is easy to verify that the number of packets to be transferred, in one transmission cycle inside each cluster, is given by

$$M_s^{sh} = Ln_{sc}(2 - P_c) \tag{4.24}$$

Dividing (4.24) by the number of channels per each cluster, we obtain the value of $T_s^{sh}$ as

$$T_s^{sh} = \frac{Ln_{sc}(2 - P_c)}{n_\lambda} \tag{4.25}$$

Through a similar approach, it is possible to modify (4.8) and then derive the following expression:

$$T_s^{mh} = L\frac{1 - P_c}{N - n_{sc}}\frac{n_{sc}^2}{p}\sum_{\text{shufflenet}} hg_h \tag{4.26}$$

which, in turn, extends the validity of (4.9) to the present case.

Combining (4.23), (4.25), and (4.26), also keeping in mind expression (4.12), the throughput per user becomes

$$C_u = \frac{n_\lambda}{n_{sc}(2 - P_c)} \qquad \text{if } T_s^{mh} \le T_s^{sh} \tag{4.27a}$$

$$C_u = \frac{2p(p-1)(kp^k - 1)}{n_{sc}(1 - P_c)\left[kp^k(3k-1)(p-1) - 2k(p^k - 1)\right]} \qquad \text{if } T_s^{mh} \ge T_s^{sh} \tag{4.27b}$$

By imposing the optimum condition that, again, corresponds to having $T_s^{mh} = T_s^{sh}$, we finally obtain

$$n_{\lambda 0} = \frac{2p(p-1)(2 - P_c)(kp^k - 1)}{(1 - P_c)\left[kp^k(3k-1)(p-1) - 2k(p^k - 1)\right]} \tag{4.28}$$

which gives the most favorable choice for the number of channels inside each cluster, once the other network parameters have been fixed. The normalized total network throughput $C_{tot}$ and the cost figure $\chi$ can be finally obtained on the basis of (4.15) and (4.20), respectively. It is interesting to observe that the condition $T_s^{sh} \leq T_s^{mh}$ is equivalent to putting $n_\lambda \geq n_{\lambda 0}$ while, conversely, the condition $T_s^{sh} \geq T_s^{mh}$ is not different from setting $n_\lambda \leq n_{\lambda 0}$.

An interesting point could be to explore the effects of the parameters $k$ and $p$, once having fixed the number $N$ of users to connect, and having constrained the total number of wavelengths to be smaller than a given $W_a$. For the sake of clarity, we refer to a specific numerical example. Nevertheless, as the present section is mainly devoted to illustrate an analysis method, we do not assume a particularly stringent bound. Precisely, we set $W_a = 300$. An example more respectful of the technological limits will be considered in Section 4.4.3.

As regards the number of users, we assume $N = 500$. This means that, once fixed $k$ and $p$, $n_{sc}$ is chosen in such a way as to have $N$ close to 500 but, if necessary, greater. With the same spirit, the number of channels per cluster $n_\lambda$ is let equal to $n_{\lambda 0}$ if we have $W \leq W_a$, otherwise $n_\lambda$ is let equal to the greatest integer satisfying such inequality.

Now we have to face some practical problems. The first one is that only a few combinations of $p$, $k$, $n_{sc}$, and $n_{\lambda 0}$ give rise to an integer value of $N$ close to 500 and simultaneously $W \leq W_a$. The second one is that each selected combination provides a different value of the uniform traffic distribution probability $P_u$, assumed as a reference point, making the interpretation of the results more complex than before. Nevertheless, some useful support information can be also derived from the analytical expressions of $C_u$, $C_{tot}$ and $\chi$. For example, it can be easily seen that $C_u$, $C_{tot}$, together with $n_{\lambda 0}$, increase uniformly with $P_c$, while $\chi$ decreases. Moreover, since we have chosen to optimize the resources distribution, any analyzed configuration is characterized by different numbers of used channels. Therefore the variability of the throughput is also due to the different amount of resources used.

In Figures 4.17 and 4.18 we have plotted $C_u$ and $\chi$ respectively, assuming $k = 2$ and $p$ variable between 2 and 5. We have omitted, instead, to show the behavior of $C_{tot}$, as it is quite similar to that of $C_u$ except for a scale factor on the order of 500 (the value assumed for $N$). The same will also be made in the following, for the sake of brevity. Based on the results of [12], reminded before, the assumption of low values for the number of columns in the connectivity graph of the shufflenet at level 2 seems to be best solution for the particular optimization problem here proposed. This justifies the choice of a fixed $k = 2$. Nevertheless a similar analysis could be developed considering $k$ variable and $p$ fixed. In each figure we have reported four curves: one refers to the case of $P_c = P_u$, while the others assume $P_c = 0.2P_u$, $P_c = 5P_u$, and $P_c = 7P_u$, respectively. As expected, the throughput is maximized and the network cost is minimized for increasing values of $P_c$, so quantitatively demonstrating the importance of a correct clusters organization, made on the basis of the traffic characteristics. The intersection of the curves $P_c = 5P_u$ and $P_c = 7P_u$ appearing in Figure 4.18, which seems to contradict the trend, is only due to the discretization operated (for physical reasons) in the value of $n_\lambda$.

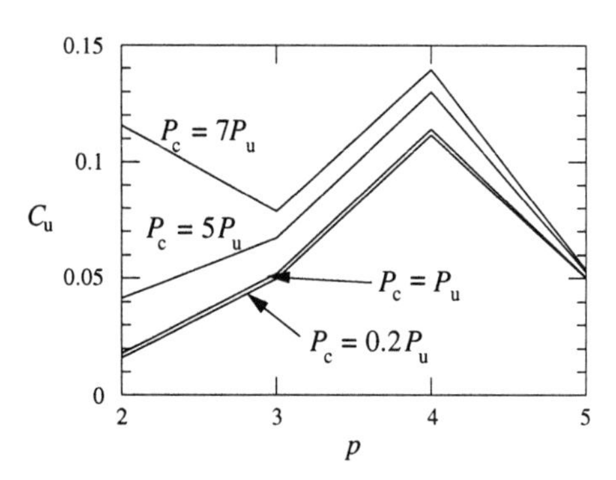


**Figure 4.17** Normalized throughput per user as a function of $p$ for different values of the intracluster connection probability.

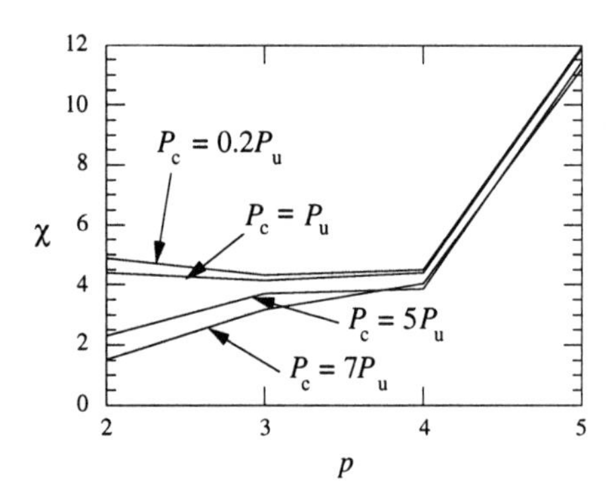


**Figure 4.18** Cost figure as a function of $p$ for different values of the intracluster connection probability.

Finally, Figures 4.19 and 4.20 show $C_u$ and $\chi$ versus $P_c$ assuming $p$ as a parameter and $k = 2$. From Figure 4.20 we can see that the configuration with $p = 5$ is everywhere characterized by the maximum cost. Since Figure 4.19 confirms that the same value ensures rather low throughputs, the assumption $p = 5$, for the specific example here considered, seems not acceptable at all. From Figure 4.20 we also derive that the choices $p = 2$, $p = 3$, or $p = 4$ lead to quite similar trends if the cost figure $\chi$ is considered. Therefore, the selection of the optimum configuration can be made on the basis of $C_u$ or, equivalently, of $C_{tot}$. $p = 3$ and $p = 4$ provide greater values but the final choice between them has to take into account the particular value of $P_c$; the two curves, in fact, intersect one another at $P_c \approx 0.8$. Such behavior can be qualitatively explained by considering that, when $p = 4$ and $P_c$ is sufficiently high, the finite number of available channels $W_a$ does not allow one to assign a number of channels per cluster $n_\lambda$ equal to $n_{\lambda 0}$. Therefore, in this region, the network with $p$ = 4 does not operate in optimum conditions and it is preferable to assume a lower value for $p$.

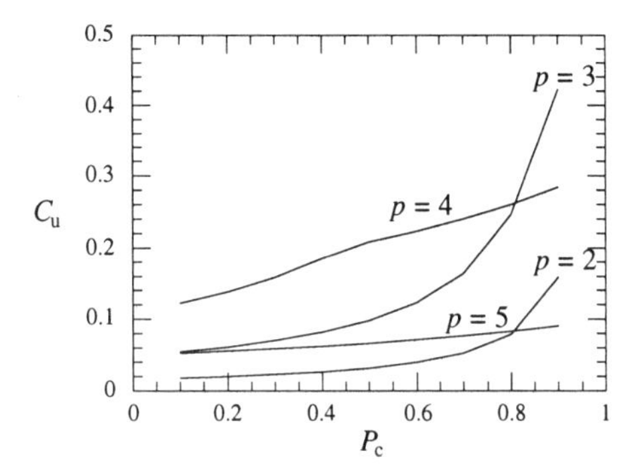


**Figure 4.19** Normalized throughput per user as a function of $P_c$ for different values of $p$.

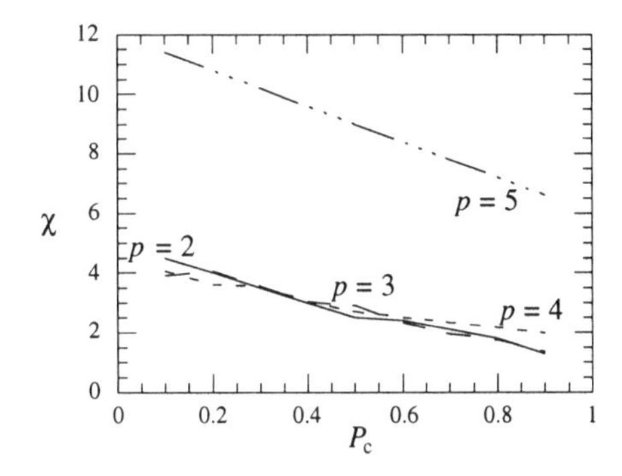


**Figure 4.20** Cost figure as a function of $P_c$ for different values of $p$.

### 4.4.3 Comparison with Shufflenet

The mathematics presented in the previous sections allows one to develop a simple comparison between a pure shufflenet (from now on denoted by SN) and a MON [13]. To make the analysis practical enough, we assume that $W_a$ = 50 channels are approximately available in both networks.

First, we wish to characterize the SN under uniform traffic conditions (Section 3.3.1). For this purpose, it should also be noted that a two-level MON reduces itself, in principle, to a SN when $n_{sc}$ = 1 is assumed. Nevertheless, the equations previously reported have to be slightly modified for the SN analysis, since they imply that at least one channel is devoted to intracluster communications, independently of the cluster size. After some algebraic manipulations, we obtain some of the results (those of interest for the present purposes) already discussed in Table 3.6, with the addition of a new column that gives the "cost figure" as defined by (4.20). Because of the need to satisfy the requirement on the number of channels, only some configurations of the SN are admissible. They are summarized, with their own parameters, in Table 4.2, where the subscript SN has been used to identify the topology.

**Table 4.2**
Performance of Some SNs with a Number of Channels on the Order of $W_a = 50$ or Less

| $k_{SN}$ | $p_{SN}$ | $N_{SN}$ | $W_{SN}$ | $C_{uSN}$ | $\chi_{SN}$ |
|---|---|---|---|---|---|
| 2 | 2 | 8 | 16 | 1 | 2 |
| 2 | 3 | 18 | 54 | 1.378 | 2.177 |
| 3 | 2 | 24 | 48 | 0.613 | 3.263 |

Passing now to consider the MON, besides $k$ and $p$, it needs a third parameter to determine univocally the virtual topology. This basically relates to the cluster size, so that $n_\lambda$ or $n_{sc}$ can be used for such a goal. Coherent with the previous analysis, $n_{sc}$ is usually preferable, but taking into account that $n_\lambda = n_{\lambda 0}$ is assumed in order to guarantee the optimization of the two-level architecture. Because of the constraint on the number of channels, only one configuration is acceptable at level 2 of the MON, precisely that characterized by $k = 2$ and $p = 2$. This means that the network considered is composed, at its upper level, by eight gateways connected through a suitable shufflenet; eight clusters are then present at level 1, each grouping $n_{sc}$ end-nodes. Since we are referring to the case of uniform traffic distribution, directly from Figure 4.15 we can view the dependence on $n_{sc}$ of the maximum normalized throughput per user, to be compared with $C_{uSN}$. A similar dependence can also be made explicit for the minimum value of the cost figure $\chi_{min} = \chi|_{n_\lambda = n_{\lambda 0}} = 1/\eta$; it is shown in Figure 4.21, limited to the range $2 \leq n_{sc} \leq 50$. This value must be compared with the values of $\chi_{SN}$ reported in Table 4.2.

Analyzing the results, one could be tempted to conclude that the performance of the MON is worse than that of the SN since, for any value of $n_{sc}$, we have $C_u < C_{uSN}$ and $\chi_{min} > \chi_{SN}$. Actually, the situation is not so simple, and further considerations must be added for a full comprehension.

For the parameter ranges here adopted ($2 \leq n_{sc} \leq 50$), continuing to see $n_{\lambda 0}$ (and also $W$, by virtue of (4.5)) as a real variable, we have that $n_{\lambda 0} \in [2.07, 2.14]$ and $W \in [32.56, 33.12]$. More physically, this means that, using 32 channels ($n_\lambda = 2$) in the whole network we can serve, for example, 400 users ($n_{sc} = 50$), at an acceptable cost. The assumption of $n_\lambda = 2$ corresponds to a slight underdimensioning of the bandwidth availability for intracluster communications. Otherwise, if an overdimensioning of the latter is preferred (three wavelengths per cluster), 40 channels are globally needed to connect the same number of end-nodes. This way the throughput of the MON is really maximum. In other words, even though it is difficult to match different networks determined by integer constraints, the comparison here shown, apparently disadvantageous for MONs, becomes interesting if the number of connectable users is taken into account, under the constraint on the number of channels. In fact, SNs using less than 50 wavelengths are not able to admit in the network more than 24 users whereas, in MONs, $N$ is an almost independent variable. Obviously, its increase is partly paid in terms of reduced throughput per user (as shown in Figure 4.15). Moreover, we have to recall that these results are obtained for a uniform distribution of traffic between nodes, so that $P_c = P_u = (n_{sc} - 1)/(N - 1)$, whereas it is evident that MONs exhibit their best performance when some unbalanced load characterizes the network.

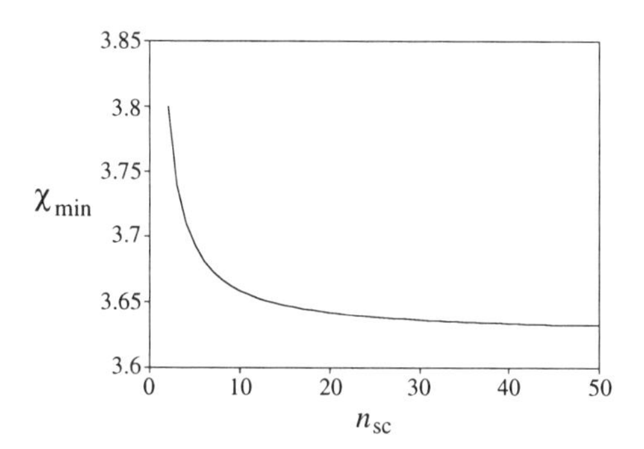


**Figure 4.21** Minimum cost figure as a function of the cluster size.

In this sense, we can now examine the behavior of the parameters considered so far, as functions of $P_c$, that is, for variable probability of intracluster connection requests, and assuming different values of $N$. Actually, the overall number of users is changed only acting on the cluster size, since the previous analysis shows that the network structure should be fixed at level 2, assuming $p = 2$ and $k = 2$, in such a way as to maintain the value of $W$ as low as required.

In Figure 4.22 we have plotted $(C_u)_{max}$, as a function of $P_c$, for different values of $n_{sc}$ and $n_\lambda = n_{\lambda 0}$. The curves have the same behavior as that for $p = 2$ in Figure 4.19, which, however, concerned a greater number of users per cluster. On the other hand a sort of "saturation effect" appears, by virtue of which, for high values of $n_{sc}$, the curves tend to overlap each other. According to (4.28), the value of $n_{\lambda 0}$ depends on $P_c$; therefore it changes, point by point, in Figure 4.22. Correspondingly, the number of channels $W$ varies as well, increasing with $P_c$, as is shown explicitly in Figure 4.23. On the right of $P_c \approx 0.7$ we have $W > 50$, and the number of channels is no longer compatible with the design constraints. The curve of $\chi_{min}$ as a function of $P_c$ is unique and given by

$$\chi_{min} = \left(1 + \frac{p}{n_{\lambda 0}}\right)(2 - P_c) \tag{4.29}$$

This curve is plotted in Figure 4.24.

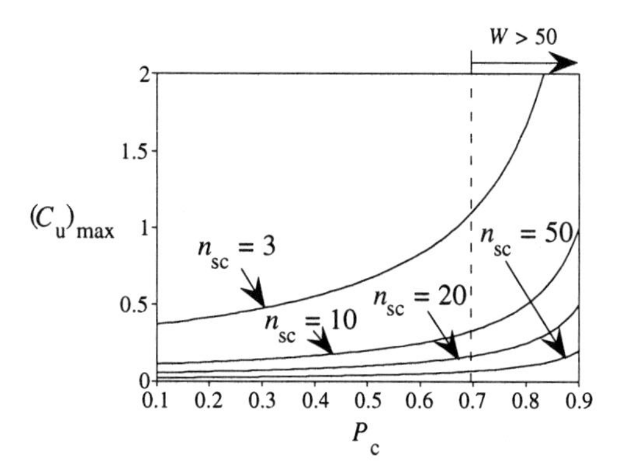


**Figure 4.22** Maximum normalized throughput, in the case of nonuniform traffic distribution, for different numbers of connectable users.

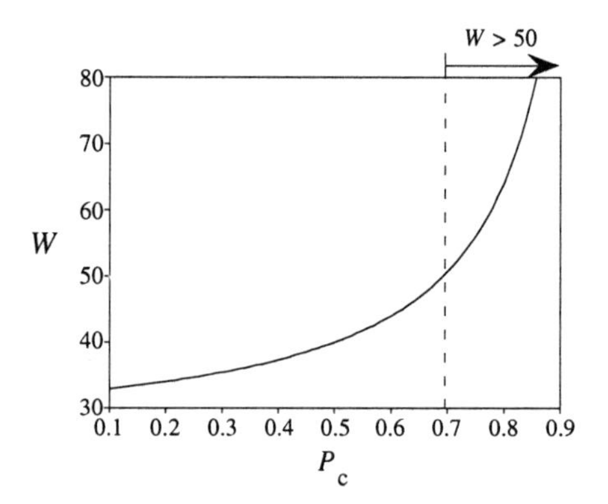


**Figure 4.23** Number of channels as a function of the intracluster connection probability.

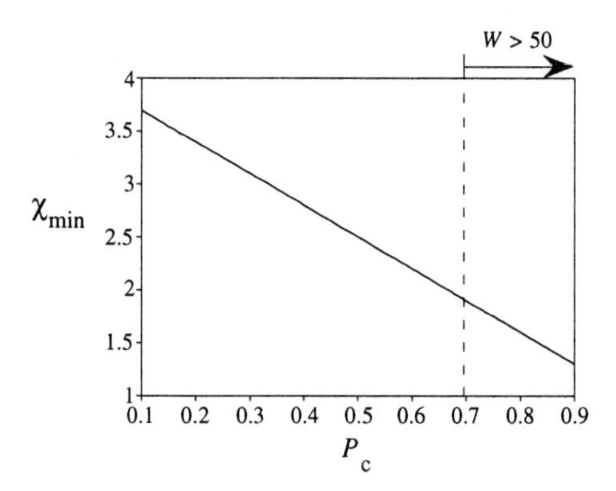


**Figure 4.24** Minimum cost figure as a function of the intracluster connection probability.

Figures 4.22 and 4.24 confirm (as expected) that the performance of the MON improves for higher values of $P_c$. The curve of the cost figure, in particular, demonstrates that the increase in the total network throughput is greater than that in the number of channels. In conclusion, the MON seems to be a very efficient solution for connecting a large number of users, each of them linked for most of the time with a specific subset of other users, and more rarely with all the others.

### 4.4.4 Multiple Hierarchical Levels

The analysis developed in the previous sections can easily be extended to a MON with any number of levels [14]. The virtual topology of a network organized on $n$ levels is schematically depicted in Figure 4.25: each ellipse represents an elementary subnetwork, where transmission can be performed using a single-hop or a multihop strategy. As before, source and destination nodes (end-nodes) are all located at level 1, whereas the upper virtual levels collect the nodes playing the role of gateways between adjacent levels. In practice, with the usual notation, every user is connected directly to $(n_{sc} - 1)$ end-nodes through a subnetwork at level 1, and is connected to the remaining $(N - n_{sc})$ users through the subnetworks at levels 2 to $n$.

**Figure 4.25** Virtual structure of a MON with $n$ hierarchical levels.

Generalizing (4.6), we can say that, in the case of uniform traffic distribution, the normalized throughput per user is given by

$$C_u = \frac{N-1}{\max\left[T_{jh}^{(1)}, T_{jh}^{(2)}, \ldots, T_{jh}^{(n)}\right]} \tag{4.30}$$

where $T_{jh}^{(i)}$ ($j$ = s for single-hop subnetworks or $j$ = m for multihop subnetworks) represents the duration of the transmission cycle, in time slots, as it would be imposed by the $i$th level individually. Using the fixed assignment WDM/TDM transmission protocol already discussed for the two-level MON, it is easy to find

$$T_{\mathrm{sh}}{}^{(i)} = \left[\left(n_{\mathrm{sc}_i} - 1\right)n_{\mathrm{uc}_{i-1}} + 2\left(1 - \delta_{in}\right)\left(N - n_{\mathrm{uc}_i}\right)\right]\frac{n_{\mathrm{sc}_i} n_{\mathrm{uc}_{i-1}}}{n_{\lambda_i}} \tag{4.31a}$$

$$T_{\mathrm{mh}}{}^{(i)} = n_{\mathrm{uc}_{i-1}}^2 k_i \left[\frac{p_i^{k_i-1}\left(3k_i - 1\right)}{2} - \frac{p_i^{k_i-1} - 1}{p_i - 1}\right] + \left(1 - \delta_{in}\right) n_{\mathrm{sc}_1} n_{\mathrm{uc}_i} k_i \, p_i^{k_i-1}\left(n_{c_i} - 1\right) \tag{4.31b}$$

In these expressions, $n_{\mathrm{c_i}}$ represents the number of subnetworks at level $i$, $n_{\mathrm{sc}_i}$ the number of nodes in each subnetwork at the same level, $n_{\mathrm{uc_i}}$ the number of users that can be connected through level $i$, and $n_{\lambda_i}$ the number of wavelengths used for internal links in the case of single-hop subnetworks; dually, $k_i$ and $p_i$, the first denoting the number of columns in the connectivity graph and the second the number of wavelengths for transmission or reception by each node, describe the generic multihop subnetwork at level $i$, again assumed to be of the shufflenet type. Finally $\delta_{in}$ is the Kronecker delta function ($\delta_{in} = 1$ if $i = n$ or $\delta_{in} = 0$ if $i \neq n$). Its introduction is necessary to take into account that the $n$th level is atypical, in some respects, since it has no further levels above it. Obviously we have $n_{\mathrm{sc}_1} = n_{\mathrm{sc}}$ whereas, conventionally, we have put $n_{\mathrm{uc}_0} = 1$.

Equation (4.31a) is basically obtained dividing the number of permissions to transmit granted inside each single-hop subnetwork, in a transmission cycle, by the number of wavelengths available. In (4.31b), instead, differently from (4.9), it has been taken into account that in each shufflenet the traffic can be unbalanced between the various WDM channels [15] (see also Section 3.4.4). Such an assumption has been made in order to compare the results of the analysis with those of a simulation, where the shortest path routing algorithm, with random choice of don't care paths, has been used. The conclusions of this comparison will be discussed below. On the other hand, it will focus the case of nonuniform traffic distribution, which is more important in practice.

Therefore, with reference to a generic source, let us denote by $P_{ci}$ the probability that an emitted packet is sent to the users connected via the $i$th level. Generalizing (4.23), we can write

$$C_{\mathrm{u}} = \frac{L}{\max\left[T_{j\mathrm{h}}{}^{(1)}, T_{j\mathrm{h}}{}^{(2)}, \ldots, T_{j\mathrm{h}}{}^{(n)}\right]} \tag{4.32}$$

while the arguments of function max[·,·] are here expressed as

$$T_{\mathrm{sh}}{}^{(i)} = L\frac{n_{\mathrm{sc}_i} n_{\mathrm{uc}_{i-1}}\left[P_{ci} + 2\left(1 - \delta_{in}\right)\left(1 - \sum_{j=1}^{i} P_{cj}\right)\right]}{n_{\lambda_i}} \tag{4.33a}$$

$$T_{\mathrm{mh}}{}^{(i)} = L\frac{P_{ci}}{\left(k_i p_i^{k_i} - 1\right)} n_{\mathrm{uc}_{i-1}} k_i \left[\frac{p_i^{k_i-1}(3k_i - 1)}{2} - \frac{p_i^{k_i-1} - 1}{p_i - 1}\right]$$
$$+ \left(1 - \delta_{in}\right) L \frac{1 - \sum_{j=1}^{i} P_{cj}}{N - n_{\mathrm{uc}_i}} k_i p_i^{k_i-1}\left(n_{\mathrm{c}_i} - 1\right) n_{\mathrm{sc}_1} n_{\mathrm{uc}_i} \tag{4.33b}$$

As mentioned above, the reliability of the analytical model can be tested by comparing its results, in some typical situations, with those obtained by a numerical simulation of the network behavior [14]. A significant example is constituted by a three-level MON ($n = 3$), operating under overload conditions, which means that, in each transmission cycle, any user has a packet to send to any other user. In this way, the assumption of the fixed bandwidth assignment protocol is well justified.

The proposed scenario seems significant enough to allow one to draw general conclusions about the model efficiency. More precisely, the network logical organization is as follows [15]:

- Level 1: single-hop;
- Level 2: multihop;
- Level 3: multihop.

This structure seems reasonable, and applicable in many practical contexts. Three different choices of the parameters describing levels 2 and 3, have been considered in [14]; more precisely:

1. $k_2 = k_3 = p_2 = p_3 = 2$;
2. $k_2 = k_3 = p_3 = 2, p_2 = 3$;
3. $k_2 = k_3 = p_2 = 2, p_3 = 3$.

On the contrary, $n_{\mathrm{sc}} = 6$ and $n_{\lambda_1} = n_\lambda = 2$ have been set everywhere, for the parameters describing level 1. Since

$$N = n_{\mathrm{sc}} k_2 p_2^{k_2} k_3 p_3^{k_3} \tag{4.34}$$

all three structures connect a very large number of users, being $N = 384$ for (1) and $N = 864$ for (2) and (3), respectively.

The results obtained are reported in Figures 4.26 to 4.28, where the normalized total network throughput (defined by (4.15)) is plotted as a function of $P_{\mathrm{c}2}$, assuming $P_{\mathrm{c}1}$ as a parameter: continuous curves refer to the analytical expressions, whereas dots represent the simulation results for a discrete number of points. For each pair ($P_{\mathrm{c}1}$, $P_{\mathrm{c}2}$) the probability of connection through the third level can be obviously computed as $P_{\mathrm{c}3} = 1 - P_{\mathrm{c}1} - P_{\mathrm{c}2}$. As evident from Figures 4.26 to 4.28, the agreement between theory and simulation is very good everywhere; the simulated results are slightly lower than the analytical ones, but this is partly expected because of some unavoidable nonidealities in reproducing the transmission protocol.

In accordance with (4.32), a change in the curves' concavity corresponds to the transition from a situation in which the throughput is limited by the $i$th level to a situation in which it is limited by the $j$th level, with $j \neq i$. In the considered cases, $j < i$, which means that the network performance may be first limited by the multihop subnetwork at level 3, then by those at level 2, and finally, by the single-hop subnetworks at level 1. Actually, the effect of the intermediate level is missing in some cases (e.g., in all the curves of Figure 4.27) while, in some others, level 1 is never the bottleneck (the curves are broken off where $P_{c3} = 0$).

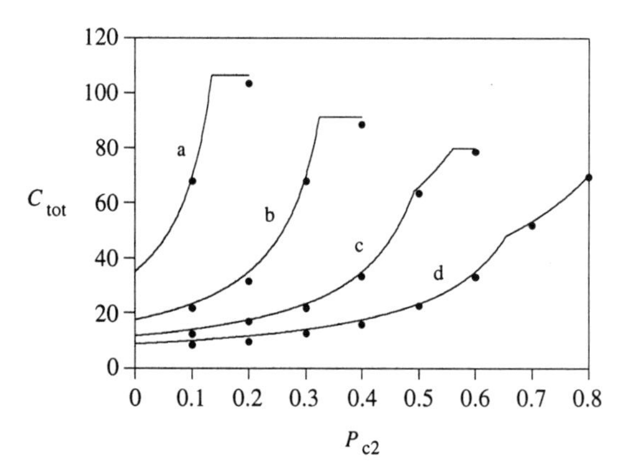


**Figure 4.26** Comparison between theory and simulation for network (1): (a) $P_{c1} = 0.8$, (b) $P_{c1} = 0.6$, (c) $P_{c1} = 0.4$, (d) $P_{c1} = 0.2$.

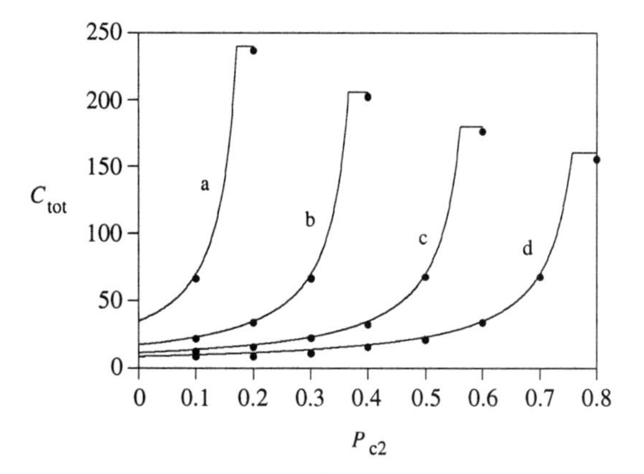


**Figure 4.27** Comparison between theory and simulation for network (2): (a) $P_{c1} = 0.8$, (b) $P_{c1} = 0.6$, (c) $P_{c1} = 0.4$, (d) $P_{c1} = 0.2$.

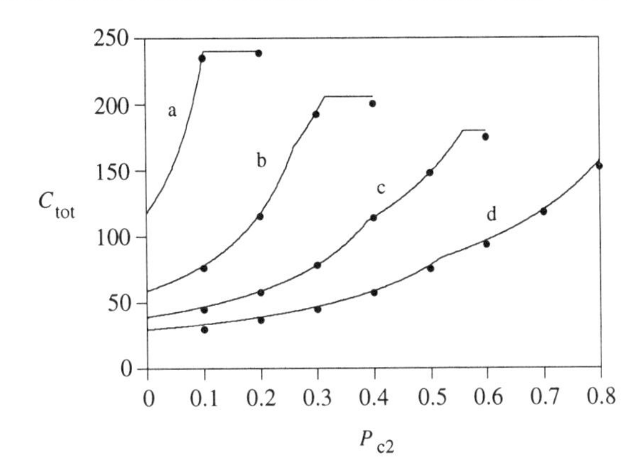


**Figure 4.28** Comparison between theory and simulation for network (3): (a) $P_{c1} = 0.8$, (b) $P_{c1} = 0.6$, (c) $P_{c1} = 0.4$, (d) $P_{c1} = 0.2$.

## REFERENCES


[1] Kaminow, I. P., C. R. Doerr, C. Dragone, T. Koch, U. Koren, A. A. M. Saleh, A. J. Kirby, C. M. Özveren, B. Schofield, R. E. Thomas, R. A. Barry, D. M. Castagnozzi, V. W. S. Chan, B. R. Hemenway, Jr., D. Marquis, S. A. Parikh, M. L. Stevens, E. A. Swanson, S. G. Finn, and R. G. Gallager, "A wideband all-optical WDM network," *IEEE J. Select. Areas Commun.*, Vol. 14, No. 5, June 1996, pp. 780–799.

[2] Ganz, A., and Y. Gao, "Traffic scheduling in multiple WDM star systems," *Proc. ICC '92*, Chicago, IL, June 1992, Paper 349.4, pp. 1468–1472.

[3] Ganz, A., and Y. Gao, "Time-wavelength assignment algorithms for high performance WDM star based systems," *IEEE Trans. Commun.*, Vol. 42, No. 2/3/4, Feb./Mar./Apr. 1994, pp. 1827–1836.

[4] Jonsson, M., and B. Svensson, "On inter-cluster communication in a time-deterministic WDM star network," to be presented at *INFOCOM '97*, Kobe, Japan, April 1997.

[5] Stern, T. E., K. Bala, S. Jiang, and J. Sharoni, "Linear lightwave networks: Performance issues," *IEEE/OSA J. Lightwave Technol.*, Vol. 11, No. 5/6, May/June 1993, pp. 937–950.

[6] Gerla, M., M. Kovačevíc, and J. Bannister, "Optical tree topologies: Access control and wavelength assignment," *Computer Networks and ISDN Systems*, Vol. 26, March 1994, pp. 965–983.

[7] Borella, A., F. Chiaraluce, M. Gerla, M. Kovačevíc, and F. Meschini, "An analytical approach to the performance evaluation of MONETs," *Proc. Second Int. Conf. on Computer Commun. and Networks ($IC^3N$)*, San Diego, CA, June 1993, pp. 257–261.

[8] Mukherjee, B., D. Banerjee, S. Ramamurthy, and A. Mukherjee, "Some principles for designing a wide-area WDM optical network," *IEEE/ACM Trans. Networking*, Vol. 4, No. 5, Oct. 1996, pp. 684–696.

[9] Wagner, R. E., R. C. Alferness, A. A. Saleh, and M. S. Goodman, "MONET: Multiwavelength optical networking," *IEEE/OSA J. Lightwave Technol.*, Vol. 14, No. 6, June 1996, pp. 1349–1355.

[10] Chlamtac, I., and A. Ganz, "Channel allocation protocols in frequency-time controlled high-speed networks," *IEEE Trans. Commun.*, Vol. 36, No. 4, April 1988, pp. 430–440.

[11] Borella, A., F. Chiaraluce, and F. Meschini, "Optimization of multilevel optical networks", *Proc. Third Int. Conf. on Computer Commun. and Networks (IC$^3$N)*, San Francisco, CA, Sept. 1994, pp. 401–408.

[12] Borella, A., F. Chiaraluce, and F. Meschini, "Channel utilization in multilevel optical networks," *Proc. EFOC&N '94*, Heidelberg, Germany, June 21-24, 1994, Papers on ATM and Networks, pp. 226–229.

[13] Borella, A., F. Chiaraluce, and F. Meschini, "Throughput/cost trade-off in passive optical networks: A comparison between shufflenets and MONETs," *Proc. Eur. Symp. on Advanced Networks and Services*, Amsterdam, Netherlands, 20-23 March, 1995, SPIE Vol. 2450, pp. 419–427.

[14] Borella, A., and F. Chiaraluce, "A general method for the analysis of combined single-hop and multi-hop connection modes in MONETs," *Proc. Melecon '96*, Bari, Italy, May 13-16 1996, pp. 1004–1007.

[15] Borella, A., F. Chiaraluce, and F. Meschini, "Analysis of hierarchical optical networks," *Proc. EFOC&N '95*, Brighton, England, June 27-30, 1995, Papers on ATM, Networks and LANs, pp. 56–59.

# *About the Authors*

**Andrea Borella** was born in Bologna, Italy in 1962. In 1987 he received the "Laurea in Ingegneria Elettronica" (summa cum laude) from the University of Bologna. Between 1987 and 1988 he was at the Radioastronomy Institute of the Italian National Research Council in Bologna. In 1990, after serving in the army, he joined the engineering faculty of the University of Ancona (Italy) where he received, in 1993, his Ph.D. in electronic engineering. At present he is visiting professor in digital communications at the Department of Electronics and Automatics of the University of Ancona. In recent years he has also been involved in some applied research programs, carried out by private industrial companies, as consultant or project leader. His research activity essentially concerns performance evaluation of high-speed networks: topologies, architectures, protocols, and routing techniques; traffic modeling and network simulation. He is coauthor of about 50 papers.

**Giovanni Cancellieri** was born in Florence, Italy in 1952. He received the "Laurea in Ingegneria Elettronica" in 1976, and the "Laurea in Fisica" in 1978, both from the University of Bologna. After a period of about 10 years, during which he was engaged by the Fondazione G. Marconi, he joined the University of Ancona, first as associate professor of microwaves, and then as full professor of electrical communications. From 1989 to 1995 he was director of the Department of Electronics and Automatics at the same university. His main research interests are in the field of fiber optics, telecommunication systems, multimedia, and teleeducation. He is coauthor of about 100 national and international scientific papers and of seven books. He is a member of the Italian Electric and Electronic Engineering Association (AEI) and of the European Association for Telematic Applications (EATA).

**Franco Chiaraluce** was born in Ancona, Italy in 1960. He received the "Laurea in Ingegneria Elettronica" (summa cum laude) from the University of Ancona in 1985. Since 1987 he has been Technical Manager in the Department of Electronics and Automatics of the University of Ancona, being actively engaged in the fields of integrated optics and optical fiber communications. A significant part of this research work was developed in cooperation with industry or in the framework of European joint projects.

At present his main research interests involve various aspects of optical communication systems theory and design, especially as regards optical networks and nonlinear devices. He is coauthor of more than 80 papers and a book on single-mode optical fiber measurements. He is a member of IEEE and AEI.

# Index

# The Artech House Optoelectronics Library

Brian Culshaw and Alan Rogers, *Series Editors*

*Amorphous and Microcrystalline Semiconductor Devices, Volume II: Materials and Device Physics*, Jerzy Kanicki, editor

*Bistabilities and Nonlinearities in Laser Diodes*, Hitoshi Kawaguchi

*Chemical and Biochemical Sensing With Optical Fibers and Waveguides*, Gilbert Boisdé and Alan Harmer

*Coherent and Nonlinear Lightwave Communications*, Milorad Cvijetic

*Coherent Lightwave Communication Systems*, Shiro Ryu

*Elliptical Fiber Waveguides*, R. B. Dyott

*Field Theory of Acousto-Optic Signal Processing Devices*, Craig Scott

*Frequency Stabilization of Semiconductor Laser Diodes*, Tetsuhiko Ikegami, Shoichi Sudo, Yoshihisa Sakai

*Fundamentals of Multiaccess Optical Fiber Networks*, Denis J. G. Mestdagh

*Germanate Glasses: Structure, Spectroscopy, and Properties*, Alfred Margaryan and Michael A. Piliavin

*Handbook of Distributed Feedback Laser Diodes*, Geert Morthier and Patrick Vankwikelberge

*Helmet-Mounted Displays and Sights*, Mordekhai Velger

*High-Power Optically Activated Solid-State Switches*, Arye Rosen and Fred Zutavern, editors

*Highly Coherent Semiconductor Lasers*, Motoichi Ohtsu

*Iddq Testing for CMOS VLSI*, Rochit Rajsuman

*Integrated Optics: Design and Modeling*, Reinhard März

*Introduction to Lightwave Communication Systems*, Rajappa Papannareddy

*Introduction to Glass Integrated Optics*, S. Iraj Najafi

*Introduction to Radiometry and Photometry*, William Ross McCluney

*Introduction to Semiconductor Integrated Optics*, Hans P. Zappe

*Laser Communications in Space*, Stephen G. Lambert and William L. Casey

*Optical Document Security, Second Edition,* Rudolf L. van Renesse, editor

*Optical FDM Network Technologies,* Kiyoshi Nosu

*Optical Fiber Amplifiers: Design and System Applications,* Anders Bjarklev

*Optical Fiber Amplifiers: Materials, Devices, and Applications,* Shoichi Sudo, editor

*Optical Fiber Communication Systems*, Leonid Kazovsky, Sergio Benedetto, Alan Willner

*Optical Fiber Sensors, Volume Two: Systems and Applicatons,* John Dakin and Brian Culshaw, editors

*Optical Fiber Sensors, Volume Three: Components and Subsystems,* John Dakin and Brian Culshaw, editors

*Optical Fiber Sensors, Volume Four: Applications, Analysis, and Future Trends,* John Dakin and Brian Culshaw, editors

*Optical Interconnection: Foundations and Applications,* Christopher Tocci and H. John Caulfield

*Optical Measurement Techniques and Applications,* Pramod Rastogi

*Optical Network Theory*, Yitzhak Weissman

*Optoelectronic Techniques for Microwave and Millimeter-Wave Engineering*, William M. Robertson

*Reliability and Degradation of LEDs and Semiconductor Lasers,* Mitsuo Fukuda

*Reliability and Degradation of III-V Optical Devices,* Osamu Ueda

*Semiconductor Raman Laser,* Ken Suto and Jun-ichi Nishizawa

*Semiconductors for Solar Cells*, Hans Joachim Möller

*Smart Structures and Materials*, Brian Culshaw

*Ultrafast Diode Lasers: Fundamentals and Applications*, Peter Vasil'ev

*Wavelength Division Multiple Access Optical Networks,* Andrea Borella, Giovanni Cancellieri, and Franco Chiaraluce

For further information on these and other Artech House titles, including previously considered out-of-print books now available through our In-Print-Forever™ (IPF™) program, contact:

Artech House
685 Canton Street
Norwood, MA 02062
781-769-9750
Fax: 781-769-6334
Telex: 951-659
email: artech@artech-house.com

Artech House
Portland House, Stag Place
London SW1E 5XA England
+44 (0) 171-973-8077
Fax: +44 (0) 171-630-0166
Telex: 951-659
email: artech-uk@artech-house.com

Find us on the World Wide Web at:
www.artech-house.com